Tracking Adult Literacy and Numeracy Skills

Routledge Research in Education

1. Learning Communities in Education
Edited by John Retallick, Barry Cocklin and Kennece Coombe

2. Teachers and the State
International Perspectives
Mike Bottery and Nigel Wright

3. Education and Psychology in Interaction
Working with Uncertainty in Inter-Connected Fields
Brahm Norwich

4. Education, Social Justice and Inter-Agency Working
Joined up or Fractured Policy?
Sheila Riddell and Lyn Tett

5. Markets for Schooling
An Economic Analysis
Nick Adnett and Peter Davies

6. The Future of Physical Education
Building a New Pedagogy
Edited by Anthony Laker

7. Migration, Education and Change
Edited by Sigrid Luchtenberg

8. Manufacturing Citizenship
Education and Nationalism in Europe, South Asia and China
Edited by Véronique Bénéï

9. Spatial Theories of Education
Policy and Geography Matters
Edited by Kalervo N. Gulson and Colin Symes

10. Balancing Dilemmas in Assessment and Learning in Contemporary Education
Edited by Anton Havnes and Liz McDowell

11. Policy Discourses, Gender, and Education
Constructing Women's Status
Elizabeth J. Allan

12. Improving Teacher Education through Action Research
Edited by Ming-Fai Hui and David L. Grossman

13. The Politics of Structural Education Reform
Keith A. Nitta

14. Political Approaches to Educational Administration and Leadership
Edited by Eugenie A. Samier with Adam G. Stanley

15. Structure and Agency in the Neoliberal University
Edited by Joyce E. Canaan and Wesley Shumar

16. Postmodern Picturebooks
Play, Parody, and Self-Referentiality
Edited by Lawrence R. Sipe and Sylvia Pantaleo

17. Play, Creativity and Digital Cultures
Edited By Rebekah Willet, Muriel Robinson and Jackie Marsh

18. Education and Neoliberal Globalization
Carlos Alberto Torres

19. Tracking Adult Literacy and Numeracy Skills
Findings from Longitudinal Research
Edited by Stephen Reder and John Bynner

Tracking Adult Literacy and Numeracy Skills

Findings from Longitudinal Research

Edited by
Stephen Reder and
John Bynner

Routledge
Taylor & Francis Group
New York London

First published 2009
by Routledge
711 Third Avenue, New York, NY 10017

Simultaneously published in the UK
by Routledge
2 Park Square, Milton Park, Abingdon, Oxon OX14 4RN

Routledge is an imprint of the Taylor & Francis Group, an informa business

First published in paperback 2011

Typeset in Sabon by IBT Global.

Library of Congress Cataloging in Publication Data

Tracking adult literacy and numeracy skills : findings from longitudinal research / edited by Stephen Reder & John Bynner. — 1st ed.
p. cm. — (Routledge research in education ; 19)
Includes bibliographical references and index.
ISBN 978-0-415-95858-5
1. Functional literacy—English-speaking countries—Longitudinal studies. 2. Numeracy—English-speaking countries—Longitudinal studies. 3. Adult education—English-speaking countries—Longitudinal studies. I. Reder, Stephen M. II. Bynner, J. M.
LC149.7.T73 2009
374—dc22
2008017897

ISBN13: 978-0-415-95858-5 (hbk)
ISBN13: 978-0-415-80788-3 (pbk)
ISBN13: 978-0-203-88888-9 (ebk)

Contents

Figures

Tables

Acknowledgments

The editors wish to acknowledge the support, encouragement, and collaboration of their many colleagues at the National Center for the Study of Adult Learning and Literacy (NCSALL) and at the National Research and Development Centre for Adult Literacy and Numeracy (NRDC). We have been very fortunate to have benefited from the stimulating intellectual environment provided by these centers and the consortia of their participating institutions. We are indebted and very grateful for the expert assistance of Melyssa Sharp in the preparation of the manuscript. Our deep gratitude and appreciation extends to the many thousands of adult learners who have participated in the longitudinal research studies described in this volume. Without their cooperation, neither the research nor this volume would have been possible. We hope that they will benefit from the improved understanding, policies, and programs that the research is intended to bring about.

Introduction

The Need for Longitudinal Studies in Adult Literacy and Numeracy Education

Stephen Reder and John Bynner

INTRODUCTION

Understanding the origins of poor literacy and numeracy skills in adulthood and their consequences for individuals in adult life is of increasing importance in modern industrialized societies where an ever higher premium is placed on proficiency in these basic skills. The origins of this concern lie in the technological transformation of employment and globalizing pressures, which have been a dominant feature of their economies since the 1970s. For this reason there have been major initiatives in all societies to raise skills levels for which the basic skills of literacy and numeracy supply the foundations.

Countries differ in the prevalence of these problems, with English-speaking nations, especially those in North America and the United Kingdom, having particularly high proportions of their adult populations with poor literacy and numeracy skills. Such poor skills may not only be a challenge for the individual in functioning effectively in adult life but are also seen as constituting a major economic cost in terms of lost productivity and international competitiveness. Apart from their national economic consequences, low levels of skills may frequently lead to marginalization of individuals in the modern labor market and society more widely, bringing about low levels of civic participation, high costs in terms of state benefits, and various forms of "social exclusion" as manifested in drugs, crime, and poor physical and mental health.

This book examines these issues, bringing together the results of longitudinal studies carried out in North America and the United Kingdom that greatly extend our understanding of them. The chapters authored by leading experts in the field increase focus on "what works" in raising skills levels and what the social and economic returns—private and public—are to skill improvement.

RESEARCH STRATEGY

Many fundamental research questions in adult education involve change over time:

- how adults learn
- how participation in programs influences adults' acquisition of knowledge and skills
- how adults' educational development interacts with their social and economic performance
- how adults' private and social situation in the family and community may enhance or obstruct their progress in learning

Much of the contextual aspect of adults' skill acquisition reflects a history of disadvantage extending back to early childhood, a history that may also be compounded, particularly in the case of immigrants, by the challenges of mastering a second language. The main point to emphasize is that adults with poor skills are not a homogenous population, but show considerable diversity in the features of their lives and in their past experiences and achievements. Some research has been conducted on all these aspects of skills acquisition using one-off descriptive surveys or as studies deploying the research tools of social biographical enquiry and ethnography.

Longitudinal studies can add methodological rigor and analytic tools for investigating changes in individuals, measuring changes over time in representative samples of a defined population. These supply meaningful profiles of a population and changes in it over time that can reveal aspects of likely causal processes. For example, the role of educational experience set against that of family and community life may point to particular features of learning careers that underpin success or failure. The representative sample offers opportunities for generalization from individuals to groups and the wider population of which they are members. These notions can point to changes in education provision and support provided to learners that can be hypothesized to rectify past problems and prevent their repetition in the next generation. Longitudinal research that is more qualitatively oriented in the sense of analyzing biographies of "individuals in context" can amplify the picture in complementary fashion through revealing the kinds of processes in individual lives that steer educational pathways in one direction or another.

Sometimes the longitudinal data in adult educational research take the form of a series of cross-sectional snapshots taken at different historical moments, as in some of the national and international literacy surveys that have been repeated every five or ten years. Such "continuous" surveys form a "time series" that provides the means for monitoring changes in the population as a whole. The complementary longitudinal survey of individuals can supply a basis for causal inference. Such studies may involve samples of populations of all ages, such as in the household or family *panel study*, or of groups of learners of a particular age—children, young people, adults—who form the sample for investigation. Another version is the birth cohort study that follows individuals over an extended period of their lives beginning at birth. These latter kinds of study involve data collection often at extended intervals updating the individual life record on each new occasion. Other studies investigating, for example, progression in response to educational provision will monitor the individuals through interview and assessment fairly frequently so that relatively small elements of incremental achievement can be observed. The end product of all such studies will be some form of longitudinal data set that supplies information in the form of measures taken of a sample of the same individuals from a specified population at different times.

Although there is a growing number of longitudinal studies in adult basic education that have been completed or are in progress, or of individuals in general population samples whose educational achievement has been recorded, no systematic compilation of findings and methods has yet to be undertaken. The book helps to bridge this gap, bringing together findings from a variety of longitudinal studies that were originally presented in a conference at the National Academy of Sciences Conference Center in Woods Hole, Massachusetts. The focus of the conference was principally on research designs and methods from a comparative perspective drawing on studies, many of which were ongoing. The book's chapters move the focus to research findings and include some additional chapters from authors not at Woods Hole but whose longitudinal research is essential to the topics covered in the book. Many authors conducted the research through the auspices of the major U.S. and UK research centers with which they were affiliated—in the United States the National Center for the Study of Adult Learning and Literacy (NCSALL) and in England, the National Research and Development Centre for Adult Literacy and Numeracy (NRDC). The Woods Hole conference was jointly organized by the two centers.

The research reported embraces a range of methodological approaches from quantitative (statistical modeling) to qualitative (biographical) investigations. Others draw on administrative records as core sources of longitudinal data. One UK study uses experimental methodology to assess the effectiveness of a program with the aim of establishing unequivocally cause and effect. Triangulating findings from different methodological perspectives and research designs and across countries produces convergence on key conclusions on the role of basic skills in the modern life course and the most effective ways of enhancing them. The organization of the book is structured to reflect the wide range of perspectives and findings bearing on a number of key topics: literacy and numeracy skills, classroom and teacher studies, program impact, social and economic outcomes, the influence of family and community contexts on individually measured change.

THE MACRO CONTEXT

Although the main focus of the research reported is on individuals' needs for basic skills in the context of their everyday lives and their response to programs directed at enhancing those skills, the wider context of these needs supplies an important backcloth to understanding them. This is particularly the case in relation to employment but also relates directly or indirectly to many aspects of family and community life. The growth of IT-based service industry at the expense of manufacturing and the decline of heavy industry and unskilled manufacturing work have had the effect of restructuring the work force with a major shift towards "white-collar" and computer-based employment. Much of the effects of these changes began in the 1970s, reaching a peak in the early 1980s, when a worldwide recession added to the pressure on jobs. The outcome was a new employment structure that increasingly demanded literacy, numeracy, and IT skills of its employees and, particularly in the case of the United Kingdom, certified skills and qualifications. The consequence was large numbers of young people finding difficulty in gaining employment and many adults losing the employment that they had.

The counterpart of these processes is *globalization,* where investment moves cross-nationally to wherever the skills sought can be matched at lowest cost and increasing pressures are placed on local industries and communities (Ashton & Green, 1996; Rifkin, 1995). Overall, the effect is increasing polarization of the labor force, in which opportunities and good prospects are available to those who have the resources to take advantage of them. At the same time, marginalized employment in the form of casual jobs and spells out of the labor force occur and especially for women having and caring for children, become the main prospects for many. In this scenario, basic skills are increasingly recognized as having high priority for individuals and for societies as a whole, because without them educational prospects and life chances more generally can be severely curtailed.

Thus, the world in which every adult, depending on the state of the local economy, could expect to move relatively effortlessly from school into some kind of stable job now becomes more problematic with numeracy and literacy as prerequisites for employment. The consequent pressure for upskilling impacts on every area of people's lives, where written competence in reading, writing, and arithmetic are taken for granted in diverse areas, ranging from understanding regulations and instructions to filling in tax forms. Thus, although much of the focus of the research reported in this book is on individual lives, basic skills performance, and outcomes, this needs to be set against the wider set of social and economic trends occurring worldwide in which the premium placed on these forms of competence and capability is increasingly raised.

INTERNATIONAL DIFFERENCES IN EDUCATION SYSTEMS AND TERMINOLOGY

Although awareness of the need to raise literacy and numeracy levels under modern conditions has existed at least since the 1960s, one of the major drivers of current concerns was the International Adult Literacy Survey (IALS), which was modeled after the National Adult Literacy Survey (NALS) first conducted in the United States in 1992. The IALS was carried out by Statistics Canada on behalf of the Organisation for Economic Co-operation and Development (OECD) between 1994 and 1998 on a representative sample of adults aged between sixteen and sixty-five in fourteen countries (Murray et al., 1998). IALS showed marked differences on a standardized assessment of three dimensions of literacy ("prose," "document," "quantitative") among countries, with English-speaking countries faring particularly badly, especially the United States and the United Kingdom. The term *quantitative literacy,* employed by Statistics Canada, is not in common use in the United Kingdom, where "numeracy," perceived as a quite different attribute from literacy, tends to be preferred.

The IALS graded population samples in the participating countries in terms of four literacy levels. This revealed such differences as 7 percent of the adult population performing at the lowest level of prose literacy (Level 1) in Sweden compared with over 20 percent in the United Kingdom and the United States. Such differences can be interpreted in a number of ways. Perhaps the English language has complexities that make it particularly susceptible to poor performance among substantial numbers who have, for example, missed out on part of their earlier education. Another explanation relates to the education system itself, suggesting that those countries with relatively small numbers at the lowest performance levels, such as the Nordic countries, supply their students with a much better grounding in the basic skills, which pays off throughout the whole of their educational careers. Another explanation relates to differentiation in the societies themselves, with larger proportions in such countries as the United States and the United Kingdom engaged in unskilled casual work to which poor basic skills are not an impediment or growing up in families in which less transmission of good basic skills from parents to children tends to occur. In countries such as England, the problem is compounded through an education system that implicitly accepts that a substantial proportion of children will "fail." Scotland, on the other hand, adopts a more flexible and community-based approach, keeping opportunities open for longer, through modularized assessment, which may account for its slightly higher proficiency level (Bynner & Parsons, 2006)

The various studies reported in this book are limited to North America and Great Britain, with the exception of one chapter that encompassed the wider sphere of OECD countries. Our main focus, however, is on differences and similarities in the identified role of basic skills in the life course in the United Kingdom and the United States and what, if anything, in educational experience, makes a difference.

Such extensive examination of basic skills in just two industrialized countries sharing a common language can reveal both communalities in skills building in different societal contexts while also pointing to particular features of contrasting educational systems that work for or against skill acquisition. The U.S. education system is comprehensive in the sense that every family expects their child to enter kindergarten at the age of five and elementary school at the age of six, moving through the twelve grades of education, from elementary school to secondary school (high school), with graduation after grade 12, for the majority of students at age eighteen. Post-eighteen, equipped with a high school diploma, young people are encouraged to move into two- and four-year college and university degrees. Others leave education to enter the labor market, often before graduating from the stage they are at.

For a large number of young people who fail to graduate from high school, however, there are major obstacles to progression in the labor market and a consequent desire among a large proportion of those who drop out to return to "school" at some stage in their lives. A major opportunity for adults to do this comes through passing the General Education Development (GED) tests, preparation for which is offered through various Adult Basic Education and Adult Secondary Education courses operated by public schools, community colleges, and other agencies. The GED has historically functioned as the equivalent of high school graduation—though as one of our chapters shows, its equivalence in terms of labor market value may be questionable. Nevertheless, the GED provides a focus for adult education in the United States. Immigrants can also use this route to improve their prospects in the labor market and to gain access to higher education. In fact, the majority of those engaged in adult education classes in the United States as reported in the book were doing so for this reason.

Adult education (literacy, numeracy, and ESOL) programs in the United States are currently funded by a combination of federal and state programs. Federal support is provided under the Workforce Investment Act (WIA) of 1998, which coordinates federal programs for job training, vocational rehabilitation, and adult education and literacy. The Adult Education and Family Literacy Act (AEFLA), also known as Title II of WIA, authorizes federal appropriations to the states based on formulas involving the educational characteristics of their populations. States must match these federal funds with funds of their own to operate adult literacy and numeracy classes. A National Reporting System has been established under WIA that requires the states to collect and report data from local programs about participants' skill levels and improvements. The standards put into place by national policy and legislation here are thus specified as system *outputs*. Programs and states are accountable for the numbers of students served and their skill gains. There are no national *input* standards for the system, such as would occur if there were national curriculum or teacher-qualification standards within the adult-education system.

The situation in the United Kingdom could hardly be more different. Unlike the standard route to high school graduation through passing grade 12 of the general U.S. educational system, in the United Kingdom (but with significant variation between England and Wales, as opposed to Scotland) the equivalent to graduation are the school-leaving examinations, which are taken at 16, to gain the General Certificate of Secondary Education (GCSE). (We note in the United States a recent trend towards requiring students to pass state-specific competency examinations in order to obtain a standard high school diploma.) Even before this age, at thirteen to fourteen, young people will be directed into different curricula based on their prospects of achieving the goal of further progression in education, requiring a minimum of five GCSEs, from any number up to eight or nine, at grade A–C (the full range of grades is A–F)—designated in the *National Qualifications Framework* as "Level 2." Those who succeed in gaining five GCSE's at grade A–C are likely to continue onto the upper stage of secondary education (the "sixth form"), or to enter "Further Education colleges" or "Sixth Form colleges" devoted exclusively either to the higher level university entry-level qualification, A Level, or if on the alternative "vocational" route, to gain vocational qualifications with a view to entry into a skilled job.

The expectation is that those who succeed in A Levels (a minimum of two achieved) and those that do exceptionally well in vocational qualifications at the equivalent level (Level 3) will have the option to continue onto higher education to take a three-year university degree. In labor market terms, the degree carries particular value for earnings returns across the life course, although the lower level qualification 'A Levels' will also produce returns. Vocational qualifications below Level 3 are of somewhat mixed value depending on which occupational route the young person is on.

For those who leave school at sixteen and do not enter college to take full-time courses, other options available are apprenticeship, which involves employment coupled with one or two days a week doing vocational education, usually in a college. Those who do not want to register for the two-year apprenticeship, which is itself stratified between a general one-year and advanced one-year, may succeed in getting a job where training is offered by an employer or take part in a government training scheme. The others are likely to drift into casual unskilled work with periods out of the labor market altogether. Such young people constitute the problem category of much concern in British educational policy circles, young people not in education, employment and training for six months or more from the age of sixteen to eighteen (NEET).

A common feature of young people who leave the education system early at sixteen, entering casual unskilled work or leaving the labor market early (often in the case of young women, to start a family), is poor basic skills. In the light of evidence, especially from the IALS survey and a series of studies conducted for the government-established Basic Skills Agency, when the new Labour government won office in 1997, one of its early priorities was to reduce the number of adults with poor basic skills. A government committee was established under the chairmanship of Lord Moser to review the problem and suggest solutions (DfEE, 1999). The consequence two years later was the initiation of a major government initiative to raise skills levels with a particular target those at the lowest level of literacy and numeracy skill, *Skills for Life*.

Prior to Moser there had been various literacy campaigns using volunteers to help adults who had reading difficulties to master this basic skill (Hamilton & Hillier, 2006). *Skills for Life* adopted the quite different approach of professionalizing tuition through raising the skills level of tutors to that of a trained teacher, that is, a three-year degree, and supplying a standard curriculum with targets for adult learners to achieve. Initially the target was Level 2—the equivalent of a GCSE, A–C in literacy terms. Subsequently there was recognition of the need to target also those who were at a far lower level than this. The national qualifications framework in fact offered three levels at which targets were to be set: Level 2 (GCSE A–C equivalent), Level 1 (elementary qualifications), Entry Level (prequalification level). Within Entry Level itself there were three further divisions reflecting the incremental steps in progression that needed to be achieved: Entry Level 3 (the highest), Entry Level 2, and Entry Level 1 or below (the lowest). A national survey of adults, known as the "Skills for Life Survey of Basic Skills Needs," identified 15 percent of the population at Entry Level, 30 percent at Level 1, and 55 percent at Level 2 or above.

Skills for Life provision is offered in a range of contexts including colleges of the kind discussed earlier, not far removed from the American community colleges in what they are attempting to do. The traditional adult education institutes, where nonvocational courses were offered and the early basics skills education was delivered, are now largely absorbed into the Further Education colleges within the vocationally oriented *Skills for Life* program. There are also various forms of community provision, especially in Scotland. In England these can include local community centers and venues like libraries ("Learn Direct") and job centers ("Job Centre Plus"), where unemployed people can seek advice and be helped to find employment. Some of this community provision that may not fit strictly the *Skills for Life* progression criteria for funding may be supported by the Local Council with a view to linking the learner to the national program later. Other locations for basic skills provision include the workplace, with a major campaign directed at employers, to raise the skills levels of their employees, and other more specialized venues, such as prisons and young offender institutions.

Taken overall, the main distinction then between the U.S. and UK systems relates to the open nature of the American system, with extended education expected up to age eighteen for all and continuation to college for most who complete high school. In contrast, the British system assumes that many young people will leave school at the statutory minimum age, sixteen, if work is available, or to do vocational qualifications, apprenticeships, and so on. Despite the Level 2 target, *Skills for Life* provision is therefore likely to be taken up typically by adults who have left the system at the earliest stage, many of whom have little or no accreditation and a large proportion of whom have very poor literacy or numeracy. In the United States, on the other hand, there is a national GED testing system but no national curriculum or teacher qualifications for adult literacy and numeracy education.

METHODOLOGICAL CONCEPTS AND ISSUES IN LONGITUDINAL STUDIES

As the reader peruses the chapters in this volume, it will become clear that conducting quality longitudinal studies of adult learners can be a daunting task. The researchers' accounts indicate they encountered many challenging problems in designing, implementing, and analyzing data in their longitudinal studies. It may be helpful to orient the reader here to a few of the key methodological problems and concepts that will arise in the chapters. There is not space, of course, to provide full coverage and technical background for these issues—for that, we will refer interested readers to other sources for more complete information about these topics.

We will mention three recurrent problems or issues that often arise in longitudinal studies: (1) sample attrition and missing data; (2) constructing measures and scales that are longitudinally stable; and (3) methods for analyzing change in longitudinal data. A few general references that are not highly technical but provide broad coverage of these and other important issues are Magnusson, Bergman, Rudinger, and Torestad's (1994) *Problems and Methods in Longitudinal Research: Stability and Change;* Saldaña's (2003) *Longitudinal Qualitative Research: Analyzing Change Through Time*; and Singer and Willett's (2003) *Applied Longitudinal Data Analysis: Modeling Change and Event Occurrence*. Additional references will be suggested below for in-depth treatment of particular methodological issues. Our discussion here is focused on quantitatively oriented longitudinal studies because most of the chapters in this volume are primarily quantitative in nature, though several of the studies employ mixed methods. There clearly needs to be more qualitative longitudinal research in this field. The Saldaña (2003) reference provides additional material about qualitative approaches in longitudinal studies in general.

SAMPLE ATTRITION AND MISSING DATA

A recurrent problem in longitudinal studies that follow adult learners over time is that they lose track of and/or fail to complete follow-up interviews and assessments with many of the individuals in their original sample. This problem is often referred to as sample attrition or lack of sample retention. All of the studies reported in this volume encountered this problem to a greater or lesser extent. Sample retention rates vary widely and tend to increase with longer intervals between initial and follow-up interviews and with increasing numbers of follow-up interviews being administered to participants. One of the longitudinal studies had such sample-attrition problems that it was prematurely terminated by its funding agency. In other studies, researchers were able to complete their studies but cautioned about the validity of their findings given the relatively high rates of sample attrition they experienced.

There are many causes, of course, of sample attrition. Sometimes studies lose contact with their participants and are unable to locate them for follow-up interviews. Other participants die, become seriously ill or unwilling to continue participating in the study for various reasons. In some studies, participants initially are recruited while attending a basic skills program but leave the program before follow-up interviews are conducted, and their motivation decreases for continuing in the study. With careful attention, many of these problems can be overcome or at least ameliorated, as attested to by the low rates of sample attrition in some of the reported studies. A practical guide is provided by Strawn, Lopez, and Setzler (2000, 2007).

Sample attrition can be methodologically problematic for several reasons. First, the reduced sample size can become too small for reliable statistical estimation of important outcomes. This is primarily more a function of the absolute size of the reduced sample than of the percentage of the original sample lost to attrition. Sufficiently large initial samples may have adequate follow-up sample sizes even after substantial attrition. Often a more important issue is the potential *bias* introduced by sample attrition. To what extent are individuals who remain in the sample after attrition representative of the original population? To address this concern, researchers often report not only the percentage sample attrition but statistically test whether key characteristics of individuals retained in the sample differ from those of the original sample (e.g., gender, education, literacy level). We are often asked what is a reasonable sample retention rate for a longitudinal study in adult education. There is no simple guideline to provide here. When follow-up sample sizes are sufficiently large, the important question has to do with attrition bias that may be present and the analytical capacity to deal with such bias.

Even when evidence of attrition bias is present, statistical modeling techniques may be available to satisfactorily work with the longitudinal data at hand. The seminal work of Little and Rubin (1987) provides systematic theory and methods for evaluating the mechanisms producing missing data and analytical methods that can validly compensate for missing data under certain assumptions about the mechanisms producing the missing data. In the case of longitudinal studies, missing data can result from sample attrition or from other factors that make certain data items missing even when subjects are interviewed (e.g., a respondent refuses to answer a particular question). Little and Rubin (as refined by Little, 1995) classify missingness as *Missing Completely at Random (MCAR)*, *Covariate-Dependent Dropout (CDD)*, *Missing at Random (MAR)*, or *informatively missing*. Without going into technical details here, we note that analytical techniques can compensate for missingness under the first three types of missingness, called *ignorable nonresponse* by Laird (1988). To use these techniques validly, however, researchers must first demonstrate that their missing data are indeed ignorable. Usually the easiest ignorability condition to satisfy is MAR. Technical algorithms for doing this with longitudinal data sets are provided by Diggle, Heagerty, Liang, and Zeger (2002, Chapter 13).

When missingness is nonignorable, serious methodological problems arise in trying to draw generalizable conclusions from the longitudinal data that are present. A common example of nonignorable missingness in longitudinal studies of adult education occurs when adults attending basic skills programs are followed over time to study attendance patterns and their impact on learning outcomes. In many such studies, students who drop out of the programs also tend to drop out of the longitudinal research project (Beder, 1999). This makes the missingness of follow-up data "informative" and thus nonignorable with respect to the dependent variables of interest. In such cases, complex statistical models may be needed that jointly model the nonresponse (i.e., missingness) of participants together with the distribution of the dependent variables.

CONSTRUCTING MEASURES AND SCALES THAT ARE LONGITUDINALLY STABLE

Many longitudinal studies in this volume seek to measure changes in complex constructs such as adults' literacy and numeracy proficiencies, literacy and numeracy practices, attitudes towards learning and programs, and so forth. Developing psychometrically sound measures and scales for such constructs can be challenging even for cross-sectional studies in which measurements are taken only on single occasions. Additional challenges are posed in developing such measures for longitudinal studies in which a measure is to be repeatedly gathered for purposes of assessing change in the constructs over time.

An important concern in constructing scales appropriate for measuring longitudinal change is that they have a *longitudinally stable measurement component*. Traditional psychometric criteria for constructing scales from numerous individual questionnaire items can lead to satisfactory instruments for use on a particular occasion that are not valid as repeated measures for detecting change over time. Purcell-Gates and colleagues (2004), for example, constructed a scale of engagement in literacy practices from a large number of individual questionnaire items about individuals' frequency of use of specific literacy materials and tasks. Using item response theory (IRT) to scale these items, they found a high degree of fit of the items to an IRT scaling model and excellent psychometric properties. Yet when the same items were subsequently administered to subjects as part of a follow-up interview, the responses could not be adequately scaled with the same item parameters, and thus the scale scores on the two occasions could not be compared in order to assess change: the same construct was not being measured on the two occasions. This is an example of lack of measurement stability over time.

Even when using a standardized test with well-developed psychometric properties, such as a standardized literacy proficiency assessment, it is important to verify (rather than assume) that the instrument has a stable measurement component over time. With an IRT-based proficiency assessment, for example, the stability of item parameters over time can be judged (given sufficient data). If item parameters are not stable, then it is difficult to interpret observed changes in overall (i.e., IRT-scaled) proficiency scores.

Often factor analytic methods are used to identify latent variables (or factors) within subjects' responses to sets of questionnaire items. Structural equation models (SEMs) offer important extensions of classical confirmatory factor analysis to the longitudinal modeling context. Working with SEMs, analysts are able to specify separately the structural and measurement components of various models of interest. In the longitudinal context, analysts can statistically test important equality constraints among the loadings of variables at different time points within the measurement component. Where such equality constraints cannot be statistically rejected, the latent variables (scales) are said to have longitudinally stable measurement properties. Farrell (1994) provides some relevant theory and examples for such work with SEMs in panel studies.

METHODS FOR ANALYZING CHANGE IN LONGITUDINAL DATA

Longitudinal studies in adult education offer the potential for understanding the processes of literacy and numeracy development over the life span, the consequences of those changes for other aspects of adult life, and some of the factors and interventions that influence the changes that take place. The careful design and implementation of such longitudinal studies are paramount; the analysis of the collected data is obviously very important as well.

An important motivation for conducting longitudinal studies is to be able to identify causal relationships between certain experiences (e.g., attending a basic skills program) and observed changes (e.g., improved basic skills, increased earnings). Although correlational evidence may often suggest causal relationships, it rarely provides conclusive evidence. Within longitudinal studies, additional evidence of *temporal ordering* (such that changes in one variable precede changes in another variable) and the *inclusion of all relevant variables* in the analysis are generally required to provide persuasive evidence of causality. When key explanatory variables are omitted from an analysis of the relationship between two variables of interest, it is inappropriate to conclude that correlation implies causality. An omitted variable correlated with each of the two variables could be responsible (rather than a causal link) for their observed correlation. The difficulty, of course, is in demonstrating that all relevant variables have been included. This is why random assignment of subjects to experimentally controlled conditions is seen as the "gold standard" for identifying causal relationships: with sufficient numbers of individuals randomly assigned to various conditions, the effects of *all* omitted variables are effectively controlled.

Experimental controls, however, are often difficult to implement effectively with adults. Only one of the studies in this volume attempted to do so. Other techniques are used in data analysis to examine potential causal relationships among key variables of interest. Repeated-measures analysis of variance and regression-based methods are the most commonly used methods in the studies within this volume. These methods can be difficult to use in longitudinal studies because they typically require *balanced* data, that is, the same number of observations on each variable in the analysis. Because of the prevalence of missing values in longitudinal studies (see earlier), this requirement typically results in *listwise deletion* of data, that is, eliminating any subject from the analytic data set who does not have balanced or complete data on all relevant variables. Such listwise deletion reduces the sample size, sometimes drastically, and raises questions about the representativeness of the remaining balanced data. Alternatively, *imputation* techniques are used in some studies to create balanced data sets by statistically estimating and filling in the missing values (Rubin, 1987). One of the studies reported in this volume uses such techniques.

Other approaches to longitudinal data analysis are based on *structural equation modeling* (SEM). Structural equation models allow the researcher to specify a variety of complex relationships among both observed variables and latent variables. As noted previously, SEMs also allow researchers to separate structural components (specifying relationships among key latent variables) from measurement components in models, which is very important for confirming that the latent variables are being measured in the same way at various time points of a panel study. Another important feature of SEMs for modeling longitudinal data is their capacity to model cross-domain relationships such as the relationships between changes in basic skills and changes in other aspects of individuals' lives, such as changes in employment, in earnings, in uses of technology, and so on. There is a rich literature, numerous introductory texts, and software that are especially useful for panel data applications of SEMs (e.g., Bentler, 1995; Joreskog & Sorbom, 1993; Maruyama, 1998; Muthén & Muthén, 2006; Singer & Willett, 2003).

A widely used family of techniques for analyzing panel data is called *latent growth curve* models. Two of the studies in this volume use these techniques extensively. Growth curve models focus on the systematic properties of change over time in repeated measures such as literacy proficiency. Taking their name from the familiar graphs of children's changing heights and weights over time, growth curve models seek to fit growth curves from a selected family of functions (e.g., linear growth curves, quadratic growth curves, etc.) to *individual* changes in a measure over time. Optimal estimates of each individual's growth curve parameters can be made based on the entire longitudinal data set (not just the selected individual's data). In fitting linear growth curve models, for example, person-specific intercepts and slopes are estimated. In fitting quadratic growth curves, person-specific intercepts, slopes, and curvatures (i.e., rates of acceleration) are estimated. Techniques are also available for estimating how various time-independent and time-varying covariates influence the shapes of individuals' growth curves (i.e., the values of their growth curve parameters; Goldstein, 1995; Raudenbush & Bryk, 2002). Depending on the type of growth curve modeling involved, these individual parameter estimates are called *empirical Bayes* estimates (in multilevel models of growth curves) or factor scores (in SEM models of growth curves).

There are a number of different approaches to modeling growth curves. Two of the most commonly used ones are multilevel or hierarchical linear models (Goldstein, 1995; Raudenbush & Bryk, 2002) and SEM-based models of growth curves (Bollen & Curran, 2006). There are important overlaps between these two approaches to modeling growth curves, as well as some important differences between them (Willett & Sayer, 1994). The multilevel models do not require balanced data as do the SEM-based models. Since almost all panel data sets in practice have missing data, multilevel models are more easily applied to the full data sets that are collected. To utilize the SEM growth models with unbalanced data sets, one either has to utilize listwise deletion or imputation to create balanced data sets for analysis (see earlier). In all cases, it is advisable to test unbalanced data sets for their structure of missingness as described previously, though this seems rarely to be done in practice. SEM-based models, on the other hand, can be constructed with a broader range of error covariance structures for growth curve modeling than their multilevel counterparts can. There are thus tradeoffs between using the two approaches to growth curve modeling.

ORGANIZATION OF THE BOOK

Although most of the longitudinal studies reported in the volume focus on the impact of program participation on literacy and numeracy development, some of the studies consider skills growth and program impact within a broader focus on changes in adults' lives. In these studies, participation in programs is but one of the life experiences potentially linked to basic skills development over time. Both qualitative studies and surveys collecting broad information about adults' activities and life experiences provide lenses for examining the influence of program participation and other activities on the development of adults' basic skills, attitudes, and other attributes. A few of these studies include both program participants and nonparticipants, whose systematic comparison can deepen our understanding of the role of participation.

The longitudinal studies reported in the book are organized into four sections. The chapters in the first section consider findings from several broad longitudinal studies of adults over time. Both those participating in basic skills programs and those not participating are part of the studied populations, and many social and economic dimensions of life changes are explored in addition to education and skills progression. The second section comprises longitudinal studies specifically of students, teachers, and classrooms in adult basic skills programs. Chapters in the third section present a range of longitudinal studies that examine the impact of such instructional programs (and of the policy environment within which they operate) on adult students' learning outcomes. The fourth section includes a variety of longitudinal studies that examine the broader impact of programs on individuals' social and economic outcomes.

PART I: LITERACY AND NUMERACY DEVELOPMENT

Several longitudinal studies reported in this section look broadly at changes in adults' lives and how literacy and numeracy abilities are constitutive of those changes. Because these studies generally follow a broad segment of the adult population over time, rather than just those adults who choose to participate in basic skills programs, they provide an important lens for viewing the impact of programs and policies on basic skill development and other aspects of adults' lives.

John Bynner and Samantha Parsons write about their research based on the experiences of a large cohort that has been followed from birth through adult life in the United Kingdom. In their chapter, they describe how individual differences in literacy and numeracy abilities, already evident in measures taken during childhood, are relatively stable across the life span and exert profound influences on a range of educational, economic, and family outcomes in adult life. By comparing the sizes of the regression coefficients between basic skills and various indicators of social and economic status at different ages (while controlling for differences in many other variables), Bynner and Parsons argue that many changes in adult life are in fact mediated by earlier skills acquisition. They suggest that greater emphasis be concentrated on supporting basic skills development in the early childhood years and that additional research examine the smaller but still formative role of various life experiences between the ages of sixteen and thirty on literacy and numeracy development, especially employment opportunities and experiences.

Stephen Reder presents some findings from a long-term longitudinal study of a population of high school dropouts in the United States. Repeated measures of both *proficiencies* and *practices* in literacy and numeracy provide data that are analyzed by growth curve methodology in order to understand basic skill development in adults. Although proficiency and practices measures are correlated and both change systematically in adult life, Reder's analyses indicate these measures are subject to different dynamics of change. Participation in basic skills programs, for example, has a clear impact on measures of literacy and numeracy practices but not on proficiency measures. Pronounced effects of age were found on proficiency changes, such that younger adults tend to gain proficiency (through their early 30s), whereas older adults tend to lose proficiency over time. Reder suggests that the design and evaluation of basic skills programs be based on improvements in everyday literacy and numeracy practices rather than on just proficiency growth (the prevailing policy in the United States). He also suggests that new types of programs and policies may be needed to meet the skill development needs of our aging populations and workforces. Reder's research indicates that literacy and numeracy growth in adulthood is the byproduct of ongoing developmental processes as well as life-event-driven changes. He suggests that future research should examine the key role of workplace experiences in mediating literacy and numeracy development.

T. Scott Murray presents results from several longitudinal studies in his chapter, conducted with various populations of school-aged children and adults in Canada and New Zealand. Murray's analyses of the data from these studies commonly focus on understanding processes of skill gain and loss over the life course. Murray's analysis of a longitudinal study of skill development in New Zealand primary and secondary students is consistent with Bynner and Parson's finding that individual differences in basic skills are laid down early in childhood and remain relatively stable across development. His "synthetic cohort" analyses of the Canadian data sets from two large international adult literacy surveys attempt to reconstruct longitudinal changes that took place historically by comparing distributions of skills in different age groups at fixed points in time. These analyses indicate that individuals' skill gains and losses vary systematically with a number of factors, including age and socioeconomic status. Murray's finding that there is more skill loss in older populations is generally consistent with Reder's finding based on direct measurement of proficiency changes (note that Murray and Reder analyze data based on quite similar proficiency measures). Murray also concludes that skill losses are highly concentrated among individuals in the lower socioeconomic quartile and among individuals with initially lower levels of skill. Looking at parallel data from a number of countries, Murray concludes from these inferred patterns of skill gain/loss that literacy and numeracy skills are the single most important determinant of societal economic growth since World War II. He argues that new policies and investments to support adult literacy and numeracy development will have larger economic returns.

All of these studies find that literacy and numeracy do indeed continue to develop across the life span. A variety of contextual, economic, and cognitive factors systematically influences these developmental processes. Although much differentiation of literacy and numeracy abilities occurs during childhood, additional variance occurs during adult life. Understanding the life events and activities shaping this adult development is important for policy and program development. The longitudinal studies in the sections following consider the effects of program participation on adult basic skills, one promising place to look for leverage through increased investment and policy attention. In addition, the contributions in this first section by Bynner and Parsons, by Reder, and by Murray each point to the need to look more closely at the formative role of work in mediating the growth of literacy and numeracy in adult life.

PART II: STUDENT, TEACHER, AND CLASSROOM STUDIES

The chapters in this section look closely at variations in literacy and numeracy instruction in classrooms and tutoring programs. The authors attempt, through a variety of analytical methods, to identify particularly effective instructional techniques and approaches based on differences in learners' longitudinal outcomes. Both qualitative differences among program features and quantitative differences in student persistence in programs are considered.

Judith Alamprese reports on a longitudinal study of low-level students in Adult Basic Education programs across the United States. Students in this study were taken from 130 programs and had decoding skills in the 0–3rd grade range as measured by a standardized reading test. A number of students' reading subskills were assessed two or three times as they entered, continued to participate in, or after they left the programs. Observers carefully noted, in this "what works" study, characteristics of the classroom instruction the students encountered. Multivariate analyses were then used to identify the student characteristics and instructional factors associated with reading subskill gains. The strongest predictor of reading gains was pretest score. This finding is reminiscent of Murray, who, we recall, looked at skill gains on a proficiency measure within a broader adult population and found that gains over time were smallest among adults already in the lowest quartile of skill. Whenever such studies find strong relationships between the initial value of a measure and its subsequent gain over time, growth curve analyses can be quite helpful for better understanding of the relationships between initial status and rate of change. Alamprese discusses a range of implications of her research for the delivery of services to low-level adults, including implications for assessment, program management, and professional development of teachers.

Another "what works" study is reported in the chapter by Larry Condelli, Heide Wrigley, and Kwang Yoon. Their longitudinal study was parallel to Alamprese's but focused on instruction for adult ESL "literacy" students—ESL students who lack basic literacy skills in any language and who have minimal proficiency in English. The two "what works" studies were funded by the U.S. Department of Education to address similar questions in different components of the federally funded basic skills programs for adults, Adult Basic Education and English as a Second Language. These common questions had to do with which instructional practices "work," that is, produce more learning. Condelli and colleagues demonstrate that instructional practices are related to student learning. Using growth curve modeling, they identified three instructional strategies that are related to literacy and language learning: connection to the outside world, use of the student's native language for clarification in instruction, and varied practice and interaction. They discuss the range of specific teaching activities encompassed by these three strategies and analyze how student characteristics influence the learning process. They remind us that despite the clear relationships among these variables and learning outcomes, they cannot determine which specific instructional activities will *cause* basic reading skills to improve. They argue that only random assignment experiments contrasting the learning outcomes associated with particular instructional practices can support such causal inferences.

Although both of these studies found that the amount of time that students spend attending basic skills classes is positively associated with improved learning outcomes, neither study found that students attend the programs for substantial periods of time. A recent representative national survey of programs and learners in the United States found that students attend basic skills classes for an average of only eighty to one hundred hours (depending on the program type). John Comings and colleagues have conducted research on student attendance in adult-education classes and the factors associated with attendance. Comings, Parella, and Soricone (1999) proposed the term *persistence* to describe adult students' engagement in self-directed study or distance education between periods of attendance in face-to-face programs. Reder and Strawn (2001, 2006) extended this idea to many adults who they found self-study to improve their reading, writing, and math skills but *never* attend any basic skills classes. In his chapter in this volume, Comings reviews research on persistence with special attention to the National Center for the Study of Adult Learning and Literacy's (NCSALL) longitudinal Persistence Study. He synthesizes the research into a systematic theory of persistence and suggests ways in which programs can better support learning through three chronological phases of participation: Entrance into Services, Participation in Services, and Reengagement in Learning.

All of the chapters mentioned thus far in this section—Alamprese, Condelli et al., and Comings—acknowledge the important role that teachers play in students' learning, though none of these studies was designed to focus on teacher variables systematically. Olga Cara and Jennifer Litster examine adult-education teachers more closely in their chapter about some early findings from the NRDC Teachers Study. They focus on the longitudinal formation of teachers' attitudes towards innovations in policy and programs in adult literacy and numeracy. Understanding how teachers respond to such innovations, of course, is central to planning effective future initiatives. Cara and Litster consider the formation of teachers' attitudes towards the national *Skills for Life* strategy in England. Although the longitudinal Teachers Study was still in progress at the time their chapter was written, Cara and Litster identified a number of key findings based on the first two waves of attitudinal data. With many other variables controlled by multivariate regression, they found that positive attitudes in teachers towards *Skills for Life* are associated with participation in professional development activities, with being change agents within peer networks, and with understanding new credentialing requirements. Cara and Litster conclude that teachers' responses to innovations may depend less on *what* those changes are and more on whether they feel ownership over the changes and continued empowerment and autonomy to help improve learners' lives.

A companion study to the Teachers Study is the NRDC Learners Study. Just as the Teachers Study, described earlier, examines the impact of the British government's *Skills for Life* initiative on teachers, the Learners Study investigates its impact upon adult learners. John Vorhaus, Ursula Howard, Yvon Appleby, Ann Marie Bathmaker, and Greg Brooks write about some of the initial findings of the Learners Study, which consisted of multiple strands of data and analysis about the characteristics of learners, their participation in *Skills for Life* programs, and their achievements. A common focus across the longitudinally focused research strands was how the evolving infrastructures of *Skills for Life* related to the changing characteristics and experiences of successive cohorts of learners. Their chapter describes the methods used, descriptive summaries of the data collected, and the methodological problems encountered in attempting to capture the impact of a major government initiative upon a diverse population of adult literacy and numeracy learners.

PART III: THE IMPACT OF POLICY AND PROGRAMS

The previous section described research focusing on the processes and outcomes of learning in the context of the student-teacher relationship. In this section we continue to pursue this issue but more broadly in terms of the impact of teaching programs at different levels and the implications this has for improving policy. An important methodological issue arises that we need to return to again. The idea of impact suggests some direct causal relationship between a particular intervention/treatment as applied in this case to individual learners on skills proficiency and on a given outcome in the labor market or in the workplace, in relation to the criminal justice system or for personal development. Establishing causality is the province of research design where the randomized controlled experiment reigns supreme as supplying the most robust basis for causal inference. Only one of the studies reported here embarked on a course involving such a design, and even then the design could not be sustained. Nevertheless, the other studies provide in their different ways powerful indicators of the impact programs have, relying on statistical forms of control and multiple sources of information triangulated against each other in place of randomization.

The projects reported in this section are important in relation to the insights they give on the way major political priorities, as enshrined in this case in the *Skills for Life* program in England, were reflected in specific teaching programs and projects with a major basic skills component. They also show how these fitted into what was not a fixed or static national program for evaluation but, as is typically the case, a continually evolving policy field. For example, subsequent to the research reported here being undertaken, a major government report in Britain—the *Leitch Report*—was published on raising skills levels generally in the population in which enhancement of literacy and numeracy language was presented as the first step (HM Treasury, 2006). Connections from basic literacy and numeracy enhancement to the vocational programs of skills enhancement set out in this highly influential report, and endorsed completely by government, will themselves be likely to impact further on the way that basic skills themselves are taught. Supported strongly by research (Casey et al., 2006), the idea of "embedding" basic skills teaching in vocational education and other programs, rather than treating it as relatively independent of the rest of the curriculum, is increasingly gaining momentum. This recognizes the motivational value and effectiveness of provision that map directly into the vocational and other goals that learners are pursuing.

The three studies reported comprise one by Hilary Metcalf and Pam Meadows on the major evaluation undertaken of *Skills for Life* as a whole in raising literacy levels and achieving other outcomes based on follow-up surveys of identified *Skills for Life* learners and the general population sample used in the 2002 *Skills for Life* Survey. In their project on *Skills for Life* in the workplace, Karen Evans and colleagues report research conducted in one of the target areas for *Skills for Life,* as reflected in the high priority given in the *Moser Report* to work-based learning. Despite the high prevalence of low-skilled individuals in many workplaces, few employers, until the publication of the Moser report, had taken the problem at all seriously. Skills for Life placed major emphasis on recruiting employers to the goal of enhancing their workers' skills and provided support in terms of curricula and teaching, as appropriate, or supporting indigenous programs within workplaces. The project focused on workers in a number of workplaces, examples of which were very difficult to build up into a list because of the relative rarity of the provision. Mixed methodology was employed using quantitative methods to assess progress, employing tests developed especially for the purpose and qualitative methods to investigate workers' experience of the program in detail. The third project reported here was directed at another specialized population—young people in the criminal justice system—where again there is a high prevalence of poor literacy and numeracy reflecting the very low educational achievements of large numbers of young offenders. Young offenders are located in institutions known as the "secure estate" or are engaged as part of their sentence in community-based education and training. The project started with the aim of undertaking a randomized control trial in which the new *Skills for Life* curriculum, adapted especially for the age group of young offenders, would be offered to individuals selected at random from a total group. For a number of reasons detailed in the chapter, which are of great interest, such a design could not be implemented, and a survey without randomization supported by qualitative methods had to be used instead.

Overall, the three projects reveal both problems with the assumptions on which so much intervention through government programs is based and the complexity of experiences that have to be captured in any research attempting to understand impact. They also provide important lessons about the practical problems of maintaining strictly to the cannons of classic research design and how the necessary adjustments can reveal new insights that otherwise might be missed.

PART IV: SOCIAL AND ECONOMIC OUTCOMES IN CONTEXT

Adults normally engage in literacy and numeracy classes for a clearly defined purpose. Unless their motivation is high, they are unlikely to be persuaded to take part (even if their skills are poor) and when they do so often have great difficulty in "staying the course." Motivation is therefore a critical factor in the effectiveness of adult teaching, which may come from the effectiveness of the teaching itself and the feeling of self fulfillment it produces or, more frequently, the context of people's lives in which the decision to join a course emerges. Such contexts include the family, the community, and the workplace, and often it is changes in these situations that prompt interest in, and the decision to improve, the basic skills. Thus, a child starting school or a workplace change, including the possibility of redundancy, or taking on office in a local club, may all heighten interest in improving skills. In the absence of such motivational triggers, most adults are likely to feel that, as far as their own lives are concerned, they cope.

The set of studies reviewed here approaches these issues from a number of angles employing a range of resources and methods. Stephen Rose uses U.S. state administrative records to evaluate adult education programs focusing particularly on participation and what might explain it in terms of the different socioeconomic characteristics of students. The report by Mary Beth

Bingham on the Tennessee Longitudinal Study of Adult Literacy Programs again focuses on characteristics of students using both quantitative and qualitative methods in her longitudinal investigation of labor market and other student outcomes. The results lead her to question whether tests scores alone are adequate to determine what difference participation in a program makes. Barbara MacDonald and Patricia Scollay pursued similar questions in their longitudinal studies of a family learning program based in California, using qualitative and quantitative measures to explore the context, motivations, and outcomes for participants in terms of effects on their lives. A similar approach is adopted by Kathy Maclachlan, Lyn Tett, and Stuart Hall in Scotland, in a project initiated not by *Skills for Life,* which does not operate there, but by the Scottish parallel program. This emphasizes importantly the community base for adult education and for developing the basic skills at the heart of the local community. A final paper, by Yvon Appleby and David Barton, moves from the macro picture of the large-scale survey and "mixed methods," as reported in many studies here, to the "in-depth" examination of one particular life. Here the emphasis is on a theoretical framework for understanding ethnographically how individuals will engage in, and what they will gain from, participation in a *Skills for Life* course.

Overall, the chapters provide a rich array of approaches and findings that illuminate the contexts in which educational interventions directed at adults may succeed or fail. Ultimately the studies may all be seen as addressing the key question to be addressed by government initiatives and programs generally: What effect did the program have on participants' lives?

REFERENCES

Ashton, D., & Green, F. (1996). *Education, training and the global economy.* Cheltenham, UK: Edward Elgar.

Beder, H. (1999). *The outcomes and impacts of adult literacy education in the United States.* Report #6 of the National Center for the Study of Adult Learning and Literacy. Cambridge, MA: Harvard University Graduate School of Education.

Bentler, P. (1995). *EQS structural equations program manual.* Encino, CA: Multivariate Software.

Bollen, K. A., & Curran, P. J. (2006). *Latent curve models: A structural equation perspective.* Hoboken, NJ: John Wiley & Sons.

Bynner, J., & Parsons, S. (2006). *New light on literacy and numeracy.* NRDC Report: Institute of Education, London.

Casey, H., Cara, O., Eldred, J., et al. (2006). *You wouldn't expect a maths teacher to teach plastering . . . : Embedding literacy, language and numeracy in post-16 vocational programmes—the impact on learning and achievement.* London: NRDC.

Comings, J., Parella, A., & Soricone, L. (1999). *Persistence among adult basic education students in pre-GED classes.* NCSALL Reports #12, Cambridge, MA: NCSALL.

DfEE (Department for Education and Employment). (1999). *A fresh start: Improving adult literacy and numeracy.* Report of the Working Group, Sudbury: DfEE Publications.

Diggle, P. J., Heagerty, P., Liang, K.-Y., et al. (2002). *Analysis of longitudinal data* (2nd Edition). Oxford: Oxford University Press.

Farrell, A. D. (1994). Structural equation modeling with longitudinal data: Strategies for examining group differences and reciprocal relationships. Journal of Consulting and Clinical Psychology, 62, 477–87.

Goldstein, H. (1995). *Multilevel statistical models* (2nd ed.). New York: Halstead Press.

Hamilton, M., & Hillier, Y. (2006). *Changing Faces of Adult Literacy, Language and Numeracy.* Stoke on Trent, UK: Trentham Books.

HM Treasury. (2006). *Leitch review of skills. Prosperity for all in the global economy—world class skills,* http://www.hmtreasury.gov.uk/independent_reviews/leitch_review/review_leitch_index.cfm.

Jöreskog, K., & Sörbom, D. (1993). *LISREL 8: Structural equation modeling with the SIMPLIS command language* Hove and London: Scientific Software International.

Laird, N. M. (1988). Missing data in longitudinal studies. *Statistics in Medicine, 7,* 305–15.

Little, R. J. A. (1995). Modeling the dropout mechanism in repeated-measures studies. *Journal of the American Statistical Association, 90,* 1112–21.

Little, R. J. A., & Rubin, D. B. (1987). *Statistical analysis with missing data.* New York: John Wiley.

Magnusson, D., Bergman, L. R., Rudinger, G., et al. (Eds.) (1994). *Problems and methods in longitudinal research: Stability and change.* New York: Cambridge University Press.

Maruyama, G. M. (1998).*Basics of structural equation modeling.* Thousand Oaks, CA: Sage.

Murray, S., Kirsch, I., & Jenkins, L. (1998). *Adult literacy in OECD countries.* Washington, DC: National Center for Education Statistics, U.S. Department of Education, Office of Educational Improvement.

Muthén, L. K, & Muthén, B. O. (2006). *MPlus user's guide* (4th ed.). Los Angeles: Muthén & Muthén.

Purcell-Gates, V., Jacobson, E., & Degener, S. (2004). *Print literacy: Uniting cognitive and social practice theories.* Cambridge, MA: Harvard University Press.

Raudenbush, S. W., & Bryk, A. A. (2002). *Hierarchical linear models: Applications and data analysis methods,* 2nd ed. Thousand Oaks, CA: Sage.

Reder, S., & Strawn, C. (2001). Program participation and self-directed learning to improve basic skills. *Focus on Basics, 4*(D), 14–17.

Reder, S., & Strawn, C. (2006). Broadening the concepts of participation and program support. *Focus on Basics, 8*(C), 6–10.

Rifkin, J. (1995). *The end of work.* New York: Putnam.

Rubin, D. B. (1987). *Multiple imputation for non-response in surveys.* New York: Wiley.

Saldaña, J. (2003). *Longitudinal qualitative research: Analyzing change through time.* Lanham, MD: AltaMira Press.

Singer, J. D., & Willett, J. B. (2003). *Applied longitudinal data analysis: Modeling change and event occurrence.* Oxford & New York: Oxford University Press.

Strawn, C., Lopez, C., & Setzler, K. (2000, revised 2007). *It CAN be done: Sample retention methods used by the Longitudinal Study of Adult Learning.* Portland, OR: Portland State University. Retrieved March 24, 2008, from www.lsal.pdx.edu/instruments.html.

Willett, J., & Sayer, A. (1994). Using covariance structure analysis to detect correlates and predictors of individual change over time. *Psychological Bulletin,* 116(2), 363–80.

Part I

Literacy and Numeracy Development

1 Insights into Basic Skills from a UK Longitudinal Study

John Bynner and Samantha Parsons

BACKGROUND

The work reported here took place against the background of a major new initiative in Britain, both to understand and to tackle the problem of poor basic skills in a substantial minority of the adult population. Concerns were driven by the growing body of evidence that basic skills difficulties were a major impediment to successful functioning in modern society culminating in the work of the UK Government's Moser Committee[1] and the ongoing policy development that arose from it, *Skills for Life* (SfL). This included defining national standards for adult literacy and numeracy. These map the range of literacy and numeracy skills and capabilities that adults are assumed to need in order to function effectively in the workplace, in the family, and in the community. A separate set of standards has been produced for literacy and numeracy.

<u>**Literacy** covers the ability to</u>

- speak, listen, and respond
- read and comprehend
- write to communicate

<u>**Numeracy** covers the ability to</u>

- understand and use mathematical information
- calculate and manipulate mathematical information
- interpret results and communicate mathematical information

The national standards for adult literacy and numeracy, as set down in the National Qualifications Framework (NQF), are specified at three levels: Entry, Level 1 and Level 2. Entry is further divided into three sublevels: Entry 1, Entry 2, and Entry 3 to specify in detail the small steps required for adults to make progress in response to basic skills educational provision. Full details of the curriculum content expected to be mastered at each level are shown in Appendices 1.A.1 and 1.A.2.

An important part of the evidence considered by Moser was drawn from basic skills data collected for the UK Basic Skills Agency in what was then a twelve-year program of longitudinal research. The program was focused particularly on identifying the earlier circumstances and experiences, which were connected with later basic skills difficulties. This work was based on the 1958 and 1970 British birth-cohort studies, known, respectively, as the National Child Development Study (NCDS) and the 1970 British Cohort Study (BCS70), the second of which supplies the findings reported in this chapter. NCDS and BCS70 are longitudinal studies that follow up all babies born in Great Britain in a single week from birth in the year the study began to adulthood, with new data collected at regular intervals throughout the cohorts' lives. Much of the earlier work was based on findings from the literacy and numeracy objective assessments that were conducted on 10 percent representative subsamples of the cohorts, first at age twenty-one in BCS70 (1991) and later, at age thirty-seven in NCDS (1997).

As part of the SfL strategy, the National Research and Development Centre for Adult Literacy and Numeracy (NRDC) was established. This offered the opportunity to increase the potential of the cohort studies for basic skills research with a particular focus on the socioeconomic consequences in adulthood of poor acquisition of the basic skills. This was subsequently extended to "profiling" in terms of early circumstances and experience individuals of at the lowest (Entry) skills levels. In 2004 the latest follow-up surveys of NCDS and BCS70 funded by the Economic and Social Research Council were due to take place. With additional funding from NRDC, new literacy and numeracy assessments were completed by all BCS70 cohort members at age thirty-four. Assessment of symptoms associated with dyslexia was also included. In addition, funding from the European Social Fund (ESF) supported a study of intergenerational continuities in basic skills acquisition, through an assessment of the reading and mathematical skills of all resident natural or adopted children from a randomly selected one in two sample of cohort members.[2]

This policy thrust in basic skills assessment in the birth cohorts both maps into and is informed by the broader life-course perspective (Elder, 1998; Heinz, 1991) to which design and analysis in the studies are directed (Bynner, Butler, Ferri, Shepherd & Smith, 2000). Within this conceptual framework the acquisition of literacy and numeracy skills supports the development of personal agency in building personal resources that will enable the individual to progress along the pathways toward achievement and fulfilling outcomes in the different domains of adult life: education, employment, family, citizenship. These pathways are shaped by the culturally based institutional frameworks in place to mediate and moderate the transitions to outcomes in adult life involved, including entry to school, to work, to partnership, to parenthood, to active citizenship and are themselves changing in response to economic and social change.

A particular concern in modern industrial societies is the impact of economic transformation on the nature of employment and employability in a labor market that places an increasing premium on skills and qualifications. In these terms individuals lacking the core skills of literacy, numeracy, and increasingly information and communication technology (ICT) face the risk of "social exclusion": that is to say, adult life marked by restricted life chances typified by casual unfulfilling work and unemployment and often accompanied by psychological and physical health problems, drug and alcohol abuse (Bynner, 2004). A major aim of SfL strategy, taken forward in 2007 by the report of the government-commissioned Leitch Committee, *Prosperity for All in the Global Economy,* is to find the means of bridging the skills gap.[3]

Experience in the family and in the early stages of life can work with, or be out of step with, schooling and educational progress, as can later experience in the workplace and the community. Skills supply the basic protective resources on which successful achievement in adult life is likely to be based, and at the core of these resources lie literacy and numeracy without which progress is likely to be impeded. Thus, adults whose developmental pathways to adult life are characterized by poor literacy and poor numeracy face increasing risk of marginalization and social exclusion. The main route to this status of particular concern to British policymakers is for young people not to be in education, employment, or training (NEET) in the critical period of the late teens (Bynner & Parsons, 2002).

THE BIRTH-COHORT STUDIES

As we have made clear, the study of basic skills has been a prominent feature of the major longitudinal studies carried out in Britain for some time (for an overview, see Bynner & Joshi, 2007). The Birth Cohort Study Series began in 1946 with a one-third sample of all babies born in a particular week and followed up through adult life. The model was replicated with further birth-cohort studies, one of which is the focus of this chapter, embracing the whole sample of one week's births in 1958 and 1970 with about 17,000 individuals initially in each study. Following the collection of comprehensive information surrounding the circumstances of birth, data were collected in a series of follow-up surveys ("sweeps") at ages seven, eleven, sixteen, twenty-three, thirty-three, forty-six (1958 cohort) and at ages five, ten, sixteen, twenty-six, thirty, thirty-four (1970 cohort). There was then a gap of thirty years before the most recent national birth-cohort study began in the millennium year, this time based on a whole year's births and starting with a sample of nineteen thousand stratified to overrepresent those growing up in disadvantaged circumstances and in areas with high concentrations of ethnic minorities.

Response has held up remarkably well across the different sweeps in each of the studies (Plewis, Calderwood, Hawkes, & Nathan, 2004). Thus, in the year 2000, when the 1958 cohort had reached age forty-two and the 1970 cohort, age thirty, response rates based on eligible members of the original cohort (i.e. excluding deaths and emigrants) were 73 percent and 70 percent, respectively.

In the 2004 BCS70 survey the response rate dropped significantly through unanticipated mobility of cohort members to 9,665 participants, 63 percent. Small biases are evident, with females and the more educated slightly more likely to stay in the study than their male and less educated counterparts, but overall the participating samples remain representative of the cohort at birth.

Design

Each of the studies adopts the life-course perspective attempting to embrace holistically the course of human development with data collected relevant to the particular stage of life reached as shown in Figure 1.1.

In the early years the focus was on home circumstances, parental attitudes and behaviors, and the behavioral and cognitive development of the cohort members. Through the preschool and school years, educational achievement was brought into the picture, together with information collected about the context of schooling supplied by teachers. In the teenage years the cohort members themselves supplied the data, with, in the case of the 1970 cohort at age sixteen, extensive questionnaire coverage of all facets of their educational and leisure lives. In adulthood the emphasis moved to participation in

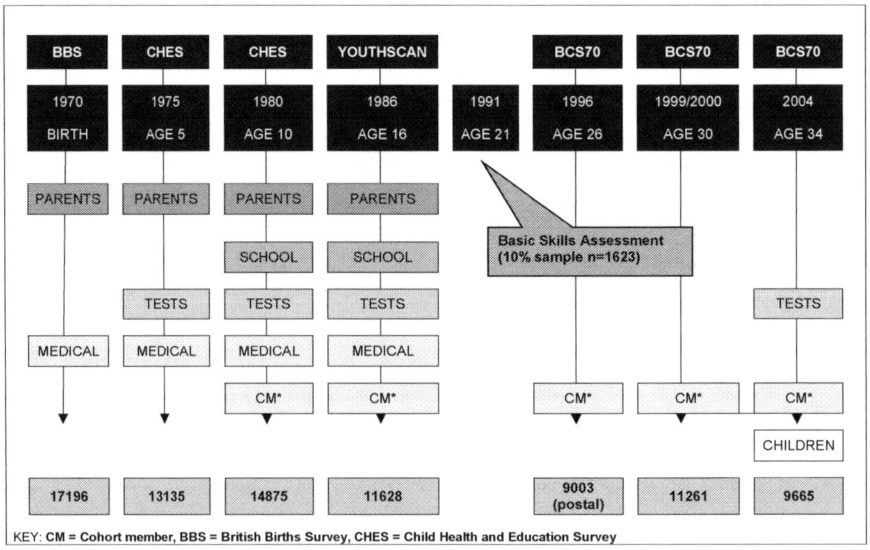

Figure 1.1 1970 cohort longitudinal design from 1970–2004.

the labor market and the other domains of adult life, including partnership, family formation, housing, and citizenship. Health, well-being and health-related behavior, have also been a recurrent theme throughout.

Although reading attainment and maths attainment assessed through educational and cognitive tests featured through the childhood years with measures taken at ages five, ten, sixteen (1970 cohort) and at ages seven, eleven, and sixteen (1958 cohort), the objective assessment of adult skills (as opposed to self-reports) did not begin until 1991, when a 10 percent sample of the 1970 cohort at age twenty-one years completed tests of functional literacy and numeracy (Bynner & Steedman, 1995; Ekinsmyth & Bynner, 1994). The exercise was repeated with a similarly designed test at age thirty-seven for a 10 percent representative sample of the 1958 cohort (Bynner & Parsons, 1997; Parsons & Bynner, 1998). As already noted, it was not until 2004, when the opportunity arose through the SfL program and the establishment of the NRDC, that a full assessment of literacy and numeracy in a whole cohort was possible—the 1970 cohort at age thirty-four.[4] The assessment, funded by NRDC, comprised forty-seven multiple-choice computer-administered test items at the different National Qualifications Framework (NQF) levels selected from the 2002–2003 National SfL Survey of Adult Basic Skills. This survey was carried out to assess the general public's basic skills problems (Williams, Clemens, Oleinikova & Tarvin, 2003). Full details of the BCS70 test are supplied in Parsons and Bynner (2005).

For purposes of continuity, the multiple-choice assessment was coupled with a second "open response" interview–administered assessment comprising test items used in the previous age twenty-one 1970 cohort study follow-up. Out of a one and a half hour interview the literacy and numeracy tests took on average twenty minutes to complete. Apart from the literacy and numeracy assessment, the survey also embraced cohort members' current employment, income, and family situation and history back to the previous survey in 2000 for employment, partnership and family formation and housing. Health and well-being and civic participation were also extensively covered.

Following the design of the 1958 cohort age thirty-three sweep, this survey also extended data collection to one-half of the cohort members' children, supplying an intergenerational data set with comparable measures of cohort members' basic skills when they were children to those of their own children (see Bynner & Parsons, 2006). This replicated in part the design of the 1958 cohort follow-up at age thirty-three, when one-third of the cohort members' children were assessed. Notably, the expansion of the sample from one-third, 1958 cohort, to one-half, 1970 cohort, proved necessary to compensate for the delayed child bearing of the more recent cohort.

Analysis approach

Embedding adult basic skills assessments in multipurpose longitudinal surveys of the birth-cohort studies kind has particular value, in enabling research to trace the origins of adult basic skills difficulties to circumstances and experiences in the early years. Through the wide range of data in all the domains of childhood and teenage and adult life, we can also extend analysis to the part the basic skills play in life-course construction as the critical mediators of the conditions of early life on subsequent pathways into and through adulthood. These possibilities have been fully exploited in a program of research lasting now over seventeen years since 1991, when the first adult basic skills assessments were made. They have now reached fruition through the comprehensive assessment of the whole 1970 cohort at age thirty-four. *New Light on Literacy and Numeracy* (Bynner & Parsons, 2006) and *Illuminating Disadvantage* (Parsons & Bynner, 2008) report the main descriptive findings.

In this chapter we exemplify the work through longitudinal analysis revisiting some of the early questions that stimulated the program in the first place. We focus on the measure of literacy proficiency as obtained from the multiple-choice test items. Aggregation of correct answers across these test items supplies a score that can be treated for the purposes of multivariate analysis as a continuous variable. At the same time the location of the items as the different levels of the SfL curriculum enables us to assess the level each cohort member has achieved. The distribution of the sample across these levels as used in the study was for men: Entry 2 and below 4 percent; Entry 3, 3 percent; Level 1, 31 percent; Level 2, 62 percent and for women: Entry 2 and below, 4 percent; Entry 3, 4 percent; Level 1, 30 percent; Level 2, 62 percent. Note that the numbers in the sample for Entry 1 were too small to assess them separately from Entry 2. Level 2 is very heterogeneous, encompassing potentially a range of proficiency from those performing at just the required literacy standard in the General Certificate of secondary Education (taken at age sixteen in England and Wales) to people with university degrees.

Our focus here is first on two potential outcomes of basic skills proficiency: employment and family formation as measured annually from age sixteen to thirty-four. Secondly, we turn to exploratory life-course analysis with the aim of elucidating factors potentially influencing, at different life-course stages, the development of basic skills proficiency as assessed at age thirty-four. To what extent can basic skills difficulties, persisting through to adulthood, be "explained" in terms of early circumstances and experiences and what are the key components of influence that might lead to poor performance? Is there evidence of a "disadvantaged career" characterized by poor basic skills and obstacles to educational achievement compounded by lack of opportunity and resources, early school leaving without any qualifications, and the prospect of further marginalization in the labor market and in family life? How does such an analysis enhance our understanding of the origins of poor basic skills manifested in adulthood and in their connection with outcomes in employment and family life? What message, if any, does this hold for policy? We start by examining two outcomes in adult life—employment and parenthood rates at different ages, then some of the relationships between key antecedent variables and literacy, and numeracy performance in adulthood. We then combine the key predictors identified from birth up to age thirty in a multivariate analysis using the statistical technique of multiple regression as an aid to understanding the relative influence of these experiences and attributes on literacy acquisition by age thirty-four.

At a general level most of the findings for numeracy replicate those for literacy, so for simplicity we present findings just for literacy. They should be viewed as illustrative rather than definitive in a program of analysis that is continuing. Our aim is to demonstrate some of the possible consequences of poor literacy in adult life and, in broad outline, the kinds of circumstances and experiences at different life-course stages associated with them. In what follows we focus on the adult literacy assessments at age thirty-four and draw into the multiple regression analysis the data for selected variables measured earlier (up to age thirty-four) in the different sweeps.

LABOR MARKET AND FAMILY OUTCOMES

Employment

The 1970 cohort-study members were one of the early cohorts of school leavers to feel the brunt of the ICT revolution beginning in the late 1970s and the economic transformation and globalization of the economy that accompanied it. Although opportunities for unskilled work were beginning to contract during the 1970s in Britain, few of those who left school at the minimum age of sixteen failed to get a job. (e.g. 62 percent of the NCDS cohort members in 1974 moved directly into jobs, including, in the case of boys, about a quarter into apprenticeships). By the time the 1970 cohort reached age sixteen (in 1986), the youth labor market was, in the words of Ashton and Maguire (1983), "vanishing." In place of jobs, young leavers were likely to face the prospect of casual work or a one- to two-year low-quality youth training scheme (YTS) as part of the national training program directed at early school leavers. Others became unemployed or, in the case of young women, often made an exit from the labor market at the earliest opportunity, to enter partnerships and have children. In this situation, staying on to gain qualifications became an increasingly significant option for assuring entry to employment. Hence, literacy and numeracy skills on which the qualification route was founded rose up the scale of importance for employability.

A striking indication of the phenomenon is shown by graphing by literacy level the proportion of young men and young women in the 1970 cohort who had spent the majority of each year from the time of their sixteenth birthday up to the end of their thirty-third year in full-time employment (Figures 1.2 and 1.3).

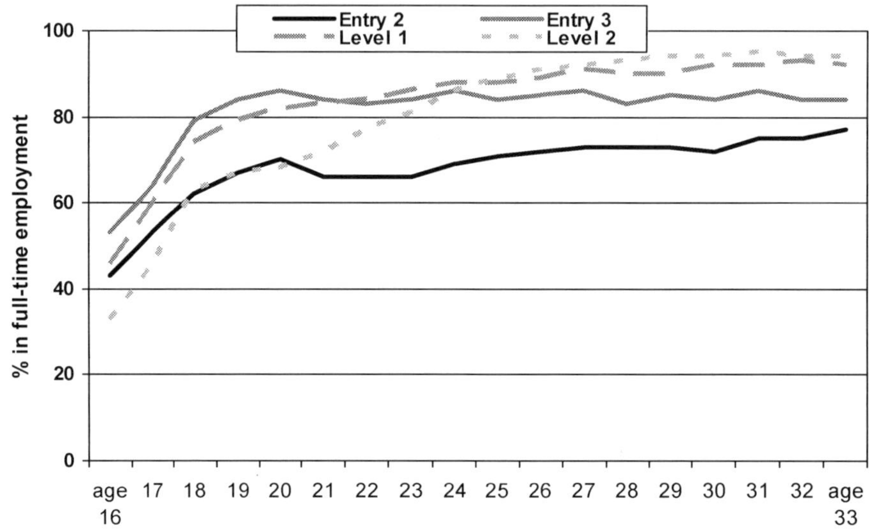

Figure 1.2 Percent of men in full-time employment by literacy level from 1986 (age 16) up to 2004 (age 33).

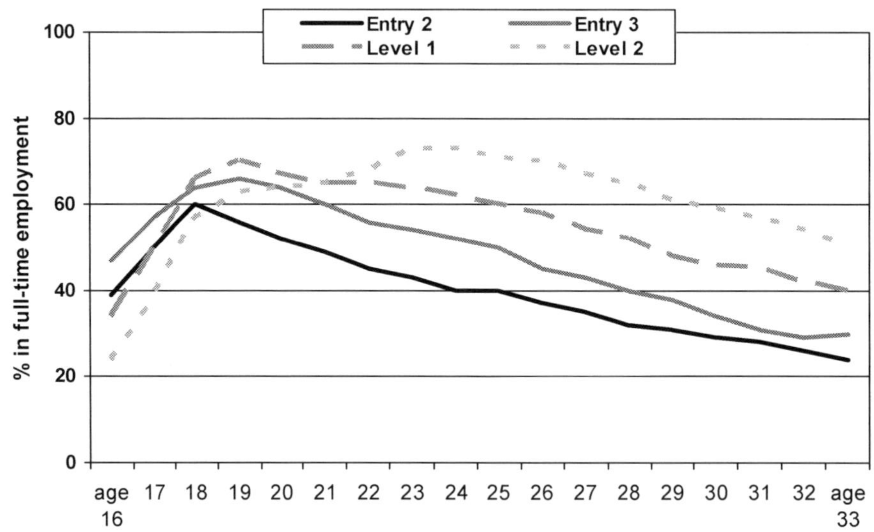

Figure 1.3 Percent of women in full-time employment by literacy level from 1986 (age 16) up to 2004 (age 33).

Although those thirty-four-year-old men with Entry literacy had been the first to leave education, with 30 percent staying on post-sixteen compared with 60 percent at Level 1, the picture for being in employment was rather different. The highest proportion in full-time employment at age seventeen was among those at Entry 3, followed by Level 1 and Entry 2, with the Level 2 group having the *fewest* in employment. By the mid-twenties the situation was quite different, with the Level 2 group overtaking the others in this respect and ending up by the late twenties with the highest employment rate, over 90 percent. In contrast, from the mid-twenties onwards, the Entry groups showed a consistently lower level of employment—with the lowest of all, Entry 2.

For women the gaps were even larger. Notably through the twenties and thirties there was a general exodus out of employment as many women at all literacy levels started families. But as for men from the early twenties onwards the employment rate was consistently lowest for those at the bottom skills level. Only a fifth of women with Entry 2 skills were still in full-time employment by age of thirty-three compared with half of the women at Level 2.

We gain further insights into the different career pathways of men and women differentiated by basic skills proficiency through charting the percentages of women and men at different literacy levels who had had a child by different ages from sixteen to thirty-four. The substantial gap between the literacy groups was again evident, especially among women—this time distinguishing between the more accelerated transition to maternity among those at Entry 2 and 3 compared with those at Level 1 and above. Around one in three of the Entry level women had become mothers by the age of twenty-one compared with one in five women who had Level 1 literacy and one in eight with Level 2 skills.

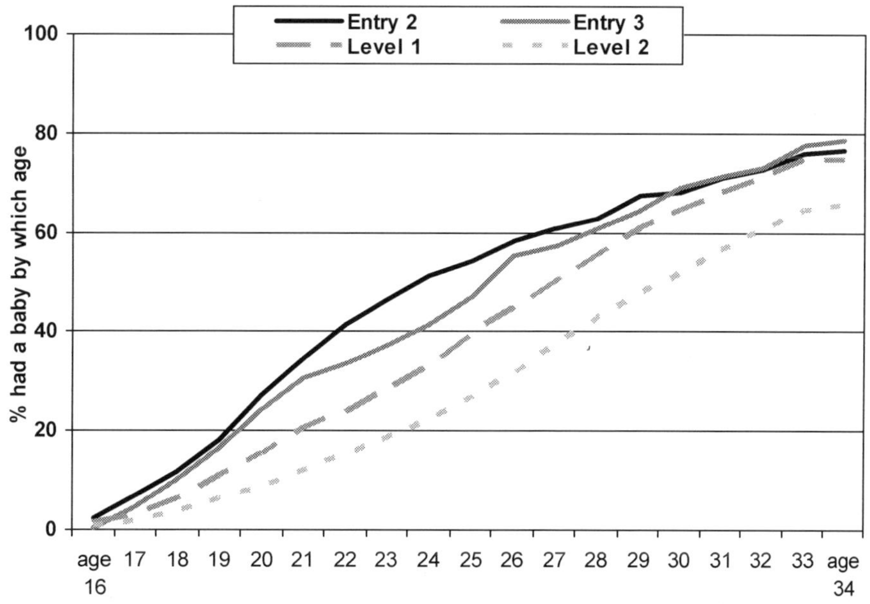

Figure 1.4 Percent of women who had a baby by age and literacy level.

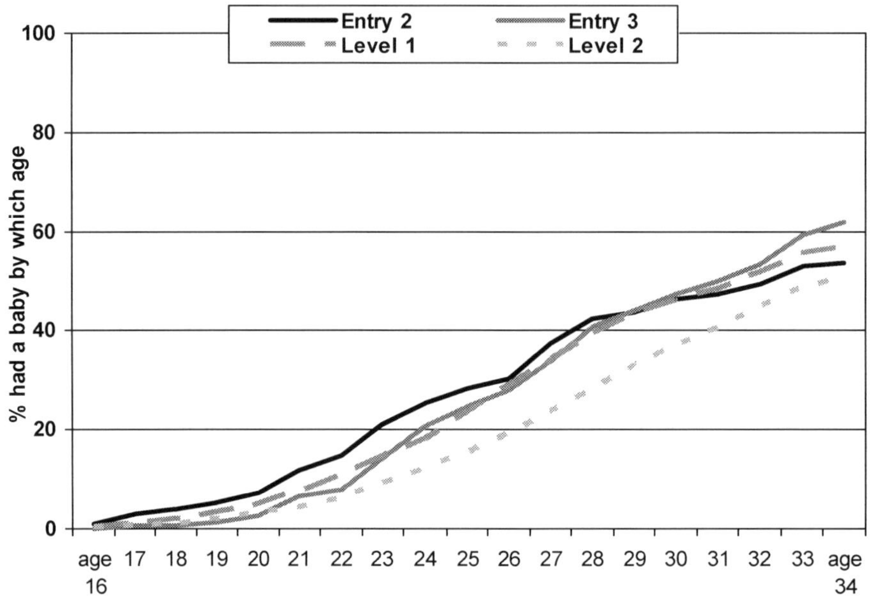

Figure 1.5 Percent of men who had a baby by age and literacy level.

For men the transition to fatherhood was much slower, with a paternity rate substantially below the maternity rate among women at the same ages (Figures 1.4 and 1.5). The gap across the skills levels was also less evident, and, if anything, there was a reversal with respect to those with Entry 2 skills over time. Although this group was more likely to have become fathers up to their mid-twenties, by age thirty-four around half of the Entry 2 men had yet to become a father compared with just under three-fifths of those at Entry 3. Level 2 men had the lowest paternity rates of all, pointing to the tendency shared with the Level 2 women to postpone the transition to parenthood.

These differences between the timing of men and women's transitions were manifested in other features of their family life. Thus, although women with poor skills tended to leave the family home early to enter into partnerships and parenthood, for men the reverse was the case. Men with the poorest literacy skills were the last to leave the family home and the last to enter into partnership.

ATTRIBUTES ASSOCIATED WITH LITERACY AT DIFFERENT LIFE COURSE STAGES

The tendencies toward exclusion statuses that these graphs reveal for men and women with poor basic skills reflect the outcomes of long-term processes to which we now turn. To what extent were the lives or people with poor basic

Table 1.1 Profile of cohort members (0–10) by their performance in the literacy assessment at age 34.

	Literacy Level			
	EL2 %	*EL3* %	*L1* %	*L2* %
Family and socioeconomic circumstances in childhood				
1970 (**age** 0)				
CM mother ever a teenage mother	29	25	19	15
CM low birthweight (< 2515 grams)	12	10	7	5
CM mother had extended education	19	20	28	43
CM father's job in 1970 RGSC I or II	9	6	14	23
CM father's job in 1970 RGSC IV or V	34	33	24	17
1975 (**age** 5)				
CM lived in an overcrowded home (1+ per room)	55	55	42	31
CM not living in an owner-occupied home	64	60	45	34
1980 (**age** 10)				
CM home in inner city/council estate	57	55	43	31
CM receives free school meals	24	22	14	10
CM ever been in care	5	2	2	2
Family educational support factors				
1975 (**age** 5)				
CM parent did not read to CM	29	26	23	16
CM parent read to CM everyday	22	31	33	44
CM did not go to preschool	40	35	27	20
CM started preschool < age 4	39	47	55	63
1980 (**age** 10)				
CM mother very interested in CM education	26	31	45	59
CM father very interested in CM education	16	19	33	47
Both parents very interested in CM education	12	16	27	41
CM parents want CM to leave education at age 16	71	60	46	30
CM parents want CM to stay on until age 18	15	17	34	51
Identification of difficulties				
1980 (**age** 10)				
CM received remedial help for reading	38	25	14	6
CM mother reports CM has some/great reading difficulties	40/16	24/6	22/3	11/1

(*continued*)

Table 1.1 Profile of Cohort Members (0–10) by Their Performance in the Literacy Assessment at Age 34 (continued)

	Literacy Level			
	EL2 %	*EL3* %	*L1* %	*L2* %
Cognitive performance: average scores (means) score range 0–10				
1975 (age 5)				
Copying Designs test	4.0	4.7	5.7	6.5
English Picture Vocabulary Test	4.6	5.2	5.9	6.0
1980 (age 10)				
Edinburgh Reading test	3.9	4.8	5.9	7.0
Friendly Maths test	4.2	5.0	5.9	6.7
Well-Being				
1980 (age 10)				
Low self-esteem	24	17	18	15
At school				
1980 (age 10)				
% children in school with high/above average ability	20	24	26	28
Average number of days missed in school term	5.2	3.9	4.5	4.0

skills marked by various forms of disadvantage? Tables 1.1 and 1.2 compare different attributes of cohort members across the four literacy levels at different life-course stages.

Early childhood

Compared with those with Level 2 literacy, cohort members at the lowest literacy levels tended to have experienced relatively poor material circumstances in their early lives and to have grown up in families offering much weaker educational support (Table 1.1). More had lived in overcrowded, social housing and, as a reflection of the parents' relatively low income, more had received free school meals. Their parents were less likely to have continued in postcompulsory education or to have gained any qualifications; fewer had read to their children or had been viewed by teachers as "very interested" in their children's education. And fewer had aspirations for their children to extend their education beyond the minimum leaving age of sixteen.

The consequence for the cohort members' own basic skills acquisition, when they were children, was reflected in their limited preschool educational experiences and relatively poor cognitive development as assessed by cognitive tests at age five and age ten. Both parents and teachers were likely to have recognized that the child's reading development after entry into school was poor compared with other children, though as many as half of those cohort members whose adult literacy performance was poor had failed to have their problems identified when they were at school.

ADULTHOOD

Leaving School and entering employment

The poor start on the pathway into and through education appeared to inhibit further progress later on (Table 1.2). Most with Entry literacy had left full-time education at the minimum age of sixteen. However, even some of those with only Entry 2 literacy did in fact have some kind of qualification by age thirty-four, suggesting that interest in learning was not entirely extinguished at school and that later opportunities appeared to have rekindled it. However, in many other respects these adults were leading disadvantaged working lives. They had the poorest employment records, spending on average of four years less in employment compared with those with Level 2 skills. Their experience of work tended to be of low-skilled jobs without training and poor or nonexistent promotion prospects. They were also more likely to have been excluded from areas of modern workplace practice, especially the use of ICT, with this form of modern impoverishment marginally more apparent for women than for men. Two-thirds of both sexes at Entry 2 did not use a computer at home or at work—over three times the proportion of those with Level 2 skills.

Family and Home Life

The disadvantaged circumstances of their childhood persisted for many of the Entry skills group into adulthood. In comparison with those with Level 2 skills, far more of the adults with the lowest literacy skills were living in rented and/or overcrowded accommodation by age thirty-four and more had experienced a spell of homelessness. Far fewer of the latter group had any savings or investments; many more were on income support and/or were living in households where none of the adults were working. Among women at Entry 2, five times as many were in nonworking households compared with those at Level 2. As we have seen, men and women's pathways to partnership and parenthood diverged. Men with the poorest skills were less likely than those with higher skills to be in a relationship or to have children by age thirty-four. In contrast, women with Entry Level 2 skills were three times as likely as women with Level 2 literacy to have been a teenage mother, twice as likely to be living as a single parent, and twice as likely to have three or more children by age thirty-four.

Table 1.2 Profile of Men and Women (16–34) by Their Performance in the Literacy Assessment at Age 34

	Literacy Level							
	EL2 %		EL3 %		L1 %		L2 %	
	m	w	m	w	m	w	m	w
The end of school and qualifications to 34								
Left full-time education at age 16	81	69	74	70	59	53	42	34
Average exam score from academic qualifications achieved at 16 (range 0–40)	2.5	3.1	3.5	3.6	5.9	6.2	9.1	9.5
No qualifications at age 34	47	41	26	17	13	10	6	4
Degree or higher at age 34	4	7	6	5	14	16	33	35
Working Life								
Average number of months in education between 16–34	10	10	9	9	17	19	30	30
Average number of months employment between 16–34	152	120	179	141	182	151	174	155
No work-related training courses from employer at 30 and 34	72	78	62	74	49	60	42	55
Average number of work-related training courses from employer at 30 and/or 34	0.8	0.8	1.8	0.8	3.2	1.9	4.1	2.3
Average number of jobs held between 16–34	3	2.7	3.4	2.9	3.4	3.4	3.4	3.6
Does not use a computer at work (age 30)	77	62	60	58	45	40	26	22
Does not use a computer at home or work (age 30)	64	62	53	7	35	42	19	22
Does not have a computer at home (age 30)	71	65	70	65	51	57	42	47
Family and home life								
Had a child when a teenager	5	18	1	17	3	11	2	6
No children at age 34	46	24	38	22	43	25	49	35
3+ children at age 34	9	25	7	19	9	14	6	11
Home owner	46	43	57	54	61	64	69	72
Overcrowded home	14	20	15	17	8	10	5	6
Experienced homelessness	10	15	5	8	6	8	5	6
No savings or investments	16	51	49	45	34	35	27	27
Income support	11	25	5	13	3	6	1	4
Nonworking household	22	31	13	16	6	10	4	6

(continued)

Table 1.2 Profile of Men and Women (16–34) by Their Performance in the Literacy Assessment at Age 34 (continued)

	Literacy Level							
	EL2 %		EL3 %		L1 %		L2 %	
	m	w	m	w	m	w	m	w
Life satisfaction and well-being								
Average Malaise score at 30 (range 0–24)	4.3	5.8	4.0	4.3	3.2	4.1	2.8	3.4
Depressed on Malaise scale (8+)	21	33	14	20	11	15	8	11

m = men; w = women.

Psychological Well-being

Finally, the Malaise scale (a measure of psychological mood aligned with depression) showed consistent differences across the literacy Levels. The proportion assessed on the scale as "depressed" (score of eight or more out of twenty-four) was two and a half times as high for men at Entry 2 compared with those at Level 2 and three times as high for women.

EXPLAINING POOR LITERACY

To trace the developmental pathways to poor acquisition of basic skills as manifested in poor performance at age thirty-four in the adult literacy test, we drew upon a range of longitudinal data to address two questions. How much of the variation in adult literacy proficiency might be attributed to earlier experience at different life-course stages? Where might the most salient associations between adult literacy and other variables measured earlier lie? To find out, we conducted a series of multiple regressions to assess the extent to which combinations of different circumstances, experiences and personal and family attributes at a number of ages in childhood—five, ten, sixteen, thiry and thirty-four—predicted performance in the literacy tests at age thirty-four. Interest lay in the way this relationship strengthened as more sets of variables were brought sequentially into the analysis and at what stage the strength of the relationship, as measured by the multiple correlation coefficient, began to level off.

The other parameter of interest was the regression coefficient, which gives the strength of the relationship between the outcome literacy score and each of the earlier predictor variables, holding constant the effects of the other predictor variables. For ease of comparison between variables, the regression coefficients were standardized to fall within the range of 0 to (+ or–) 1. Statistical significance of the estimates was established at the 5 percent, 1 percent, and .01 percent levels—e.g. an estimate significant at the 5 percent level indicated that the probability of the result occurring by chance was less than 5 percent or 1 in 20. To optimize the estimates, any information that was missing for the 9,567 BCS70 cohort members who completed the literacy and numeracy assessments in the 2004 sweep of data collection was imputed using the multiple imputation by chained equations (ICE), a program implemented in Stata 10 (Royston, 2004). The data set for analysis contained all the variables for which a statistically significant bivariate relationship with literacy test performance had been established. Five replicates of the data were created. Model estimates were averaged across these five analyses, with their standard errors calculated according to Rubin's rule (Rubin, 1987).

The predictor variables were introduced into the analysis in a series of steps beginning at birth, then at ages five, ten, sixteen, and post-sixteen up to age thirty (see Tables 1.3 and 1.4). The summary in Table 2.5 gives the final estimates for men and women and for the total sample when all variables had been entered into the analysis. For comparison, the results of the analysis for the imputed and nonimputed (full data) analyses are shown.

We start by considering the combined effects of these variables on the adult literacy outcome as illustrated by the graph of "variance explained" (multiple correlation squared) shown in Figure 1.6. The acquisition of literacy skills can be seen as rapid in response to different circumstances through the early years up to age ten. There are then signs of a marked leveling off with only small increments up to age sixteen, followed by another more modest acceleration up to thirty-four. For both men and women the graphs were very similar with the variance explained in age thirty-four literacy performance rising to 36 percent by age thirty-four for men and 32 percent for women, that is, multiple correlations of .60 and .57, respectively.

Men

Turning now to the regression estimates for men, it is notable that in the first of the five analyses (experiences and attributes present at birth), such variables as the age mother left full-time education and the socioeconomic status of the family, as assessed from father's occupation, strongly predicted the later age thirty-four literacy outcome. As other variables were introduced in the subsequent stages of the analysis, the size of the coefficients for the early variables declined, signifying that the effect of the birth circumstance variable was being largely mediated through the later variables. However, at age ten the new poverty indicator, free school meals, emerged as a (negative) predictor, and attending a school class with a large proportion of high-ability children related (positively) to the adult score. Thus, for example, at age five, cognitive scores on standardized tests became the main predictor, and at age ten, similarly, educational attainment took over as the main predictor from the earlier tests. Notably, one age five cognitive development indicator, the "copying designs" test score, actually retained its predictive value independently of other mediator variables through all the analysis stages.

Table 1.3 Predictor Variables in the Series of Multiple Regression Analyses: Predicting Literacy Score at Age 34 for Men

	Birth	*Age 5*	*Age 10*	*Age 16*	*Age 34*
***R^2* (adjusted)**	.06	.17	.34	.34	38
BIRTH					
Age mother left full-time education: 0 = no education . . . 31	.10***	.04**	.01	.00	.00
Age mother had first child: age 10–47	.07***	.03*	.01	.00	.01
Social class from father's occupation (or mother's if single): 0 = single mother, 1 = unskilled . . . 7 = professional	.12***	.03	–.02	–.02	–.03
Birthweight of child	.06***	.02	.00	.00	–.00
AGE 5					
Overcrowding (person per room ratio): range .07–6.0		–.02	–.00	–.00	.00
Home ownership: 0 = own, 1 = other		–.02**	–.01	.00	.01
Copying designs test score (standardized)		.22***	.07***	.06***	.04**
EPVT test score (standardized)		.17***	.06**	.06**	.04*
Child went to preschool: 0 = no, 1 = yes, > = age 4, 2 = yes, < age 4		.05***	.03	.04	.03
Days mother or father read to child: 0 to 7 days a week		.04**	.02	.00	.01
AGE 10					
Reading test score (standardized)			.20***	.19***	.18***
Maths test score (standardized)			.17***	.17***	.13***
Child received free school meals: 0 = no, 1 = yes			–.03*	–.04*	–.02
% of children in cm's school high / above average ability: 0—100%			.04*	.04*	.04*
Parent has interest in child's education: 1 = mum or dad little/uninterested, 2 = moderate to very interested			.03*	.03	.02
Mother identified reading difficulties: 1 = no difficulties, 2 = some, 3 = great			–.07***	–.07***	–.08***
Age parents want child to leave education: 1 = 16, 2 = 17, 3 = 18			.03*	.02	.01

(continued)

Table 1.3 Predictor Variables in the Series of Multiple Regression Analyses: Predicting Literacy Score at Age 34 for Men (continued)

	Birth	*Age 5*	*Age 10*	*Age 16*	*Age 34*
R^2 (adjusted)	.06	.17	.34	.34	38
Child lives in urban/inner city: 0 = other, 1 = inner urban/estate			–.02	–.02	–.01
Child's self-esteem: 0 = good,1 = low, 2 = very low			.02	.03	.03
Ever been in care: 0 = no, 1 = yes			–.02	–.02	–.02
Number of days missed from school in previous term: 0–150 days			.01	.01	.01
Child received remedial help/classes at school: 1 = no, 2 = yes			–.12***	–.13***	–.11***
AGE 16 (age 30 and 34 data sets)					
Exam score from GCSE/GCE/CSE exams attained at age 16 (2 points for grade A–C, grade 1cse, 1 point for lower grades): 0–40				.06***	.02
AGE 16–34					
Months in FT education 16–34: 0–211					.04*
Highest qualification by age 34: 0 = no quals . . . 4 = nvq 4–6					.13***
Months in employment 16–34: 0–216					.15***
Number of training courses from employer at 30 +/or 34: 0–111					.02
Number of babies by age 34: 0–8					.02
Number of jobs held between 16–34: 0–11					.02
Whether uses a computer at home or work at age 30: 0 = no, 1 = yes					.08***
Malaise score at age 30: 0–24					–.03*
N (100%) =	*4,573*	*4,573*	*4,573*	*4,573*	*4573*

At age ten, other family factors also emerged, such as parents' interest in education and the identification of reading difficulties. This finding is interesting in suggesting that those children who had difficulties with reading and were identified as such at an early stage in their school careers tended to retain these difficulties through to adult life. Much the same result was shown for participation in remedial classes at school,

Table 1.4 Predictor Variables in the Series of Multiple Regression Analyses: Predicting Literacy Score at Age 34 for Women

	Birth	*Age 5*	*Age 10*	*Age 16*	*Age 34*
R^2 (adjusted)	.04	.13	.32	.32	.34
BIRTH					
Age mother left full-time education: 0 = no education . . . 31	.11***	.04*	–.01	–.01	–.01
Age mother had first child: age 10–47	.07***	.01	–.01	–.01	–.01
Social class from father's occupation (or mother's if single): 0 = single mother, 1 = unskilled . . . 7 = professional	.14***	.06***	.01	.01	.00
Birthweight of child	.06***	.02	.01	.01	–.00
AGE 5					
Overcrowding (person per room ratio): range .07–6.0		–.05***	–.04**	–.04**	–.04**
Home ownership: 0 = own, 1 = other		–.07***	–.05**	–.04**	–.03*
Copying designs test score (standardized)		.20***	.05**	.04**	.04**
EPVT test score (standardiz ed)		.14***	.04*	.03*	.02
Child went to preschool: 0 = no, 1 = yes, > = age 4, 2 = yes, < age 4		.06***	.04*	.03*	–.03
Days mother or father read to child: 0 to 7 days a week		.03	–.00	–.01	–.01
AGE 10					
Reading test score (standardized)			.24***	.24***	.22***
Maths test score (standardized)			.14***	.13***	.10***
Child received free school meals: 0 = no, 1 = yes			–.01	–.01	.00
% of children in cm's school high/above average ability: 0–100%			.02*	.02	.02
Parent has interest in child's education: 1 = mum or dad little/uninterested, 2 = moderate to very interested			.01	.02	.00

(continued)

Table 1.4 Predictor Variables in the Series of Multiple Regression Analyses: Predicting Literacy Score at Age 34 for Women (continued)

	Birth	*Age 5*	*Age 10*	*Age 16*	*Age 34*
R^2 (adjusted)	.04	.13	.32	.32	.34
Mother identified reading difficulties: 1 = no difficulties, 2 = some, 3 = great			–.07***	–.07***	–.07***
Age parents want child to leave education: 1 = 16, 2 = 17, 3 = 18			.06***	.06***	.04**
Child lives in urban/inner city: 0 = other, 1 = inner urban/estate			.02	.02	.03
Child's self-esteem: 0 = good,1 = low, 2 = very low			.02	.02	.03*
Ever been in care: 0 = no, 1 = yes			–.02	–.02	–.01
Number of days missed from school in previous term: 0–150 days			–.01	–.01	–.00
Child received remedial help/classes at school: 1 = no, 2 = yes			–.12***	–.12***	–.11***
AGE 16 (age 30 and 34 data sets)					
Exam score from GCSE/GCE/CSE exams attained at age 16 (2 points for grade A–C, grade 1cse, 1 point for lower grades): 0–40	.			06***	.02
AGE 16–34					
Months in FT education 16–34: 0–211					.02
Highest qualification by age 34: 0 = no quals . . . 4 = nvq 4–6					.08***
Months in employment 16–34: 0–216					.09***
Number of training courses from employer at 30 +/or 34: 0–111					.00
Number of babies by age 34: 0–8					.01
Number of jobs held between 16–34: 0–11					–.01

(continued)

Table 1.4 Predictor Variables in the Series of Multiple Regression Analyses: Predicting Literacy Score at Age 34 for Women (continued)

	Birth	*Age 5*	*Age 10*	*Age 16*	*Age 34*
R^2 *(adjusted)*	.04	.13	.32	.32	.34
Whether uses a computer at home or work at age 30: 0 = no, 1 = yes					.06***
Malaise score at age 30: 0–24					–.07***
N (100%) =	4,994	4,994	4,994	4,994	4994

Table 1.5 Summary of Predictor Variables in the Final Multiple Regression Model Predicting Literacy Score at Age 34

	Men		*Women*		*All*	
	Complete	*Imputed*	*Complete*	*Imputed*	*Complete*	*Imputed*
R^2 *(adjusted)*	.36	.38	.32	.34	.33	.36
BIRTH						
Sex of cohort member : 1 = boy, 2 = girl					.00	.00
Age mother left full-time education: 0 = no education . . . 31	.01	.00	–.03	–.01	–.02	–.00
Age mother had first child: age 10–47)	.03	.01	–.02	–.01	.00	–.00
Social class from father's occupation (or mother's if single): 0 = single mother, 1 = unskilled . . . 7 = professional	–.02	–.03	.02	.00	.00	–.01
Birthweight of child	.03	–.00	–.01	–.00	.01	.00
AGE 5						
Overcrowding (person per room ratio): range .07–6.0	.03	.00	–.07**	–.04**	–.02	–.02*

(continued)

Table 1.5 Summary of Predictor Variables in the Final Multiple Regression Model Predicting Literacy Score at Age 34 (continued)

	Men		*Women*		*All*	
	Complete	*Imputed*	*Complete*	*Imputed*	*Complete*	*Imputed*
R^2 (adjusted)	.36	.38	.32	.34	.33	.36
Copying designs test score (standardized)	.06*	.04**	.03	.04**	.05**	.05***
EPVT test score (standardized)	.02	.04*	.01	.02	.02	.03*
Child went to pre school: 0 = no, 1 = yes, >= age 4, 2 = yes, < age 4	.01	.03	–.01	–.03	.00	.03*
Days mother or father read to child: 0 to 7 days a week	.02	.01	–.03	–.01	–.01	–.00
AGE 10						
Reading test score (standardized)	.15***	.18***	.23***	.22***	.19***	.20***
Maths test score (standardized)	.10**	.13***	.13***	.10***	.12**	.11***
Child received free school meals: 0 = no, 1 = yes	–.04	–.02	.04	.00	.01	–.01
% of children in cm's school high/above-average ability: 0–100%	.05*	.04*	.02	.02	.03*	.03**
Parent has interest in child's education: 1 = mum or dad little/uninterested, 2 = moderate to very interested	.05*	.02	.02	.00	.03*	.01

(continued)

Table 1.5 Summary of Predictor Variables in the Final Multiple Regression Model Predicting Literacy Score at Age 34 (continued)

	Men		*Women*		*All*	
	Complete	*Imputed*	*Complete*	*Imputed*	*Complete*	*Imputed*
Mother identified reading difficulties: 1 = no difficulties, 2 = some, 3 = great	–.08**	–.08***	–.10***	–.07***	–.09***	–.07***
Age parents want child to leave education: 1 = 16, 2 = 17, 3 = 18	.00	.01	.05*	.04**	.03	.02*
Child lives in urban/ inner city: 0 = other, 1 = inner urban/estate	–.05	–.01	.03	.03	–.02	.01
Child's self-esteem: 0 = good,1 = low, 2 = very low	.01	.03	.06**	.03*	.04*	.03*
Ever been in care: 0 = no, 1 = yes	–.02	–.02	–.08***	–.01	–.05***	–.01
Number of days missed from school in previous term: 0–150 days	–.01	.01	.02	–.00	.01	.00
Child received remedial help/ classes at school: 1=no, 2=yes	–.17***	–.11***	–.06*	–.11***	–.11***	–.11***
AGE 16 (age 30 and 34 data sets)						
Exam score from GCSE/GCE/ CSE exams attained at age 16 (2 points for grade A–C, grade 1cse, 1 point for lower grades): 0–40	.04	.02	.02	.02	.03	.02*

(continued)

Table 1.5 Summary of Predictor Variables in the Final Multiple Regression Model Predicting Literacy Score at Age 34 (continued)

	Men		*Women*		*All*	
	Complete	*Imputed*	*Complete*	*Imputed*	*Complete*	*Imputed*
AGE 16–34						
Months in FT education 16–34: 0–211	.03	.04*	.05	.02	.04	.02
Highest qualification by age 34: 0 = no quals . . . 4 = nvq 4–6	.11**	.13***	.06*	.08***	.08***	.10***
Months in employment 16–34: 0–216	.11***	.15***	.09**	.09***	.10***	.11***
Number of training courses from employer at age 30 +/or 34 (combined): 0–111	.04	.02	.03	.00	.04*	.01
Number of babies by age 34: 0–8	.00	.02	.02	.01	.01	.03**
Number of jobs held between 16–34: 0–11	.01	.02	.02	–.01	.01	.01
Whether uses a computer at home or work at age 30: 0 = no, 1 = yes	.06**	.08***	.03	.06***	.05**	.07***
Malaise score at age 30: 0–24	.02	–.03*	–.06**	–.07***	–.05**	–.05***
n (100%) =	*1,458*	*4,573*	*1,666*	*4,994*	*3,125*	*9,567*

where again the relationship was negative rather than positive, with the later outcome suggesting that the remedial classes failed to be of much help—at least for these children.

Labor market participation was assessed up to the age of thirty in terms of the record of full-time employment from age sixteen as derived from the job histories from which Figures 1.2 and 1.3 were derived. Notably, months

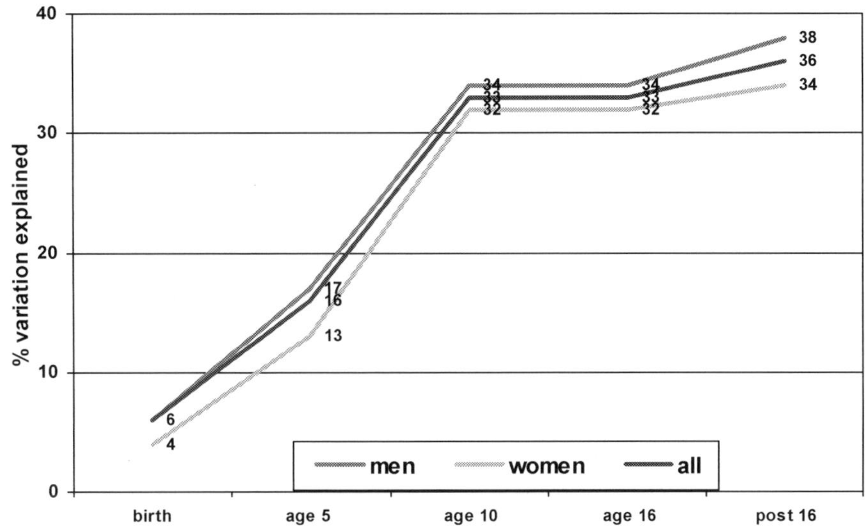

Figure 1.6 Percent of variation in literacy scores at 34 in BCS70 explained at different ages.

in employment was a positive predictor of literacy score, as was using a computer at home or at work when the cohort member was aged thirty.

Women

Much the same results were found for women, except that a wider range of circumstance variables, overcrowding and home ownership particularly, emerged as significant predictors, retaining their importance right up to age thirty-four. The "copying designs" test in this case did not retain such strong predictive value after the age of five. The results for remedial education and identification of difficulties similarly suggested the same negative effect on the later outcome as they had for men. But other predictors also proved significant: children's self-esteem and, most notably, whether the child had ever been in care. The latter had a negative relationship with adult literacy performance. In relation to participation in the labor market, as for men, months in employment between age sixteen and age thirty-four was a strong predictor of adult literacy score. Highest qualification was much more predictive of adult literacy performance for women than it was for men.

DISCUSSION AND CONCLUSIONS

The results of the analysis set against those from earlier studies demonstrate the consistency across cohorts and across ages of disadvantaged careers manifested through low skills acquisition with exclusion outcomes in education and the labor market and ramifications for family life as well. The key period when identifiable influences can be seen as playing a part in this basic skills acquisition is through the preschool and elementary school period up to age ten. From this age onwards the gain in explanation of adult skills difficulties reduces, suggesting that changes in circumstances and the environment are less likely to have a modifying influence on the basic skills set that is now in place. Rather, changes in the circumstances of individual lives, relationships formed, jobs entered, community engagement and the exercise of individual agency in place of generalizable patterns are likely to account increasingly for the levels reached and changes in them. However, our data point to a small increase in variance explained through the twenties not observed in our earlier studies. It is not unreasonable to suggest that part of this rise may be due to enhanced interest and support for raising basic skills levels in the adult population. Workplaces which at one time would have had little interest in, or certainly any feeling of responsibility for the literacy levels of their employees, are now more likely to acknowledge that there may be a job to do

As we might expect, the effect of material circumstances and family factors on later proficiency tends to be mediated by earlier skills acquisition—less so for women than for men. However, post-sixteen workplace experience shows effects that are independent of earlier cognitive development and later educational achievement, pointing perhaps to the growing importance of such experience in raising basic skills levels

Any such suggestions need to be qualified in terms of methodological considerations. The effect of imputing missing data rather than just using in the analysis data for which complete individual records were available is one such issue to address. Imputation by effectively boosting sample size increased the number of regression coefficient estimates meeting the lowest ($p < .05$) criterion of statistical significance. Some estimates also increased in size in the imputed sample, thus also achieving statistical significance. For others, however, the size reduced. Overall, we judge that multiple imputation is worthwhile and supports the most robust interpretation of the data. But the difference it makes over complete data is not substantial. Perhaps of more significance is missing variables that could have made a difference and problems of potential selection bias in the attribution of "effects" to variables that did emerge as significant in the analysis. As with all multipurpose longitudinal surveys, we are limited by the variables available for which decisions have been made on criteria over which we often have had no control. Through the multifaceted coverage of the studies, the consistency of findings across cohorts and across ages and, the increasing use of the life-course framework for variable selection, the studies are in as good if not better position than most others to claim that the critical variables are available. Nevertheless, constant development of measurement within the studies, the application of more sophisticated analytic techniques such as structural equation modeling, and new studies for replication and triangulation across all of them are needed to strengthen confidence further.

With these cautions in mind, what provisional conclusions can we draw and what are their policy implications? The main message for educational policy might appear to reinforce the argument for the highest concentration of educational effort on basic skills acquisition through the early years, ideally before primary school entry (e.g., Heckman & Masterov, 2007). The "Sure Start" program in the United Kingdom and "Early Head Start" in the United States target specifically early childhood. Family involvement in children's learning is a valuable means of giving them a better start in education. Preschool provision, in such programs, or through nursery school (UK) and kindergartens (U.S.), offers more favorable pupil/carer ratios than the elementary school, enabling more intensive forms of educational play and other forms of learning to take place. Such experience carried from, or back, into the family, with a strong "learning culture" and supported by family learning opportunities in which parents and children are assisted in learning together, supplies potentially continuous reinforcement of a learning dynamic. The result can be a positive cycle of achievement that supplies the foundation for formal schooling. The elementary school, with its large classes, offers fewer one-to-one learning opportunities of this kind, with the risk that children who have entered the school with lower levels of skills than their more favored counterparts fall increasingly behind.

Our findings suggest that what is offered in schools in the identification of difficulties and remedial opportunities frequently fails to reverse negative trends. Rather, it points to a continuing problem in the sense that those who enter remedial education in elementary school are most likely to be still showing poor literacy as adults.

Beyond the early stages of skills acquisition, the ramifications of low-skills levels carried through schooling and into the labor market can be seen as restricting opportunities with further potential adverse effects on family and community life. But the extra boost to variance explained in literacy performance over the period sixteen to thirty-four also points to factors specific to this period, such as labor market experience, that can make a difference. Clearly, there is scope for government to strengthen these.

In life-course terms, we can see the acquisition of literacy skills as playing a key part in the essential competencies and capabilities that define positive career paths and trajectories through the various stages of life. Those without them are the most likely to be found on a social exclusion path. But such a gloomy prognostication needs to be tempered with the realization that what we have demonstrated here are statistical tendencies with predictive value not inevitabilities. Clearly, substantially less than 100 percent variance explained suggests there is much room for movement in and out of disadvantaged trajectories in adulthood, in which strengthening the basic skills can play a significant part. In fact, in our analysis over 60 percent of the variance remained unexplained by age thirty-four in terms of the circumstances, experience, and early skills performance variables included in this analysis. Although some if not most of this unexplained variance may be attributed to unreliability of measurement, and missing variables, a sizable proportion almost certainly cannot. Individuals, through the agency they can exercise, can turn their lives in positive or negative directions at any stage from early childhood onwards. Supporting their skills development is one of the main means of offering movement towards positive life-course experience and fulfilling outcomes in adult life.

Such reasoning thus runs counter to the argument that the bulk of investment in the basic skills should be in early childhood. While recognizing the importance of this period to later progress, the creation of opportunities for adults to strengthen their own agency in overcoming earlier failures offers the added bonus of also potentially benefiting their children. This might be seen as supporting the case for expanding, rather than contracting, in the nature of "stopgaps" educational programs such as SfL.

APPENDIX 1.1

Table 1.A.1 Aspects Covered by the Adult Literacy Curriculum

ADULT LITERACY CURRICULUM	
Speaking and Listening: Listen and respond	
Entry 2	Listen and respond to spoken language, including straightforward information, short narratives, explanations and instructions.
Entry 3	Listen and respond to spoken language, including straightforward information and narratives, and follow straightforward explanations and instructions, both face-to-face and on the telephone.
Level 1	Listen and respond to spoken language, including information and narratives and follow explanations and instructions of varying lengths, adapting response to speaker, medium and context.
Level 2	Listen and respond to spoken language, including extended information and narratives and follow detailed explanations and multistep instructions of varying length, adapting response to speaker, medium. and context.
Speaking and Listening: Speak to communicate	
Entry 2	Speak to communicate information, feelings and opinions on familiar topics.
Entry 3	Speak to communicate information, feelings and opinions on familiar topics, using appropriate formality, both face-to-face and on the telephone.
Level 1	Speak to communicate information, ideas and opinions, adapting speech and content to take account of the listener(s) and medium.
Level 2	Speak to communicate straightforward and detailed information, ideas, and opinions clearly, adapting speech and content to take account of the listener(s), medium, purpose and situation.

(continued)

Table 1.A.1 Aspects Covered by the Adult Literacy Curriculum (continued)

ADULT LITERACY CURRICULUM	
Speaking and Listening: Engage in discussion	
Entry 2	Engage in discussion with one or more people in a familiar situation to establish shared understanding about familiar topics.
Entry 3	Engage in discussion with one or more people in a familiar situation, making relevant points and responding to what others say, to reach a shared understanding about familiar topics.
Level 1	Engage in discussion with one or more people in familiar and unfamiliar situations, making clear and relevant contributions that respond to what others say and produce a shared understanding about different topics.
Level 2	Engage in discussion with one or more people in a variety of different situations, making clear and effective contributions that produce outcomes appropriate to purpose and topic.
Reading: Read and understand	
Entry 2	Read and understand short, straightforward texts on familiar topics.
Entry 3	Read and understand short, straightforward texts on familiar topics accurately and independently.
Level 1	Read and understand straightforward texts of varying length on a variety of topics accurately and independently.
Level 2	Read and understand a range of texts of varying complexity accurately and independently.
Reading: Read and obtain information	
Entry 2	Read and obtain information from short documents, familiar sources, and signs and symbols.
Entry 3	Read and obtain information from everyday sources.
Level 1	Read and obtain information from different sources.
Level 2	Read and obtain information of varying length and detail from different sources.
Writing: Write to communicate	
Entry 2	Write to communicate information with some awareness of the intended audience.
Entry 3	Write to communicate information and opinions with some adaptation to the intended audience.
Level 1	Write to communicate information, ideas and opinions clearly, using length, format, and style appropriate to purpose and audience.
Level 2	Write to communicate information, ideas and opinions clearly and effectively, using length, format and style appropriate to purpose, content, and audience.

APPENDIX 1.2

Table 1.A.2 **Aspects Covered by the Adult Numeracy Curriculum**

ADULT NUMERACY CURRICULUM	
Understanding and using mathematical information: Read and understand	
Entry 2	Read and understand information given by numbers, symbols, simple diagrams and charts in graphical, numerical, and written material.
Entry 3	Read and understand information given by numbers, symbols, diagrams, and charts used for different purposes and in different ways in graphical, numerical and written material.
Level 1	Read and understand straightforward mathematical information used for different purposes and independently select relevant information from given graphical, numerical and written material.
Level 2	Read and understand mathematical information used for different purposes and independently select and compare relevant information from a variety of graphical, numerical and written material.
Understanding and using mathematical information: Specify and describe	
Entry 2	Specify and describe a practical problem or task using numbers, measures, and simple shapes to record essential information.
Entry 3	Specify and describe a practical problem or task using numbers, measures and diagrams to collect and record relevant information.
Level 1	Specify and describe a practical activity, problem, or task using mathematical information and language to make accurate observations and identify suitable calculations to achieve an appropriate outcome.
Level 2	Specify and describe a practical activity, problem, or task using mathematical information and language to increase understanding and select appropriate methods for carrying through a substantial activity.
Calculating and manipulating mathematical information: Generate results	
Entry 2	Generate results to a given level of accuracy using given methods and given checking procedures appropriate to the specified purpose.
Entry 3	Generate results to a given level of accuracy using given methods, measures and checking procedures appropriate to the specified purpose.
Level 1	Generate results to a given level of accuracy using methods, measures and checking procedures appropriate to the specified purpose.
Level 2	Generate results to an appropriate level of accuracy using methods, measures and checking procedures appropriate to the specified purpose.

(continued)

Table 1.A.2 Aspects Covered by the Adult Numeracy Curriculum (continued)

ADULT NUMERACY CURRICULUM	
Interpreting results and communicating mathematical information: Present and explain results	
Entry 2	Present and explain results that meet the intended purpose using appropriate numbers, simple diagrams, and symbols.
Entry 3	Present and explain results that meet the intended purpose using appropriate numbers, diagrams, charts, and symbols.
Level 1	Present and explain results that meet the intended purpose using an appropriate format to a given level of accuracy.
Level 2	Present and explain results clearly and accurately using numerical, graphical and written formats appropriate to purpose, findings and audience.

NOTES

1. A Fresh Start—Improving Literacy and Numeracy (Department for Education and Employment, 1999).
2. The details of the development of the assessment instruments for the BCS70 age 34 follow-up are supplied in an earlier report and an associated journal article. See S. Parsons and J. Bynner (2005).
3. Leitch Review of Skills, Final Report, UK Government, 2006, www.hm-treasury.gov.uk/Leitch.
4. The main survey was funded by the UK Economic and Social Research Council (ESRC).

REFERENCES

Ashton, D., & Maguire, M. (1983). *Youthaid occasional paper no.3*. London: Youthaid.

Bynner, J. (2004). Literacy, numeracy and employability. *Literacy and Numeracy Studies*, 13, 31–48.

Bynner, J., Butler, N. R., Ferri, E., et al. (2000). *The 1999–2000 survey of the National Child Development Study: Summary of contents*. CLS Working Paper No.1. London: CLS, Institute of Education

Bynner, J., & Joshi, H. (2007). *Building the evidence base from longitudinal data: The aims, content and achievements of the British birth cohort studies.*

Bynner, J., & Parsons, S. (1997). *It doesn't get any better: The impact of poor basic skills on the lives of 37 year olds*. London: Basic Skills Agency.

Bynner, J., & Parsons, S. (2002). Social exclusion and the transition from school to work: The case of young people not in education, employment or training NEET. *Journal of Vocational Behaviour*, 60, 289–309.

Bynner, J., & Parsons, S. (2006). *New light on literacy and numeracy*. NRDC Research Report. London: National Research and Development Centre for Adult Literacy and Numeracy, Institute of Education.

Bynner, J., & Steedman, J. (1995). *Difficulties with basic skills*. London: Basic Skills Agency.

Ekinsmyth, C., & Bynner, J. (1994). *The basic skills of young adults*. London: Adult Literacy and Basic Skills Unit.

Elder, G. H. (1998). The lifecourse and human development. In R. M. Lerner (ed.), *Handbook of child psychology, Volume 1: Theoretical models of human development*. New York: Wiley.

Heckman, J., & Masterov, D. V. (2007). The productivity argument for investing in young Children. *Review of Agricultural Economics*, 29(3), 446–93.

Heinz, W. R. (1991). Status passages, social risks and the life course: A conceptual framework. In W. R. Heinz (ed.), *Theoretical advances in life course research, Volume 1: Status passages and the life course*. Weinheim, Germany: Deutscher Studien Verlag.

Parsons, S. & Bynner, J. (1998). *Influences on adults basic skills*. London: Basic Skills Agency.

Parsons, S., & Bynner, J. (2005). Measuring basic skills for longitudinal study: The design and development of instruments for use with cohort members in the age 34 follow-up in the 1970 British Cohort Study (BCS70). *Literacy and Numeracy Studies*, 14, 7–30.

Parsons, S., & Bynner, J. (2008). *Illuminating disadvantage*. NRDC Research Report. London: National Research and Development Centre for Adult Literacy and Numeracy, Institute of Education.

Plewis, I., Calderwood, L., & Hawkes, D. (2004). *Changes in the NCDS and BCS70 populations and samples over time*. CLS Technical Report, London: CLS, Institute of Education.

Royston, P. (2004). Multiple imputation of missing values. *The Stata Journal*, 4(3), 227–41.

Rubin, D. B.(1987). *Multiple imputation for non-response in surveys*. New York: Wiley.

Williams, J., Clemens, S., Oleinikova, K., et al. (2003). *The SfL survey: A national needs and impact survey of literacy, numeracy and ICT skills*. Department for Education and Skills Research Report 490, UK Department for Education and Skills, London.

2 The Development of Literacy and Numeracy in Adult Life

Stephen Reder[1]

INTRODUCTION

A major function of adult education systems is to provide instruction to help adults improve their literacy and numeracy skills. The measure of these programs' success is generally taken to be the extent to which adults' skills do improve over time. The rationale for investments in adult education is often framed in terms of the favorable economic and social consequences of skill development (Kirsch, Braun, Yamamoto, & Sum, 2007; Moser, 1999; Murray, this volume). Although adult programs are frequently evaluated and sometimes held accountable in terms of their students' skill improvement, remarkably little is known about the measured development of literacy and numeracy abilities in adult life.

This chapter will explore some of the dynamics of change in literacy and numeracy across the adult life span. Although some prior research has examined change in literacy and numeracy in adult life, most of the studies used designs that limit our ability to address these issues directly. Many of the prior studies are relatively short-term follow-ups of participants in adult skills programs. In some studies, the follow-up intervals are too short to observe meaningful change in adult life. In other studies, only program participants are observed, making it difficult to understand participation and persistence patterns or the impact of program participation (Beder, 1999; Brooks, Davies, Duckett, Hutchinson, Kendall, & Wilkin, 2001). To get around some of these limitations, the Longitudinal Study of Adult Learning (LSAL) was designed as a long-term panel study to follow a target population for adult literacy and numeracy education over a relatively long period of time.

The LSAL was designed to address four major research questions about the development of literacy in adult life (Reder & Strawn, 2001a, 2001b, 2006):

- To what extent do adults' literacy abilities continue to develop after they are out of school?
- What are adult learners' patterns of participation over time in literacy training and education? In other learning contexts?

- What life experiences are associated with adult literacy development? How do formally organized basic skills programs contribute to these learning trajectories? Workplace training? Other contexts and activities?
- What are the impacts of adult literacy development on social and economic outcomes?

In following individuals' literacy and numeracy over time, LSAL utilized multiple methods of measuring literacy and numeracy abilities. Most prior longitudinal research has examined changes in *proficiency* measures alone. Proficiency measures are quite important, of course, because of the volume of evidence showing they are closely linked to educational and economic outcomes (Kirsch et al., 2007) and because they figure prominently in policy formation and program funding in many industrialized countries (Organization for Economic Cooperation and Development, 1997; Statistics Canada and Organization for Economic Cooperation and Development, 2005). LSAL included a commonly used proficiency measure. Important as they are, proficiency measures do not provide a complete picture of literacy and numeracy use and development. Individuals' literacy and numeracy *practices*, that is, their use of reading, writing, and math in everyday activities, provide another important picture of literacy and numeracy. LSAL developed longitudinally stable measures of some literacy practices and numeracy practices that are also used in the analysis of change here.[2]

This chapter will thus focus on examining some of the dynamics of change in adult literacy and numeracy across the life span. We will follow the development of literacy and numeracy through two parallel measurement lenses: proficiency and practices.[3] We will use hierarchical linear growth-curve models (Raudenbush & Bryk, 2002) for analyzing the processes of change in literacy and numeracy. We will consider how time-invariant characteristics of individuals such as gender, place of birth, and ethnicity influence the growth of literacy and numeracy in adult life. We shall also examine how various life events, modeled as time-varying predictors, influence the development of literacy and numeracy in adult life.

DESIGN OF LSAL

The LSAL was designed as a panel study, representative of a local rather than a national target population for adult literacy and numeracy education. This target population was defined as residents of the Portland, Oregon, metropolitan area, ages eighteen to forty-four, proficient but not necessarily native English speakers, high school dropouts (i.e., did not receive a high school diploma and were no longer enrolled in school), and had not received a General Educational Development (GED) or other high school equivalency credential.[4] A statistically representative sample of this population was drawn from two sampling frames: a combination of random-digit dialing and enrollment forms provided by the three major adult education programs serving the Portland metropolitan area. Sampled households were called and screened for members in the defined target population. The resulting sample contained 940 individuals and was weighted so that population statistics could be estimated from the sample data. The sample consisted of 496 individuals from the random-digit-dialing frame and 444 from the enrolled student frame. In addition to the formal sample of 940 respondents, thirty-nine pilot subjects took part in LSAL. These pilot subjects were used for instrument development, interviewer training, and for in-depth qualitative studies. .

The LSAL conducted a series of five periodic interviews and skills assessments in respondents' homes. Respondents were paid for each of these sessions, which took an average of about one and one-half hours to complete. The five sessions, or "waves," of data were collected according to the following schedule:

Wave 1: 1998–1999
Wave 2: 1999–2000
Wave 3: 2000–2001
Wave 4: 2002–2003
Wave 5: 2004–2005

Individuals were interviewed at about the same time in each wave so that there is approximately constant spacing among individuals' successive interviews and assessments (e.g., a respondent interviewed in February 1999 in Wave 1 is interviewed during February 2000 in Wave 2, February 2001 in Wave 3, etc.). Waves 1–5 have been completed and will be reported in this chapter.[5] Through Wave 5, about 90 percent of the original sample was retained in the study.

At the beginning of the study in 1998, the population had an average age of twenty-eight and was evenly divided among males and females. Approximately one-third were members of minority groups, one in ten were born outside of the United States, one-third described themselves as having a learning disability, and one in three reported having taken special education classes (designed for students with physical and learning disabilities) while they were in elementary or secondary school.

Individuals dropped out of school for a variety of reasons. The most common reasons given were that individuals were bored or did not like or did not fit in school (29 percent) or problems with academic performance (26 percent). Reasons related to employment while in school (17 percent), problems with personal relationships (15 percent), family problems (10 percent), and health or pregnancy reasons (9 percent) were also frequently reported.

Each wave of data collection consisted of an in-home interview followed by cognitive assessments. The interviews included numerous core items, repeated in each wave, regarding the individual's household and family composition, employment and educational status, social and economic status, engagement in literacy, numeracy, and information technology (IT) practices, use of learning strategies, participation in adult, continuing and postsecondary education and training, hobbies, interests, and aspirations for the future. During the Wave 1 interview, additional items gathered information about respondents' families of origin, their elementary and secondary educational experiences, their employment histories, their current household and living situation, and other baseline and background data. During subsequent interviews, this base of information about educational, social, and economic activities was updated for the intervening time period, adding to the cumulative historical profiles of the individuals and their families. Later waves of data also included onetime topical modules that looked in depth at such issues as health status and health care utilization, self-reported learning disabilities, turbulence in everyday life, and so forth.

LSAL's skills assessment includes a standardized functional literacy assessment in each wave. The functional literacy assessment used is the Document Literacy scale of the Test of Applied Literacy Skills (TALS) developed by the Educational Testing Service. Administered in a constructed response rather than multiple-choice format, the TALS assesses adults' abilities to extract and process written information in a variety of everyday document formats, such as forms, maps, tables, text displays, labels, and so forth. These written documents are processed as part of performing simulated everyday literacy tasks. Respondents are assigned proficiency scores on a 0–500 scale based on the simulated literacy tasks they are able to perform correctly using a full information scaling program.[6] The TALS instruments are similar to instruments used in the 1992 National Adult Literacy Survey (NALS), the 2003 National Assessment of Adult Literacy (NAAL), the International Adult Literacy Survey (IALS), the Adult Lifelong Learning (ALL) survey, and in numerous state-level surveys of adult literacy conducted in the United States. Two TALS Document test forms, Form A and Form B, were randomly assigned for administration to respondents in Wave 1 and were alternated thereafter for a given subject from wave to wave. The developers of the instrument analyzed LSAL's item-level responses and concluded that the responses fit the IRT model well and that the fitted item parameters were longitudinally stable.

Repeated self-reported measures of everyday reading, writing, and math practices measures were also collected. Respondents were asked in each wave about whether and how often they performed each of a set of specific reading, writing, numeracy, and computer practices in various everyday contexts (home, community, work). As described below, these numerous practices items were scaled into longitudinally stable measures of engagement in literacy and numeracy practices in everyday activities and contexts.

DEVELOPMENT OF LITERACY PROFICIENCY

Although the LSAL population was limited to adults with relatively little schooling (at the time the study began), the range of their assessed proficiencies is quite broad. Figure 2.1 displays the distribution of assessed TALS Document Literacy proficiency, distinguishing five conventionally reported levels or subranges of the 0–500 proficiency continuum.[7] The left panel of Figure 2.1 displays the distribution of these proficiency scores as assessed in Wave 1 for the LSAL population in Portland, Oregon. The right panel of Figure 2.1 displays the corresponding distribution for a matching statewide population (ages 18–44, proficient English speakers, without a high school degree or equivalent, and currently not in high school) assessed with a similar instrument in an independent survey conducted in 1990 (Oregon Progress Board, 1991).

It is clear that adults in the target population, despite having relatively little formal education, display a broad range of proficiencies. Although a small percentage of this population performed at the lowest level of proficiency (Level 1), there is also a small percentage performing at the highest proficiency levels (Level 4 and 5). Clearly, most of the population performs at the intermediate Levels 2 and 3. The similarity of the two distributions provides a measure of confidence about the overall reliability of such large-scale assessments despite being carried out in different projects, by different organizations, and using different sampling procedures and test forms.

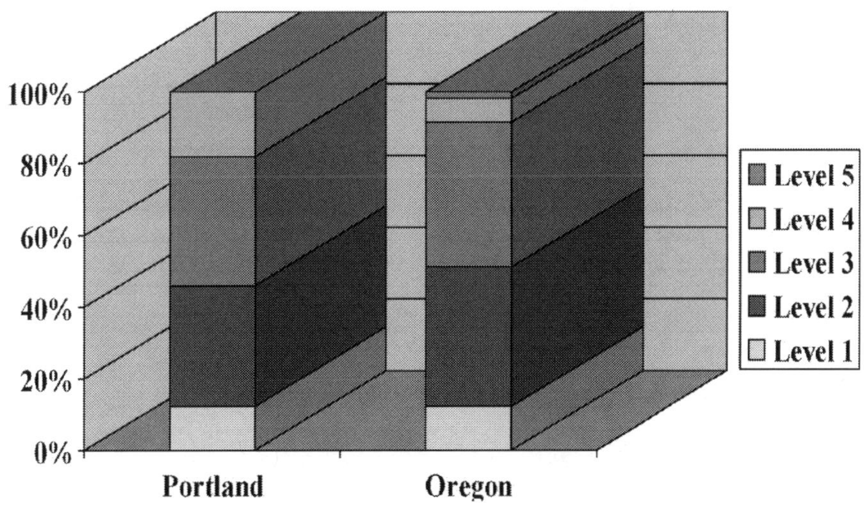

Figure 2.1 Distribution of document literacy proficiency for LSAL population of Portland and matching subpopulation of Oregon. *Source*: Longitudinal Study of Adult Learning, Wave 1, 1998 (Portland) and Oregon Literacy Survey, 1990. Author calculations.

The Oregon Literacy Survey results shown in the right panel of Figure 2.1, based on data collected in 1990, are quite similar to those of LSAL for Wave 1 collected in 1998. The similarity of these data, collected some eight years apart, suggests that there has been little overall change in levels of literacy in our target population. This is also the conclusion reached by Reder and Edmonston (2003), who used population models to predict literacy change in Oregon's adult population between 1990 and 2000. The initial NAAL report (U.S. Department of Education, 2005) also found little overall change in literacy levels between 1992 and 2003 in the adult population of the United States as a whole. But none of these studies considers *individual* change, and this is where the LSAL data become very important, as there is very little research that has directly measured changes in adult literacy proficiency over substantial periods of time using methodologies that enable individual as opposed to group or population change to be explored.

Before considering individual change within the LSAL population, let us look at the overall pattern of change. Figure 2.2 shows the mean literacy proficiency for Waves 1–5 for the 658 individuals who were assessed in each of these five years.[8] Note the time scale shown in Figure 2.2 corresponds to the constant spacing between successive assessments for any individual: the first assessment takes place at year "0", the second at year 1, the third at year 2, the fourth at year 4, and the fifth at year 6. A repeated measures ANOVA finds statistically significant differences in the proficiency of this population across the five time points (F = 3.775, df = 4, p = .005), with a significant linear trend (F = 8.879, F = 1, p = .003). The relatively small increase in population literacy level over time does not imply that literacy changes at the individual level are uniformly small (Reder & Edmonston, 2000). There can be many factors which affect the amount of individual gain (or loss) of proficiency over time,

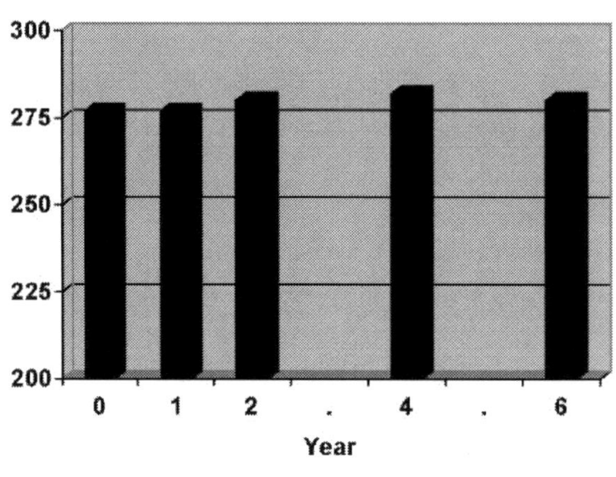

Figure 2.2 Mean document literacy proficiency assessed at Waves 1–5. Means calculated for the 629 individuals assessed at each of the five waves.

and the overall population change could be close to zero (as in the figure) if the gains of some individuals are offset by the losses in proficiency experienced by others. As we shall see in examining individual change, this is exactly what has happened in LSAL.

To examine individual change, we use the latent growth-curve methodology described in the introductory chapter of this volume. We first construct growth charts for individual adults' literacy proficiencies over time. Figure 2.3 shows such growth charts for sixteen randomly selected individuals from the LSAL sample. The literacy proficiency scores are plotted for each individual at the five time points of measurement: 0, 1, 2, 4, and 6 years after the study began. The lines drawn for each individual are the ordinary least-squares (OLS) regression lines fitted to that individual's data. Glancing at these illustrative cases, we see a broad range of overall proficiency

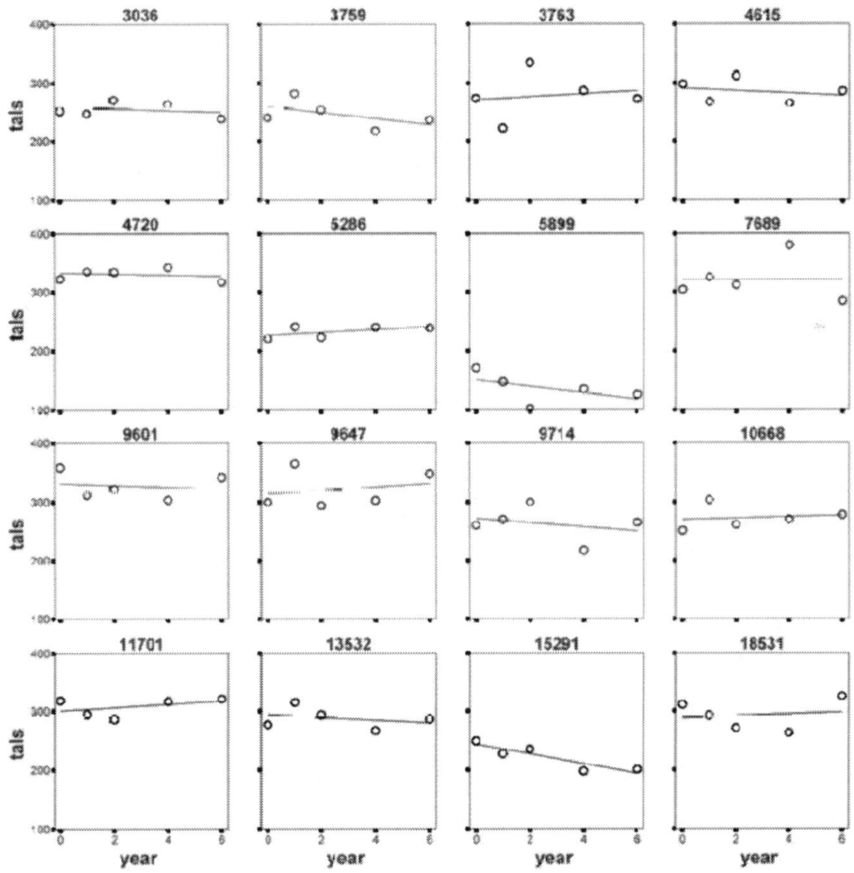

Figure 2.3 Literacy proficiency measured over time for sixteen randomly selected subjects, with OLS regression lines.

levels (as we saw above in Figure 2.1) and also a considerable range of apparent rates of change in proficiency over the six-year period. Some individuals, such as cases #5286 and #11701, display apparent positive growth (i.e., proficiency gains) over time, whereas others, such as #3759 and #9714, exhibit apparent negative growth (i.e., proficiency losses), and yet others, such as #3836 and #10668, exhibit relatively little proficiency change over time.

The empirical data plots shown in Figure 2.3 are a heuristic tool for examining some basic characteristics of observed changes in proficiency over time. These data suggest—and statistical analyses confirm—that we can use linear growth curves to model changes in proficiency over the six-year period of time. We see in Figure 2.3 an impressive variability among the individually fit ordinary least-squares (OLS) regression lines, with widely varying initial proficiency levels (the initial value or intercept of the line) as well as widely varying rates of change in proficiency over time (the slopes of the individual lines). The fluctuations of the individual data points around the regression lines in Figure 2.3 indicate that there is a fair amount of measurement error present on each assessment occasion.

We next estimate individuals' growth-curve parameters. Since we selected linear growth curves for our models, the growth curves are individual lines that can be represented by two parameters: a person-specific intercept (initial proficiency status) and a person-specific slope (rate of proficiency change).[9] We use standard statistical methods that have been developed for estimating individual growth curve parameters using multilevel models of change (Raudenbush & Bryk, 2002; Singer & Willett, 2003):[10]

Table 2.1 displays the results from statistical testing of a series of linear growth models of the LSAL proficiency data.[11] Five models are shown in the table. Model A is the "unconditional means" model, a convenient baseline. The unconditional means model assumes that each individual has his or her own proficiency level (initial status) and that no growth takes place from wave to wave. Wave-to-wave variations in assessed proficiency scores within an individual are assumed to reflect measurement error. Results shown in the table for Model A can be used to calculate the proportion of the total variance in proficiencies (598 + 1,671 = 2,269) that is attributable to between-subjects (1,671) variance. Thus, 74 percent of all the variance can be attributed to differences between persons.

Model B adds a single variable, the TALS Form (A versus B) used on the given occasion with the given respondent, to the unconditional means model. Forms were randomly assigned to respondents on Wave 1 and then alternated wave by wave thereafter. Although the two Forms were designed by the Educational Testing Service to be roughly equivalent, there is wave-to-wave within-person variance associated with the alternation of Forms, and adding the Form to the model improves the overall fit, which we test with the deviance difference test ($\chi^2 = 213$, df = 3, $p < .001$).[12]

Table 2.1 Multilevel Linear Growth Models Oof Proficiency

	A	*B*	C	*D*
Initial Status				
Intercept	277.56[a]	273.29[a]	271.44[a]	271.77[a]
Nonminority				21.33[a]
Learning disability				−16.84[a]
Highest grade				4.88[b]
U.S.-born				33.79[a]
Female				−11.62[b]
Age (Wave 1)				−0.72[b]
Linear Rate of Change				
Intercept			0.80[b]	0.91[a]
Nonminority				−0.19
Learning disability				0.22
Highest grade				0.24
U.S.-born				−1.52[c]
Female				0.75
Age (Wave 1)				−0.12[a]
Form		8.60[a]	8.53[a]	8.38[a]
Variance Components				
Level 1 within-person	598	495	468	467
Level 2:				
Initial status	1,671[a]	1,925[a]	1,925[a]	1,495[a]
Rate of change	3[a]	2[a]		
Form	265[a]	289[a]	290[a]	

Deviance	40,782	40,569	40,532	40,043
Parameters estimated	3	6	10	22

[a]p < .001; [b]p < .01; [c]p < .05.
Notes: In Model D, the predictors of Initial Status and Linear Rate of Change are centered around their grand means. Estimates made by HLM, version 6.03.

Model C is an unconditional linear growth-curve model. It adds the individual slope or rate of growth to the previous model. We term it *unconditional* because the individual growth rates, although variable within the population, are not systematically predicted by observed characteristics of individuals (e.g., their gender or age). Comparing the unconditional growth model in Column C with the unconditional means with form model in Column B, we see that the unconditional linear growth model fits the LSAL proficiency data better than the preceding baseline models ($\chi^2 = 37$, df = 4, $p < .001$). Notice that the *average* growth rate estimated is quite small, only 0.80 scale points per year. Over the six years of observation this averages about 5 scale points of growth. Although this average growth rate is significantly different than zero ($t = 2.77$, $p = .006$), it represents a modest overall rate of growth, and is generally consistent with findings of small but statistically significant gains in test scores reported by programs using short term "pre" and "post" comparisons of administrative data (Beder, 1999; Brooks et al., 2001; Rose, this volume).

Model D conditions the two individual growth-rate parameters—initial status and linear rate of change—on a number of background variables: *Nonminority* status (1 if white & non-Hispanic, 0 otherwise); *Learning disability* (1 if individual reported having a learning disability, 0 if not); *Highest grade* completed before dropping out of secondary school; *U.S.-born* (0 if foreign-born, 1 if native-born); *Female* (0 if male, 1 if female); *Age* (age at the beginning of the study). Comparing the fit of the nested unconditional growth (C) and conditional growth (D) models with the deviance difference test, we find that the conditional growth model fits better ($\chi^2 = 489$, df = 12, $p < .001$). All background variables in the conditional growth model are significant predictors of individuals' initial proficiencies. Individuals who are not members of minority groups, U.S.-born individuals, and men tend to have higher initial proficiencies. Individuals who do not have learning disabilities or who completed more years of schooling (before dropping out of secondary school) or were younger at the start of LSAL tend to have higher initial proficiencies as well. All of these results are consistent with cross-sectional findings from large proficiency surveys.

Most of these background variables, however, are not significant predictors of individuals' rates of change in proficiency over time. Only two variables—starting age and whether U.S.-born—are significant predictors of rate of change in proficiency. The negative coefficients shown in the table for these variables indicate that younger adults have higher rates of proficiency change, as do adults who were born outside the Untied States. Let's consider each of these findings in a bit more detail.[13]

According to the model, the younger the individual, the higher the rate of growth tends to be. There is a crossover point in age above which individuals' rate of change becomes negative (i.e., they tend to lose proficiency over time) and below which the rate is positive (i.e, they tend to gain proficiency over time). We can estimate the crossover age by setting the predictive equation for rate of change to 0 and solve for the starting age. This yields an estimated crossover age of about thirty-five years. Thus individuals thirty-five years of age or older tend to lose proficiency over time, whereas individuals younger than that tend to gain proficiency. According to the predictive equation for rate of change, our youngest LSAL subjects,

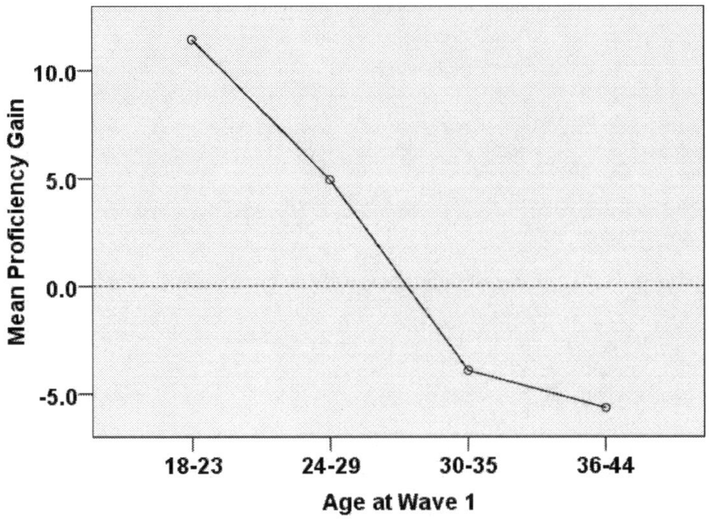

Figure 2.4 Proficiency gains between Waves 1 and 5 by age at Wave 1.

age eighteen at the time the study began, would have the highest average rate of change, estimated to be 2.05 scale points per year (equivalent to a gain of 12 points over the 6 years), whereas our oldest LSAL subjects, age forty-four at the beginning, would experience the lowest rate, averaging a *loss* of 1.24 scale points per year (equivalent to a loss of about 8 points over the 6 years of observation).

Figure 2.4 displays average proficiency changes between Waves 1 and 5 for different LSAL age subpopulations. As expected from the results of the growth-curve modeling, we see the largest positive gains in the youngest group (ages 18–23 at Wave 1), with progressively smaller gains in progressively older groups. The older groups show an average loss in proficiency.

These effects of age on proficiency growth are consistent with cross-sectional analyses of a wide range of other adult literacy and numeracy assessments. Figures 2.5 and 2.6 show the average proficiencies for adults in various age groups on large-scale national assessments conducted in the United States and the United Kingdom, respectively.[14] The Document Literacy proficiency measure used in each of these national surveys is the same as the one used in LSAL. These data are corrected for differences in education and disabilities among the age groups, in order to highlight the cross-sectional effects of age more clearly. The inverted-U shape in these age cross sections is consistent with the longitudinal results from LSAL. The slope of the line segment connecting successive data points corresponds roughly to the rate of change in proficiency in that age range. The figures show, for both the United States and the United Kingdom, that the youngest group has the highest rate of change, with successively

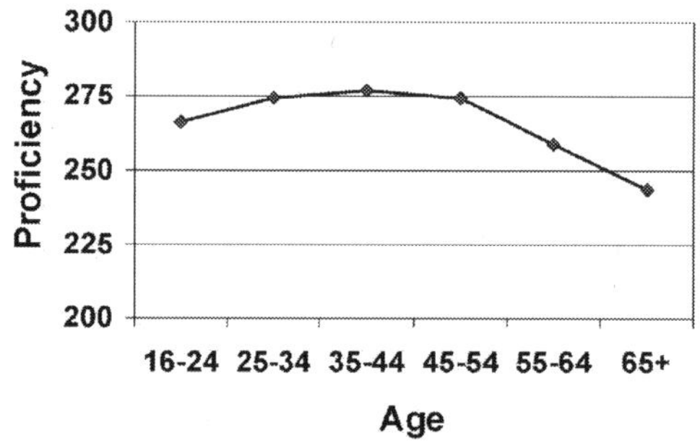

Figure 2.5 Mean document literacy proficiency by age for adults in the United States, regression-corrected for education and disabilities. *Source*: National Adult Literacy Survey, 1992. Author calculations.

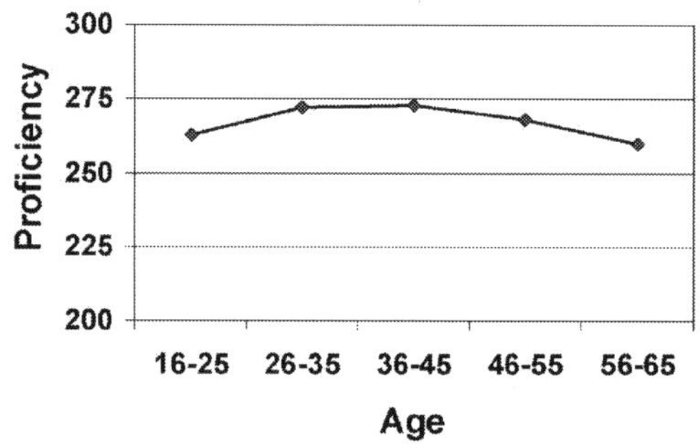

Figure 2.6 Mean document literacy proficiency by age for adults in the United Kingdom, regression-corrected for education and disabilities. *Source*: International Adult Literacy Survey, 1995. Author calculations.

older groups showing less positive slopes. A midlife group shows a relatively flat contour and the older groups show progressively more negative slopes (i.e., proficiency loss). Similar age-dependent rates of proficiency growth are evident in analyses of reading and math test scores of students in adult education courses (Reder & Kulongoski, 2006). "Synthetic cohort" studies of change over time in age cross sections of proficiency data in numerous countries are also consistent with this general picture that proficiency changes in adults are age-related (Murray, this volume; Willms & Murray, 2007).

Although the LSAL findings about age and proficiency change are consistent with cross-sectional analyses, they greatly extend previous findings. Inferring developmental change from cross-sectional age differences potentially confounds the maturational effects of aging with the period or cohort effects that may be correlated with age, as other studies of aging and cognitive change have demonstrated (e.g., Schaie, 1996). Within-person growth-curve analyses confirm the maturational nature of age-based cross-sectional differences in literacy and numeracy seen in large-scale assessments and surveys.

There is good reason to suppose that cognitive factors that change with age may directly influence adults' performance on functional literacy tasks such as those assessed in LSAL and these other studies. Longitudinal studies of adults' performance in laboratory tasks show a range of age effects on different types of tasks. Performance on some types of tasks systematically declines with age after early adulthood, whereas performance on other types of tasks shows quite a different age pattern, with performance progressively improving across much of the adult life span, leveling off and finally declining only at some point past middle age (Salthouse, 1996; Schaie, 1996). Psychologists since the 1920s have tried to account for these life-span developmental patterns in terms of two types of underlying cognitive abilities, one type that declines with age throughout adulthood (cognition variously labeled "fluid," "type A," "cognitive mechanics," and "process" by different authors) and another component that increases with age through adolescence and young adulthood (cognition variously labeled "crystallized," "type B," "cognitive pragmatics," and "product"). For present purposes, we'll label them fluid and crystallized abilities (Salthouse, 1996). Although such a framework has the potential to predict a range of age effects in the functional literacy tasks being considered here, most of its research is based on the performance of laboratory tasks rather than everyday functional literacy activities. When abilities to perform more "everyday" type of tasks are assessed, systematic decreases in performance with age have rarely been reported before the age of sixty (Schaie, 1994; Willis & Schaie, 1986).

Laboratory studies of cognitive change across the life span (e.g., Salthouse, 1996) provide some clues about the underlying cognitive determinants of these changes in adult literacy proficiency. It is possible, for example, that most functional literacy tasks draw on both fluid and crystallized abilities, with the former decreasing over the adult life span while the latter are increasing, together influencing overall task performance to yield the inverted-U shape age contours seen in the cross-sectional and longitudinal data.

Our growth-curve models indicate that the rate of proficiency change is considerably higher for immigrant than for native-born adults in the United States. This holds when we control for age in the models. Performance on the TALS proficiency assessment, in which individuals are asked to perform simulated "everyday" tasks—such as finding items on a map, extracting information from train schedules, following directions on medication labels—draws upon both cultural knowledge as well as cognitive skills. Immigrant adults, even those with conversational proficiency in English, may be continuing to acquire advanced cultural and linguistic knowledge across the life span that native-born adults already possess. To the extent that the TALS proficiency depends upon the acquisition of that linguistic and cultural knowledge, we would expect immigrants' rate of proficiency gain to be higher than that of native-born adults (holding other variables constant).[15]

Figure 2.7 further illustrates how age and country of birth influence estimated individual rates of proficiency growth in adulthood. The rate of proficiency growth in scale points per year over the six-year observation period is plotted against the individual's age at the beginning of the study. Each point in the figure is a respondent. Native-born adults are plotted with circles, whereas foreign-born adults are plotted with triangles. These are model-based *empirical Bayes* estimates (Morris, 1983) of individual slope parameters, which are similar to but more efficient than OLS estimates and are the best linear unbiased predictors (Singer & Willett, 2003). The slope parameters clearly decline with increasing age, as we saw before. The dotted horizontal line is drawn at a the zero slope level, so that points above the line are individuals showing positive growth and those below the line are showing negative growth (i.e., loss) of proficiency. We also see immigrants' higher rates of proficiency growth in Figure 2.7, where they seem to be a qualitatively distinct subgroup. Their heightened rate of proficiency growth is evident at all ages.

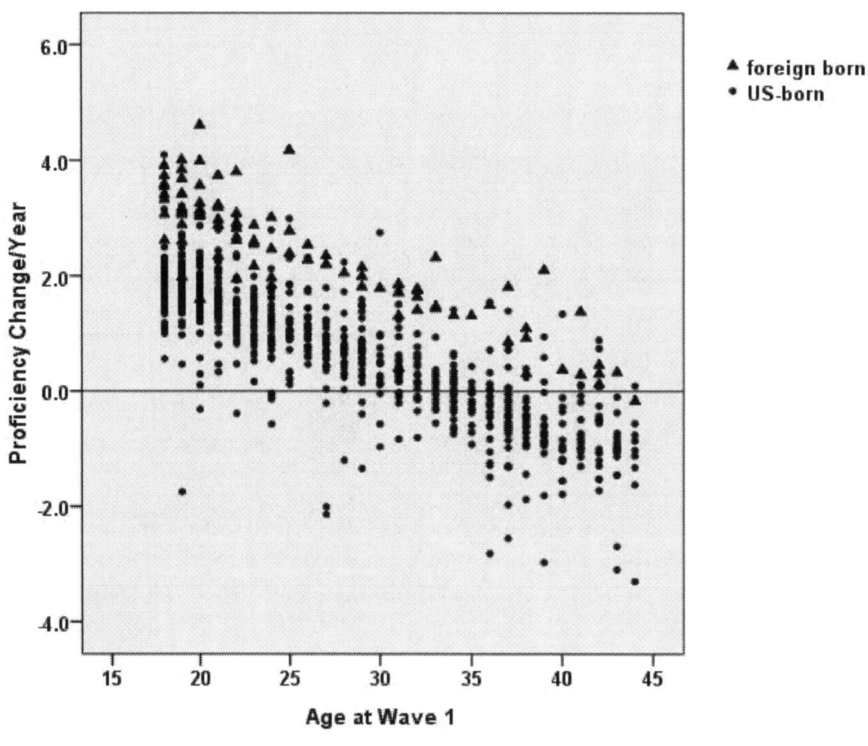

Figure 2.7 Empirical Bayes estimates of individuals' rate of proficiency change per year for Waves 1–5, by their age at Wave 1. Circles are U.S.-born; triangles are foreign-born.

In addition to understanding the effects of time-invariant demographic predictors, we are interested in the time-dependent effects of key life events that occur at different times in adults' lives. Many scholars and practitioners of adult literacy and numeracy development believe that key life events—such as the birth of a child, a qualitative change in employment, participation in a program (to mention but a few of the myriad possibilities)—have great impact on the processes of literacy and numeracy development in adult life (Beder, 1991). We explored the influence on proficiency development of a few of the key life events systematically recorded in LSAL: participation in an adult literacy/numeracy course; self-directed learning (self-study) to improve basic skills or prepare for the GED tests (Reder & Strawn, 2001, 2006); receipt of the formal GED credential. The event of obtaining a GED credential can occur at most one time, whereas the events of participating in a program and self-study can occur in multiple time periods for individual respondents. Further information about these events is available in Reder and Strawn (2006).

The effects of these events are modeled as *person-centered* time-varying predictors of proficiency, added on to the growth-curve predictions; the effects of the occurrence of a particular event at one time are measured against the effects of nonoccurrence of that event at other times for the same individual. So each person is, in effect, his or her own control. No statistically significant effects of these time-varying predictors were found on proficiency development. Receipt of a GED during a given period of time had a marginal, nearly significant positive effect on the proficiency assessed at the next wave ($t = 1.849$, $p = .064$).

DEVELOPMENT OF LITERACY AND NUMERACY PRACTICES

LSAL collected numerous measures of individuals' literacy and numeracy practices—in home, community, and workplace settings—each time they were interviewed. Respondents reported how often they engaged in various literacy practices, such as writing a note, reading fiction, reading the news section of the newspaper, doing math for a bank statement, and so forth. LSAL asked two questions about each practice. Respondents were first asked if they ever engaged in the practice, for example, "Do you ever read the news section of the newspaper?" (yes or no). If respondents answered yes, they were asked a second question about their frequency of engaging in that practice (e.g., "How often do you read the news section of the newspaper?"), answering on a five-point scale ranging from 1 ("rarely") to 5 ("every day"). Answers to the pair of questions for each practice were combined so that those indicating they had never engaged in a practice received a score of 0 for never. Thus, the possible range of scores for each practice was from 0 ("never") to 5 ("every day").

Confirmatory factor analysis was carried out on these data.[16] Confirmatory factor analysis is a highly sensitive technique that provides information about the degree to which items measuring the same concept are closely related as well as information about whether there are multiple concepts measured by the set of items. Using this analytic approach, we eliminated items that tended to be poorly associated with the other items, and we discovered that two central concepts were being assessed by these questions, which are glossed as *literacy practices* and *numeracy practices*.

After finalizing this confirmatory factor model with the Wave 1 data, we conducted a series of analyses to ensure that the measurement properties of the scales were consistent across waves of the study. It is essential to establish consistent measurement properties across waves in order to conduct valid longitudinal analyses of change. Without consistent measurement properties, it is impossible to distinguish changes in literacy practices from changes in the measurement properties of the scale. Purcell-Gates and colleagues (2004) also investigated literacy practices using similar types of interview items. They were able to construct a highly reliable scale using item response theory scaling methods but their scale was not longitudinally stable from the first interview to the second interview, so they were not able to measure change in literacy practices with their scale. LSAL's results, demonstrating that the measurement properties of its scales are stable across waves, thus provide a solid measurement foundation for longitudinal analyses.

It is important to point out, however, that in order to obtain longitudinally stable measurement properties, it was necessary to reduce the number of items used per scale, so that the resulting scales are measured with considerable error (in terms of classical psychometric criteria) on any one occasion.[17] For repeated-measures applications such as LSAL, however, such measurement error may be acceptable if it is stable over time. Also note that the particular items comprising these longitudinally stable scales were relatively broadly specified practices (e.g., "How often do you read fiction?") rather than more narrowly specified practices. Further research and development are needed to clarify the extent to which longitudinally stable measurement of practices may require such broad specification of context.

Growth in literacy practices

Growth-curve models, similar to those constructed for repeated measures of proficiency, were developed for these longitudinally stable literacy and numeracy practices measures. Table 2.2 displays modeling results for the literacy practices.

Model A is the unconditional means model, from which we can calculate that 53 percent of the total variance among literacy practices scores is between persons (as opposed to between occasions or measurement error). This is a much lower percentage than we saw for the literacy proficiency measure (where 74 percent of all variance is between persons). This indicates that there is more variation from occasion to occasion over time in literacy practices measures than in literacy proficiency measures.

Table 2.2 Multilevel Linear Growth Models of Engagement in Literacy Practices

	A	*B*	*C*	*D*
Initial Status				
Intercept	2.225[a]	2.141[a]	2.105[a]	2.096[a]
Nonminority			0.228	0.307[c]
Learning disability			0.009	–0.036
Highest grade			0.052	0.041
U.S.-born			0.097	0.131
Female			0.735[a]	0.636[a]
Age (Wave 1)			0.010	0.008
Linear Rate of Change				
Intercept		0.036[b]	0.040[b]	0.030[b]
Nonminority			0.008	0.013
Learning disability			–0.012	–0.022
Highest grade			–0.000	–0.001
U.S.-born			0.015	–0.014
Female			–0.066[b]	–0.057[c]
Age (Wave 1)			–0.003[c]	–0.003[b]
Time-varying Predictors				
Program participation				0.331[a]
Self-Study				0.253[b]
Obtain GED				0.017
Variance Components				
Level 1within-person	0.798	0.738	0.737	0.696
Level 2:				
Initial status	0.885[a]	0.999[a]	0.839[a]	0.902[a]
Rate of change		0.009[a]	0.007[a]	0.006[a]
==============				
Deviance	12,603	12,517	12,385	12,101
Parameters estimated	3	6	18	21

[a]$p < .001$; [b]$p < .01$; [c]$p < .05$.

Notes: In Models C & D, the predictors of Initial Status and Linear Rate of Change are centered around their grand means. In Model D, time-varying predictors are centered around means of individuals. Estimates made by HLM, version 6.03.

Model B is the unconditional linear growth model for literacy practices. The difference in deviance between Models B and A, 12,603–12,517 = 86, is highly significant with three degrees of freedom ($\chi^2 = 86$, df = 3, $p < .001$), indicating that the unconditional linear growth model fits the literacy practices data better than the unconditional means model. The mean rate of growth is significant, indicating that, on the average, individuals' engagement in literacy practices increases over time. The variance components associated with both the initial status and linear growth rate are significant, although the variance component for the rate of change is much smaller than for the initial status, as we found for the proficiency growth curves as well.

Model C conditions the growth-curve parameters—the initial status and linear rate of change—on individuals' background and demographic variables. The deviance difference test indicates that Model C fits the data better than Model B does ($\chi^2 = 132$, df = 12, $p < .001$). Women and individuals who are not members of minority groups have higher initial statuses on this measure of engagement in literacy practices. The initial status does not vary significantly with other background variables as it did in the proficiency growth curves. The rate of change is lower for females, indicating that the gender difference in initial status, favoring women at the beginning of the study, diminishes over time. Age also influences the rate of change in this measure of literacy practices, such that older individuals gain less proficiency over time than do their younger counterparts. Further analysis indicates that the effects of age on change in literacy practices is far less pronounced than it is for literacy proficiency, as indicated, for example, by the nonsignificant effects of age on initial status of literacy practices.

Model D adds time-varying predictors to the growth-curve models for literacy practices. These time-dependent predictors, centered within individuals so that each person serves as his or her own control over time, significantly improve the fit of the model according to a deviance difference test ($\chi^2 = 284$, df = 3. $p < .001$). Both program participation and self-study have significant, positive time-dependent effects on the growth curves, while obtaining a GED does not have a significant effect. It is quite interesting that program participation and self-study activities have demonstrable effects on the development of literacy practices but not on literacy proficiency. We will consider this more fully below.

Growth in numeracy practices

Table 2.3 shows the results for the corresponding growth models of engagement in numeracy practices. Model A is the unconditional means model. Half (50 percent) of the variance in the numeracy practices scores is between subjects, comparable to the 53 percent figure we saw earlier for literacy practices but much lower than the 74 percent we obtained for the proficiency scores. Model B is the unconditional growth model, which fits the numeracy practices data better than Model A ($\chi^2 = 57$, df = 3, $p < .001$). The mean growth rate

Table 2.3 Multilevel Linear Growth Models of Engagement in Numeracy Practices

	A	*B*	C	*D*
Initial Status				
Intercept	2.579[a]	2.474[a]	2.434[a]	2.461[a]
Nonminority			0.216	0.416[c]
Learning disability			–0.059	–0.138
Highest grade			0.298[a]	0.266[a]
U.S.-born			0.240	0.292
Female			0.265	0.068
Age (Wave 1)			0.029[b]	0.023[c]
Linear Rate of Change				
Intercept		0.018[c]	0.049[b]	0.042[b]
Nonminority			–0.006	–0.053
Learning disability			–0.036	–0.007
Highest grade			–0.035[c]	–0.026
U.S.-born			–0.035	–0.055
Female			0.019	0.038
Age (Wave 1)			–0.004	–0.002
Time-varying Predictors				
Program participation				0.197[c]
Self-Study				0.167[c]
Obtain GED				–0.167
Variance Components				
Level 1within-person	1.194	1.095	1.078	1.057
Level 2:				
Initial status	1.175[a]	1.328[a]	1.056[a]	1.127[a]
Rate of change	0.016[a]	0.014[a]	0.010[a]	
=============				
Deviance	14,200	14,143	13,972	13,732
Parameters estimated	3	6	18	21

[a]p < .001; [b]p < .01; [c]p < .05.

Notes: In Models C & D, the predictors of Initial Status and Linear Rate of Change are centered around their grand means. In Model D, time-varying predictors are centered around means of individuals. Estimates made by HLM, version 6.03.

is significantly positive, as are the variance components associated with the initial status and growth rate parameters of the linear growth curve models.

Model C conditions the growth rate parameters on individuals' background and demographic characteristics. This conditional growth model fits the numeracy practices data significantly better according to a deviance difference test ($\chi^2 = 161$, df = 12, $p < .001$).

Education and age are positively associated with the initial status, such that older individuals and those who completed more years of schooling (before dropping out) have higher initial levels of engagement in numeracy practices. There are no significant gender or minority-group differences in numeracy practices, differences that we found in the growth of literacy practices. The only significant predictor of rate of change is education, which is negative, indicating that the higher initial status associated with more years of schooling completed is offset by a slightly lower rate of change associated with increasing years of schooling completed. The non-significant effect of age on rate of change, coupled with the positive effect of age on initial status, is consistent with the overall positive growth rate in Model B for engagement in numeracy practices. Individuals' engagement in literacy practices tends to increase steadily across the life span

The effects of adding time-varying predictors associated with self-study, program participation, and GED receipt to the growth-curve model are shown in Model D. The deviance difference test indicates that Model D fits the numeracy practices data significantly better than Model C does ($\chi^2 = 240$, df = 3, $p < .001$). As found for literacy practices, both program participation and self-study add positively to engagement in numeracy practices, whereas receipt of the GED credential has no significant effect on the growth of numeracy practices.

DISCUSSION

These results deepen our understanding of the development of literacy and numeracy in adult life. First, and foremost, is the central finding that literacy and numeracy do indeed continue to develop among adults after they leave school. Although the experiences of adult literacy students and teachers and a wealth of administrative data from programs (e.g., Rose, this volume) indicate that adults going through basic skills programs undergo at least short-term proficiency gains (follow-ups are generally limited to short-term intervals), LSAL provides longer-term data about proficiency changes among target adults who attend programs and those who do not. We also see systematic changes in literacy and numeracy in measures based on literacy and numeracy practices.

Growth-curve modeling indicates that these changes, however measured, are heterogeneous: some adults' proficiencies or engagement in literacy or numeracy practices increase over time, while others' decrease, while still others' change relatively little. This heterogeneity of change is structured with respect to both demographic and background characteristics as well as with respect to life events and experiences, in ways that vary with the particular measure of literacy or numeracy used.

There are important similarities and differences in the changes observed among the various measures or dimensions of literacy and numeracy examined. To begin with, as one would expect, the measures are positively correlated. Individuals with relatively high levels of literacy proficiency also have relatively high levels of engagement in literacy and numeracy practices. The overall levels (initial statuses) of proficiency and of practices are positively correlated. Empirical Bayes estimates (EBs) of initial status in proficiency are positively correlated with those of literacy practices ($r = 0.237$, $p < .001$) and with those of numeracy practices ($r = 0.266$, $p < .001$). When these correlations are disattenuated for the measurement error present in both the proficiency and practices measures, the underlying correlations are somewhat stronger: 0.308 and 0.352, respectively. The initial statuses of the literacy and numeracy practices measures are also positively correlated ($r = 0.409$, $p < .001$), which, when disattenuated from measurement error, yields an underlying correlation of 0.580.

Although there is moderate correlation among individuals' *levels* of proficiency and practices, their dynamics of change over time are quite different. We saw that different background characteristics influence rates of change over time in the various measures. Furthermore, we saw that life events and activities affect the growth curves of the measures in different ways. Participation in skills programs, for example, does not affect the development of proficiency but does affect engagement in literacy and numeracy practices. Overall, there is some correlation between individuals' rates of change (or slopes) for proficiency and practices. The correlation between the EBs of proficiency and reading practices *slopes* is not significant ($r = 0.054$, $p > .05$). The corresponding correlation between proficiency slopes and math practices slopes is only marginally significant ($r = 0.065$, $p = .045$). When disattenuated, this correlation rises to 0.421.

Although many background variables influence the level or initial status of the proficiency measure, only two variables influence the slope or rate of change of proficiency, age, and place of birth. The rate of proficiency change is negatively related to the individual's age at the start of the period. This finding, we saw, is closely related to the "inverted-U" cross section of proficiency by age seen in many national and international surveys, showing literacy lowest among young and old adults and at intermediate levels among middle-aged adults. The loss of proficiency over time among older adults—and according to our findings, this decline may start as early as thirty-five years of age—has far-ranging implications. Traditionally there have been two sources of new adults with basic skills needs in a society: youth leaving school without the skills they need and immigrants arriving without the skills they need. Our findings suggest that there may be a third source: older adults who have lost skills they once had. With our graying workforces and societies, there may be increasing need for adult educators to develop new types of programs that focus on skill retention among older adults.

Our finding that younger adults tend to gain proficiency over time also carries important implications for adult education. Recall that we also found that participation in basic skills programs does not impact proficiency development according to our growth-curve analyses. Taken together, these findings suggest that administrative reports of proficiency gains among program participants, often taken as evidence of program impact, may instead reflect proficiency gains being experienced by young adults regardless of their participation in programs. Since administrative data such as these typically do not include nonparticipating adults as controls, it is easy to misinterpret gains observed among program participants.

It is important to emphasize that we are *not* suggesting that programs are generally ineffective. We are suggesting instead that proficiency gains may not be the appropriate measure of program impact. Other dimensions of growth, such as changing levels of engagement in everyday literacy and numeracy practices, may be better indicators of program impact and effectiveness. Our growth-curve models found strong, immediate connections between program participation and measures of literacy and numeracy practices. Using quite a different methodology, Purcell-Gates, Degener, Jacobson, and Soler (2000) also found a direct impact of adult literacy programs on students' literacy practices. Rather than holding programs accountable for their students' short-term proficiency gains, the LSAL findings suggest that changes in learners' engagement in literacy and numeracy practices would be a more effective way to assess program impact

Considerable effort, of course, would need to be invested in developing valid and reliable measures of literacy and numeracy practices. Some of the initial progress made in analysis of the LSAL practices data can inform such efforts. The traditional psychometric criterion for developing scales with high reliability may not be an adequate approach. For longitudinal analyses of the LSAL practices data, it was necessary to build scales with longitudinally stable measurement properties. This resulted in constructing two scales—one for literacy practices and one for numeracy practices—comprised of very few items. These scales of course had low Cronbach alpha values (due to the small number of items) but were very useful analytically as repeated measures. It is quite interesting that the items comprising these longitudinally stable scales were very broad in terms of practices (e.g., how often one reads fiction) rather than highly specific. Classic psychometrically constructed scales, on the other hand, tend to be comprised of many more items that are highly specific in nature. Purcell-Gates et al. (2004), for example, had more than thirty items in their IRT-constructed scale of engagement in literacy practices, but found that the resulting scale did not have stable item parameters when fitted to data at two different points in time. It thus could not be used for longitudinal analyses (Purcell-Gates, personal communication, 2003). Further research and development will help us understand how best to measure engagement in literacy and numeracy practices for use in longitudinal research.

It may well be that further research will demonstrate that proficiency gains are a long-term outcome of participation in programs. The models we tested in this chapter looked at relatively short-term effects of participation events on growth curves. We found no short-term effects of such events (including program participation, self-study, or receipt of the GED) on proficiency development, but we did find clear short-term effects on measures of engagement in literacy and numeracy practices. Since we found positive correlations between rates of growth in proficiency and practices, it may be that these short-term impacts on practices will eventually mediate longer-term effects on proficiency development. This is predicted by practice engagement theory (Reder, 1994). Further modeling of the LSAL data will allow us to examine these possibilities more carefully.

Several of the growth-curve models we examined included time-varying covariates as markers of the occurrence of specific life events and activities. These event indicators are added to the underlying growth-curve processes and statistically tested to see if such events as program participation, self-study, and receipt of the GED credential influence various measures of literacy and numeracy development. For measures of engagement in literacy and numeracy practices, both program participation and self-study activities have positive, time-specific effects on growth curves.

It is important to note that these findings are based on models that combine developmentally oriented growth curves with time-varying life-history covariates. The richness of these models can help us understand literacy and numeracy development in new ways. These models, for example, extend our understanding of previous findings about differences in rates of improved skills and GED attainment among groups that self-study or participate in basic skills programs (Reder, 2007; Reder & Strawn, 2006). These types of models for literacy and numeracy development, combining growth processes and life histories, could be made much richer by incorporating other key life events such as the birth of a child, changes in family or household composition, changes in health or employment status, new uses of technology, and so forth. Plans are underway to develop this approach further in continuing analyses of the LSAL data.

NOTES

1. This research was supported through the National Center for the Study of Adult Learning and Literacy (NCSALL) under Award Number R309B960002 from the U.S. Department of Education. Thanks to Clare Strawn, my colleague and collaborator on the Longitudinal Study of Adult Learning (LSAL), for her many contributions to this work. Thanks to Erik Simensen and Cynthia Lopez for their leadership and management of the LSAL interviewers and trackers too numerous to mention individually, without all of whose tireless efforts this work would not be possible. All of us who work on the LSAL project are humbled by the enormous contributions of the nearly 1,000 LSAL respondents who have given so generously to the project by opening up

their lives, their homes, and in many cases, their hearts. Our hope is that the project will be not only *about* them but also *for* them.

2. LSAL also asked respondents, in each wave other than the first, about any changes they noticed in their reading, writing, and math skills since the previous interview. For each skill domain, an initial question probed whether there were any changes since the previous interview. If any change was noted for a skill domain (e.g., reading), follow-up questions asked about whether there had been a change in the individual's reading ability since the previous interview (*better*, *same*, or *worse*), a change since the previous interview in how often they read (*more often*, *the same*, or *less often*), or a change since the previous interview in the kinds of materials they read (*same* kinds or *different* kinds of materials). If the individual reported reading different kinds of materials, specific examples were requested. Parallel sets of questions were asked about changes in writing and math. These data provide a third and very interesting lens for studying the development of literacy and numeracy, but limited space prevents their inclusion here.
3. Although LSAL also collected some data about literacy *subskills* (e.g., vocabulary measures, reading fluency measures, phonological processing measures), these are not included in this chapter. Chapters by Condelli et al. and by Alamprese in this volume consider changes in literacy subskills in adult education students.
4. These characteristics were determined at the time that the LSAL sampling took place; individuals initially sampled from the defined population have been followed over time even though they may subsequently have moved from the Portland area, received a GED, and so on.
5. A sixth wave was collected during 2006–2007 but is not included in this chapter.
6. Thanks to Irwin Kirsch of the Educational Testing Service for providing this scoring program.
7. The levels of this scale shown here, corresponding to the levels reported for the 1992 NALS data, differed from the levels used in reporting the 2003 NAAL data.
8. Approximately 90 percent of the subjects were interviewed on any given wave in waves 2–5; subjects who failed to be interviewed in one wave might be interviewed in a subsequent wave, however. Analysis of the "in" and "out" pattern of subjects over waves indicates that these LSAL data are *missing at random* according to Little and Rubin's (1987) criteria and thus all data can be utilized in many analyses rather than just the smaller subsample of cases interviewed at *each* occasion. Nevertheless, certain analyses such as these repeated measures ANOVAs require the use of complete or "balanced" panel data.
9. Although the OLS regression lines such as those shown in Figure 2.3 provide reasonable initial estimates of these individual growth-curve parameters, they are usually not the best estimates of individual growth-curve parameters. Because we have reason to suppose that these estimates will reflect both *within-person change* and *between-person differences in change*, a multilevel statistical model for change is often appropriate (Bryk & Raudenbush, 1987; Rogosa & Willett, 1985).
10. These methods, in contrast to OLS methods, do not consider each individual's data in isolation from others' data. Instead, both the person-specific intercept and the person-specific slope are assumed to be random variables with known distributions in the sampled population. Statistical models can be formulated and tested about how other individual characteristics and variables affect their intercepts and slopes, that is, their growth trajectories for literacy proficiency.
11. The HLM program, Version 6.03, was used to estimate these models.

12. For readers unfamiliar with the deviance difference test, this compares the relative goodness of fit of two nested models to the same data. The closeness of fit for each model is measured by its so-called likelihood "deviance." Comparing Model B to Model A, we see a difference in deviance of 40,782—40,569 = 213 that is gained by having three additional parameters estimated in Model B than in Model A. The difference in deviance has a chi-square distribution with degrees of freedom equal to the difference in number of parameters estimated.
13. The small but statistically significant variance components associated with individuals' rates of proficiency growth (marked in the table as "Level 2: Rate of change" in Models C, D & E) suggest that a simpler model might be appropriate in which individuals' rate of proficiency growth does not vary randomly but deterministically depending on age and place of birth. Attempts to fit that model to these data indicate that the current form of the model fits better.
14. These figures are based on author calculations using nationally representative household data collected by the National Adult Literacy Survey (NALS) in the United States in 1992 and in the United Kingdom by the 1996 International Adult Literacy Survey (IALS).
15. We explored this idea further by adding oral vocabulary scores assessed in Wave 2 to our growth-curve models to see how they would interact with place of birth in predicting the rate of proficiency growth. With oral vocabulary controlled, foreign-born adults still have significantly higher rates of proficiency growth than U.S.-born adults.
16. Jason Newsom conducted the scaling analyses of the literacy and numeracy practices.
17. The literacy practices scale was comprised of three items and the numeracy practices scale of two items.

REFERENCES

Beder, H. (1991). *Adult literacy: Issues for policy and practice.* Malabar FL: Krieger Publishing Company.

Beder, H. (1999). *The outcomes and impacts of adult literacy education in the United States.* Cambridge, MA: National Center for the Study of Adult Learning and Literacy, Report #6 (http://ncsall.gse.harvard.edu/research/report6.pdf).

Brooks, G., Davies, R., Duckett, L., et al. (2001). *Progress in adult literacy: Do learners learn?* London: The Basic Skills Agency.

Bryk, A. S., & Raudenbush, S. W. (1987). Application of hierarchical linear models to assessing change. *Psychological Bulletin*, 101(1), 147–158.

Kirsch, I. S., Braun, H., Yamamoto, K., & Sum, A. (2007). *America's perfect storm: Three forces shaping our nation's future.* Princeton: Educational Testing Service.

Kirsch, I. S., Jungeblut, A., Jenkins, L., et al. (1993). *Adult literacy in America: A first look at the results of the National Adult Literacy Survey.* Washington, DC: U.S. Department of Education, National Center for Education Statistics.

Little, R. J. A., & Rubin, D. B. (1987). *Statistical analysis with missing data.* New York: Wiley.

Morris, C. N. (1983). Parametric empirical Bayes inference: Theory and applications. *Journal of the American Statistical Association*, 78, 47–65.

Moser, C. (1999). *A fresh start—improving literacy and numeracy.* London: Department for Education and Employment.

Oregon Progress Board (1991). *The Oregon Literacy Survey: Measuring adults' functional skills.* Salem OR: Author.

Organization for Economic Cooperation and Development. (1997). Literacy skills for the knowledge society: Further results from the International Adult Literacy Survey. Paris: Author.

Purcell-Gates, V., Degener, S., Jacobson, E., et al. (2000). *Affecting change in literacy practices of adult learners: Impact of two dimensions of instruction*. Cambridge, MA: Harvard Graduate School of Education, National Center for the Study of Adult Learning and Literacy.

Purcell-Gates, V., Jacobson, E., & Degener, S. (2004). *Print literacy: Uniting cognitive and social practice theories*. Cambridge, MA: Harvard University Press.

Raudenbush, S. W., & Bryk, A. S. (2002). *Hierarchical linear modeling: Applications and data analysis methods* (2nd ed.). Thousand Oaks, CA: Sage.

Reder, S. (2007). Giving literacy away, again: New concepts of promising practice. In A. Belzer (ed.), *Toward defining and improving quality in adult basic education: Issues and challenges* (pp. 255–76). Mahwah, NJ: Lawrence Erlbaum Associates.

Reder, S. (1994). Practice engagement theory: A sociocultural approach to literacy across languages and cultures. In R. M. Weber, B. Ferdman, & A. Ramirez (eds.), *Literacy across languages and cultures*. Albany, NY: State University of New York Press.

Reder, S., & Edmonston, B. (2003). *Oregon Literacy Project: Final report*. Salem, OR: Department of Community Colleges and Workforce Development.

Reder, S., & Edmonston, B. (2000). *Demographic change and literacy development in a decade*. Washington, DC: National Center for Education Statistics.

Reder, S., & Kulongoski, K. (2006). *Using statewide administrative data for research, evaluation and policy development in adult education* Paper presented at the annual meeting of the Commission on Adult Basic Education (COABE), Houston, TX, April 27–29, 2006.

Reder, S., & Strawn, C. (2006). Broadening the concepts of participation and program support. Focus on Basics, 8(C), 6–10.

Reder, S. & Strawn, C. (2001a). The K-12 school experiences of high school dropouts. Focus on Basics, 4(D), 12–13.

Reder, S. & Strawn, C. (2001b). *Program participation and self-directed learning to improve basic skills*. Focus on Basics, 4(D), 14–17.

Rogosa, D. R., & Willett, B. (1985). *Understanding correlates of change by modeling individual differences in growth*. Psychometrika, 50, 203–228.

Salthouse, T. A. (1996). The processing speed theory of adult age differences in cognition. *Psychological Review*, 103(3), 403–28.

Schaie, K. W. (1994). The course of adult intellectual development. *American Psychologist*, 49(3), 304–13.

Schaie, K. W. (1996). *Intellectual development in adulthood*. Cambridge, MA: Cambridge University Press.

Singer, J., & Willett, J. (2003). *Applied longitudinal data analysis: Modeling change and event occurrence*. New York: Oxford University Press.

Statistics Canada and Organization for Economic Cooperation and Development (2005). Learning a living: First results of the Adult Literacy and Life Skills Survey. Ottawa and Paris: Author.

U.S. Department of Education. (2005). *A first look at the literacy of America's adults in the 21st Century*. Washington, DC: Author.

Willis, S. L., & Schaie, K. W. (1986). Practical intelligence in later adulthood. In R. J. Sternberg & R. K. Wagner (eds.), *Nature and origins of competence in the everyday world*. Cambridge: Cambridge University Press.

Willms, D., & Murray, T. S. (2007). *Gaining and losing literacy skills over the lifecourse*. Ottawa: Statistics Canada.

3 Longitudinal Research Related to the Acquisition and Maintenance of Literacy

T. Scott Murray

This chapter describes three linked elements of longitudinal research that are being undertaken by Statistics Canada to explore the acquisition, maintenance, and loss of literacy skill and related implications for public policy. The three studies are:

> The Organization for Economic Cooperation and Development's (OECD) 2000 Programme for International Student Assessment and Youth in Transition longitudinal study (PISA/YITS) being conducted by Statistics Canada.
>
> A synthetic cohort analysis of the International Adult Literacy Survey (IALS) and the Adult Literacy and Life Skills Survey (ALL) data sets for Canada.
>
> The addition of IALS/ALL literacy and numeracy measures to New Zealand Center for Educational Research (NZCER's) Competent Learners study.

The three studies share a common goal of understanding processes of skill gain and loss over the life course. All three studies focus on literacy, the ability to understand and apply information derived from print to solve everyday problems. Taken together, the studies in question represent an investment of a significant amount of public resources. It is reasonable to ask, therefore, why Canada's policymakers worry about literacy.

In a word, "change." Change is a defining feature of modern life. Technologies change, the organization of work changes, terms of trade change, communities change, and social roles change as individuals negotiate the life course. Change is unavoidable. It obliges individuals, social institutions, and nations to adapt. Individuals and institutions that have the ability to adapt are resilient—they survive and have the chance to flourish.

Individuals and institutions that lack the ability to adapt become vulnerable and dependent. The ability to adapt to change depends, among other things, upon the skills that individuals, social institutions, and nations possess. Literacy has been recognized as an important economic and social asset for at least 3,200 years (Statistics Canada & HRDC, 1995),[1] Thus, it is in Canadians' collective self-interest to understand how literacy will influence their lives in the future and how they need to respond as parents, learners, citizens, and employees.

Literacy has been identified as a critical tool for achieving balanced and equitable social and economic development (Murray, Binkley, & Clermont, 2005; Rychen & Salganik, 2001, 2003;). Literacy—its level and distribution—has played a central role in generating the economic wealth that allows Canadians to enjoy one of the highest standards of living in the world (Coulombe & Tremblay, 2006a, 2006b; Coulombe, Tremblay, & Marchand, 2004). It would seem that literacy will become more important in the emerging global knowledge economy, a fact that argues for the attention of policymakers (Canadian Council on Learning, 2007).

Low literacy levels are also responsible for creating most of the social inequality in many of the outcomes that Canadians value most—their physical health, their ability to access learning, their ability to participate socially and democratically, and to enjoy stable, high-wage employment. Canada's markets—its labor markets, health systems, learning systems, and political systems—are savagely efficient at identifying and rewarding skill, to the point that one has to worry about the consequences for those with low skill levels (Macracken & Murray, 2008; Maxwell & Teplova, 2008).

Literacy also plays a central role in setting the levels of taxes and what is received from the activities of governments. The demand for, and the cost of, providing public goods and services, including health and educational services, would be much reduced if literacy levels were higher.

Thus, worry about the economics of literacy is pragmatic. Like it or not, economics dominates the public policy process in Canada. Thus, to be successful, advocacy for public investments in literacy must move beyond moral imperatives to economic ones.

Understanding the economics of literacy becomes an imperative for those who want to change the world for the better and begs answers to a series of simple questions, including:

- What is the level and distribution of adult literacy skill?
- How does the level and distribution of skill compare to that of competitors?
- How does the level and distribution of literacy skill influence individual outcomes?
- How does the level of literacy skill influence Canada's performance at the macro level, including relative macroeconomic performance, population health, and citizen engagement?
- How will the literacy skill profile evolve over the medium term?

- Would Canada's quality of life be improved if literacy skill levels were higher?
- If so, what role could or should governments play in encouraging such improvement, including funding remedial programming for adults?

Answering these questions depends upon access to information about the stock of literacy skill, the quality of the current flow of literacy skill, an understanding of how literacy is acquired, and how it is gained and lost over the life course. The three studies in question were put in place to respond directly to these questions.

The NZ study offers an incredibly rich variable set, including repeated cognitive measures since birth, that allow an examination of key mediators of performance during the period when literacy is generally acquired. It offers insight into how early childhood experience and the initial education system serve to amplify (or attenuate) social inequality in levels of acquired literacy. As such, it provides policymakers with a sense of where and how they might intervene to both raise average proficiency levels and to reduce prevailing levels of social inequality in literacy outcomes.

The first PISA/YITS cohort was inducted in 2000 at age fifteen. The PISA part of the study provides the means to monitor changes in the quality of skill over time in repeated cohorts of students. The YITS longitudinal follow-up lacks early covariates but has a large sample size and will retest to explore skill gain and loss between fifteen and twenty-four years of age—where the cross-sectional estimates suggest the most skill gains and losses occur in adulthood. The final study is not a true longitudinal study but rather a synthetic cohort study aimed at reconstructing average longitudinal trajectories for different population subgroups from IALS and ALL data sets. Policymakers believe that Canada is exposed economically. Relatively small birth cohorts will constrain the ability to realize skill gains through educational reform. High levels of immigration necessitate large expenditures to ensure rapid integration and to contain skill-based social inequality in labor market and other outcomes. The globalization of markets for goods and services have leveled the economic playing field at a time when educational investment and reform is rapidly increasing the global stock of literacy skill and allowing developing economies to compete on both price and quality. Among nations, Canada is perhaps one of the most exposed to such shifts in the terms of trade because of its dependence on international trade. A deep understanding of how the various skill flows will transform Canada's stock of literacy skill would require longitudinal data systems that would take decades to yield results. The synthetic cohort approach provides estimates of the average skill gain and loss experienced by different population subgroups—close enough, we think, to inform policy options and priorities.

Each of the three studies is described below:

THE 2000 PISA/YITS, BACKGROUND, AND RATIONALE FOR THE STUDY

The Youth in Transition Survey (YITS) was designed to examine the patterns of, and influences on, major transitions in young people's lives, particularly with respect to education, training, and work. Content includes measurement of major transitions in young people's lives, including virtually all formal educational experiences and most informal learning experiences, achievement, aspirations and expectations, and employment experiences. The implementation plan encompasses a longitudinal survey of each of two cohorts, ages fifteen and eighteen to twenty, to be surveyed every two years.

PISA/YITS is one project, which consists of two parallel survey programs: the Programme for International Student Assessment (PISA) and the Youth in Transition Survey (YITS). PISA is an international assessment of the skills and knowledge of fifteen-year-olds which aims to assess whether students approaching the end of compulsory education have acquired the knowledge and skills that are essential for full participation in society.

YITS is designed to examine the patterns of, and influences on, major transitions in young people's lives, particularly with respect to education, training, and work. Human Resources and Skills Development Canada and Statistics Canada have been developing the YITS in consultation with provincial and territorial ministries and departments of labor and education. Analysis of the cross-sectional proficiency estimates from the IALS and ALL studies by age have established that average literacy skill level continues to rise after the age of normal secondary graduation. The rise in the average is, however, accompanied by an increase in the variability of proficiency. Individuals in the top skill quartile appear to gain skills and individuals in the bottom skill quartile appear to lose skills. Without longitudinal data it is impossible to isolate the underlying processes that mediate skill change—hence the need for a large sample longitudinal cohort that tracks the evolution of skill over the crucial sixteen- to twenty-six-year age range and that provides a rich array of covariates to explain observed skill gain (or loss) and offer insight for policy.

Methodology

By retesting these respondents in PISA/YITS Cycle 5 in the winter of 2009, at the age of twenty-four, we will be able to: compute skill gain (or loss) measures for each sampled individual, analyze the factors that are associated with literacy skill gains and losses, and establish how tight those relationships are.

Skill loss is highly problematic, as it erodes the return on educational investments, reduces rates of productivity growth, and creates higher levels of social inequality in other outcomes, such as health and social engagement. The study will look explicitly at how skill conditions access and persistence in postsecondary education, access to employment, and the quality of employment, including access to adult education and training and high levels of skill use. This information is crucial for public policy—we expect it to reveal that skill loss is the product of either a demand deficiency or of displacement associated with the adoption of Information and Communications Technology (ICTs). The former would constitute a market failure, the latter a market success. Government intervention is required for both, but the nature of the intervention would be quite different.

The research population is a longitudinal cohort of fifteen-year-olds living in Canada's provinces. Initially inducted in 2000, and reinterviewed biannually until reaching the age of twenty-four in 2009, when they will be retested using a variant of the same literacy instruments.

The fifteen-year-old respondents to the PISA 2000 participated in both PISA and YITS. Starting in 2002, they are to be followed up longitudinally by YITS every two years. The survey population for PISA 2000 (15-year-olds) comprises persons who were born in 1984 and were attending any form of schooling in the ten provinces of Canada. Schools on Indian reserves were excluded, as were various types of schools for which it would be infeasible to administer the survey, such as home schooling and special-needs schools. These exclusions represent less than 4 percent of fifteen-year-olds in Canada.

The sample design for the reading cohort (15-year-olds) entails two-stage probability sampling, with a stratified sample of 1,200 schools selected at the first stage and a sample of eligible students selected within each sampled school. The initial student sample size for the reading cohort which was conducted in 2000 was 38,000.

The 2000 PISA focused on reading literacy but also provided measures of mathematics literacy (numeracy) and science literacy. The longitudinal content includes measurement of major transitions in young people's lives, including virtually all formal educational experiences, most labor market experiences, achievement, aspirations, and expectations.

Data are collected directly from survey respondents. The original PISA assessment was administered in classrooms with parent and teacher questionnaires administered separately. Collection for Cycle 2 took place from mid-February to mid-June 2002 using computer-assisted telephone interview (CATI). The response rate for the reading cohort (15-year-olds) was 90.5 percent. The combined response rate for both cohorts in Cycle 2 was 88.1 percent. The need to administer a literacy test in 2009 will likely necessitate a personal visit. The test itself will be Web-based and adaptive and so will yield a reliable point estimate of proficiency. The team is currently investigating the possibility of having the respondent do some of this over the Web.

The PISA reading test used a BIB design—one that sacrifices the reliability of individual proficiency estimates for good coverage of the content domain. This means that the initial estimate is inherently noisy but unbiased, whereas the second estimate will be relatively precise. We do not think this will interfere with the analysis.

FINDINGS AND EMERGING CONCLUSIONS

Analyses of the first few cycles of PISA/YITS have confirmed that literacy skill has a significant impact on downstream outcomes with lower skilled individuals doing much worse (Statistics Canada, HRDC and CMEC, 2001; Statistics Canada, HRSDC and CMEC, 2004). The following charts show the impact that literacy skill is having on secondary completion and postsecondary participation (Knighton & Bussière, 2006). Given the strong relationship between these outcomes and measures of early labor market success, we expect that time will serve to amplify skill-based inequalities.

These findings argue strongly for government policies designed to reduce social inequality in literacy skill in the education system. Given that these gaps manifest themselves early in life, generally before the age of school entry, the policy response would necessarily include early childhood education programs. Primary education systems would also need to devote energy and resources to diagnosing and remediating students

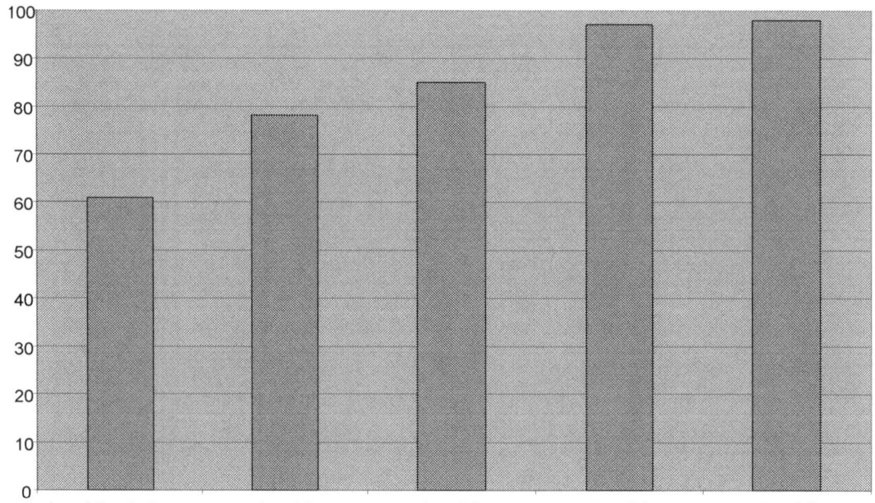

Figure 3.1 Percent of youth who completed high school by age 19, by reading proficiency level at age 15.

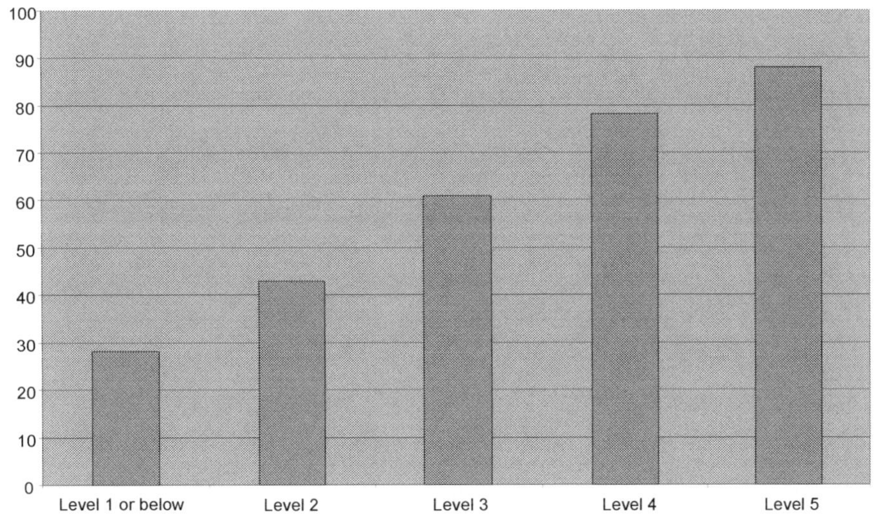

Figure 3.2 Postsecondary participation rates by reading proficiency level at age 15.

with low literacy achievement if they are to avoid amplifying the level of inequality in literacy skill observed at school entry.

A SYNTHETIC COHORT ANALYSIS OF THE IALS AND ALL DATA SETS FOR CANADA: BACKGROUND AND RATIONALE FOR THE STUDY

Comparison of the distributions of adult prose literacy and document literacy from IALS and ALL for those countries that have both data sets has identified the presence of significant change (see Statistics Canada & OECD, 2000; Statistics Canada & OECD, 2005). This study attempts, through a synthetic cohort analysis, to decompose skill change in the population and to relate the level and direction of change to background variables that true longitudinal research has shown to be important mediators of skill gain and loss.

Methodology

Comparison of IALS and ALL estimates allows one to identify net skill change for different population subgroups. This analysis used a Hierarchical Linear Model (HLM) to decompose average skill change into its components. Isolating skill loss demographically and occupationally will tell policymakers how to intervene, whether in the supply side, the demand side, or to improve the efficiency of the markets that mediate skill supply and demand.

Key research questions include:

- What is the overall level of skill gain and loss observed between 1994 and 2003?
- How does the level of average skill gain and loss differ by population subgroup?
- What factors appear to influence the magnitude of average skill gain and loss of different groups?

The IALS and ALL studies surveyed the civilian, non-institutional adult population of Canada living in the ten provinces of Canada in 1994 and observed again in 2003. HLM was used to isolate the influence of covariates on skill gain and loss in a way that allowed group trajectories to be reconstructed. The population, sixteen to sixty-five years of age in 1994, is compared to the population aged twenty-five to seventy-four in 2003. Immigrants to Canada between 1994 and 2003 are excluded from the 2003 data set. Out-migration and death in the 1994–2003 period remain unobserved. Estimates from IALS for 1994 will be based upon roughly 6,900 respondents aged sixteen and up, from ALL for 2003 some 16,000 respondents aged twenty-five and up.

IALS and ALL provide some nine hundred variables divided into three groups: demographic variables, variables thought of as determinants, variables thought of as outcomes related to health, the labor market, social and educational domains. The distinction between determinants and outcomes is somewhat blurred because what is an outcome at one time point becomes a determinant at later time points. Both IALS and ALL employed a personal interview in which a trained interviewer administered a background questionnaire and a test booklet.

The synthetic cohort approach is flawed in that it assumes that the relationships have been stable over the age range of the subjects and thus ignores period effects. The policy issues at play are thought to be sufficiently pressing, and the cost of inaction too high, that one cannot wait for real longitudinal data sets to yield an answer. Analyses of the macroeconomic impact of literacy skill differences simply backcast from the IALS, and ALL data suggest that skill has been far more important to GDP and productivity growth than conventionally assumed. The synthetic cohort approach is judged to be "good enough for government work" and will allow for several of the key assumptions in the growth modeling to be tested.

FINDINGS AND EMERGING CONCLUSIONS

Simple comparisons of average skill for different population subgroups at the two time points have already revealed the basic structure underlying skill gain and loss. This evidence fits with what is known from the available longitudinal data sets, particularly the work on numeracy by Bynner and Parsons (this volume) using the British birth cohorts. The analysis (Statistics Canada & OECD, 2005); Willms & Murray, 2007) has revealed several important facts, including that:

- Skill loss is concentrated in adults from lower socioeconomic backgrounds.
- Skill loss and gain are evident over the entire age range but is largest during prime working age.
- Provinces differ greatly in the magnitude of net skill gain and loss.
- A range of characteristics appears to influence the magnitude of observed skill gain or loss including age, education, and the intensity of skill use on the job.

The following chart, drawn from Willms and Murray (2007), shows that significant skill loss occurred between 1994 and 2003 in Canada over the entire life course.

The findings carry important implications for policy. First, skill loss is real and large enough to have offset skill gain associated with higher levels of initial education (up a full year between 1994 and 2003), higher levels

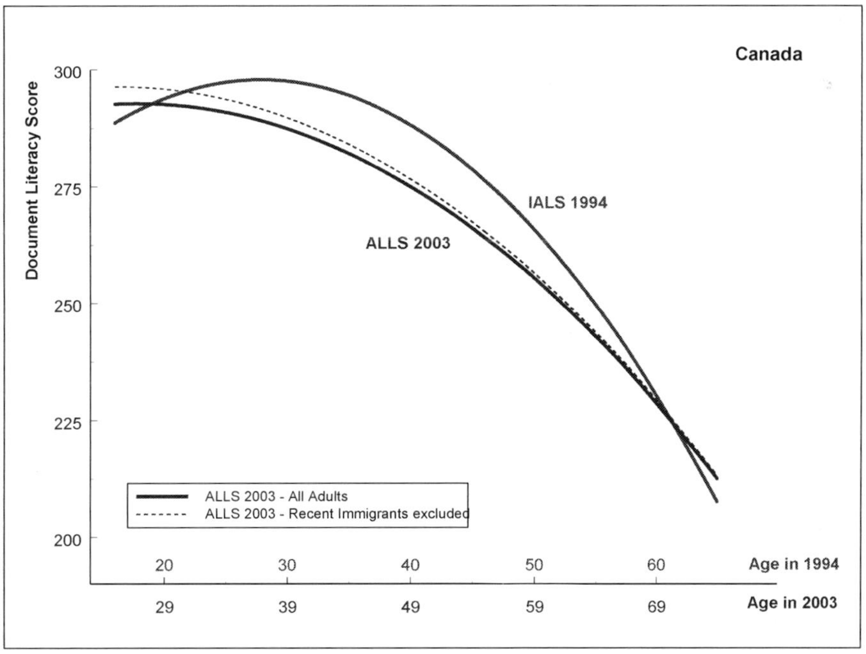

Figure 3.3 Skill loss in Canada across the life course, 1994–2003.

of adult education and training (up 15 percent), stable quality flowing out of the secondary system (unchanged PISA scores), and rising knowledge intensity of jobs. Thus, skill loss is eroding the potential return on both public and private investments in education. Skill loss is also likely reducing overall levels of productivity growth and, thus, depriving the economy of a competitive edge.

Second, the fact that skill loss is concentrated in individuals from lower socioeconomic backgrounds reveals one of the key mechanisms by which social inequality in valued outcomes is generated and suggests that investments to reduce social inequality in literacy skill would help to reduce current levels of wage, incomes, and employment inequality. This is particularly true for immigrants, who will constitute the primary source of net labor force growth over the coming decades.

Finally, the very existence of skill loss raises the questions on the demand side of public policy. Thought of in economic terms, skill loss can be taken as prima facie evidence of inadequate social and economic demand for skill. The concentration of skill loss in industries and occupations with low indices of skill use is troubling. In many cases these are capital-intense industries that have been the beneficiaries of government subsidies and other public supports. This suggests a need to look carefully at government policies that might unwittingly be reducing the economic demand for skill use by substituting capital for skilled labor. It also argues for government to focus on the broad array of policy instruments that might be used to increase levels of skill demand over the short and medium term. They might also consider reducing investments in postsecondary education in areas where skill demand is thought to be inadequate.

Governments need to pay attention to more than the supply side/skill formation. They also need to pay attention to the demand side and understand the incentive structure underlying skill gain and loss, particularly at work.

Analysis of IALS and ALL data suggests considerable variation in skill use by firms in the same industries. Ideally this would be obtained through longitudinal tracking of representative samples of workers within firms so that one can observe heterogeneity in firm skill demand conditioned on firm characteristics. We are also working to see if we can use the output of the synthetic cohort analyses to isolate the growth effects of skill gain and loss on long-term GDP trajectories. This would provide an estimate of the economic cost associated with insufficient demand for skill.

THE COMPETENT LEARNERS STUDY: BACKGROUND AND RATIONALE FOR THE STUDY

Statistics Canada has been cooperating with the New Zealand Center for Educational Research on a piece of longitudinal research executed in the context of their Competent Learners study. The Competent Learners study is a continuation of the Competent Children project. This project is a longitudinal study of a sample of New Zealand young people, who have been followed from their final early childhood education centers in the Wellington region. The main aim of the project is to chart the development of competencies in the context of home and educational experiences which may account for differences in patterns of development and young people's performance. Reports from the study and associated papers are available on the NZCER Web site, www.nzcer.org.nz. The project is funded by the Ministry of Education, with some additional funding from NZCER.

Methodology

We have helped the NZ research team by providing a set of test items that will allow us to estimate document literacy and numeracy scores for each of the children at age sixteen on the NALS/IALS and ALL proficiency scales, respectively.

This adds value from the NZ perspective because it provides a clear link to the NZ IALS data set. This linkage provides evidence of the relationship of literacy proficiency to real-world labor market, educational, health, and social outcomes and provides a way to test if the observed relationships are fundamental or are second- or third-order effects of deeper developmental processes.

Part of the ALL development process involved work aimed at refining the theoretical basis of the assessment. This work, known as the DeSeCo project, was managed by the Swiss in cooperation with the OECD, Statistics Canada, and NCES. It attempted to define the competencies that should be measured within the context of an international comparative skill assessment. The ALL development process attempted to reflect the DeSeCo work in three ways:

- It attempted to develop measures in eight cognitive domains.
- It added attitudinal measures thought to be important to the application of skills.
- It adopted a design that allowed for an exploration of a partial interskill covariance matrix.

Development only yielded reliable measures in four of the eight domains—prose literacy, document literacy, numeracy, and problem solving. Development efforts failed to yield measures of teamwork, practical intelligence, ICT skill, and speaking/listening.[2] Thus, the ALL study is limited in its ability to explore the full interskill covariance matrix as specified in DeSeCo. The Competent Learners study will let us explore essentially the full matrix to establish dependencies and variability. This will help to inform developmental priorities for the next round of adult skill assessments.

The study also provides three measures that are of particular interest to our research team. The first is the measure of IQ based upon the administration of Raven's matrices. Even though the NALS/IALS/ALL tests are not timed, and differences in performance can be explained almost completely with variables that are related to the mechanics of using print, critics claim that our results are simply a reflection of IQ. The current study affords us an opportunity to explore the relationship empirically.

The second is the measures of soft skills. Much is being made about the need to develop these skills. We believe that this will distort adult education priorities away from fundamentals like reading and math. The data suggest that soft skills are fixed early in life. These skills only matter to employers because, used to labor surpluses, they take the hard skills for granted. This explains much of strong relationship between literacy and labor market success.

The third is a measure of oral language proficiency. Many researchers have argued that the IALS and ALL literacy measures for immigrants simply reflect a lack of oral language proficiency in the language(s) of the test. The NZ database will allow this hypothesis to be tested directly.

At age fourteen, NCZER has data for 475 young people. These data include results from reading comprehension, writing, mathematics, and pattern completion tasks, and ratings from core-subject teachers (English, mathematics, and science) in relation to social and attitudinal competencies. The latter comprise curiosity, perseverance, self-management, self-efficacy, social skills with peers, social skills with adults, and communication.

NCZER has just reported the first results and analysis of the material gathered from late 2002 and during 2003, when the sample were aged fourteen years (Wylie, Ferral, Hodgen, & Thompson, 2003; Wylie & Hodgen, 2007). They describe their competency levels and relationships between the competencies. They then analyze the relationship between the young people's current competency levels and four social characteristics: gender, family income levels, maternal qualification levels, and ethnicity. To do so, they compare performance above and below the median, followed by modeling, to see which of these characteristics contributes significantly to performance.

Finally, they report an analysis of the predictability of current levels of performance in relation to earlier levels of performance. To answer this question, they analyzed the data at a number of different levels. Overall trends were modeled using structural equation modeling. Quartile groupings were used to describe differences in patterns for high and low performers.

A first set of analyses of the age sixteen data have just been released. The analyses employed indices of four attitudinal competencies estimated from the detailed batteries of questions: thinking and learning, focused and responsible, social skills, social difficulties.

The Competent Children/Competent Learners sample was originally chosen in relation to the main focus of the first phase of the study, which was the role of early childhood education experiences and quality. This meant the units for sampling were early childhood education types, other than ngä köhanga reo, rather than social characteristics. This, and the fact that the sample was chosen from the Wellington region, has resulted in a sample which is not nationally representative in terms of social characteristics. The sample has higher proportions of young people from high-income families, and those whose mothers have trade or tertiary level qualifications than the national average, and lower proportions of Mäori and Pasifika young people.

Our interest lies in the ability to relate our adult skill measures to a wide range of repeated measures that are available for each child. These include cognitive measures of IQ, math, reading comprehension, writing, and logical problem-solving skills, social and attitudinal measures of curiosity, perseverance, self-management, self-efficacy, social skills with peers, and social skills with adults and communication.

The study involves conducting personal interviews with the children themselves, their parents, and key teachers using a set of questionnaires. The protocol also involves the administration of direct assessments of a range of cognitive domains and the collection of information about the child from administrative records held by schools.

In their most recent publication the NZ researchers have used a range of advanced analysis techniques, including structural equation modeling, to explore the interrelationships between competencies at age fourteen, gradients and ANOVA to explore the relationships between competency and social characteristics, path analyses to explore the forward and backward predictability of competency at different ages, and an analysis of individual cognitive trajectories to explore interquartile mobility.

Data analysis at age sixteen is currently underway. Feedback from the interviewers, scoring of the test, and statistical analyses suggest that the literacy and numeracy measures worked as intended. As with any longitudinal study, there is some concern about representativeness and sample attrition bias. Five hundred forty-nine children have been part of the Competent Children/Competent Learners study. The characteristics of the 475 who took part at age fourteen and those who have departed the study (74, or 13 percent), including those who took part at age twelve but not at age fourteen (21, or 4 percent), have been compared to see if there were any differences in social characteristics and age-eight competency scores. There were no differences in social terms. However, those who have left the study had slightly lower mathematics and logical problem-solving scores at age eight. Those who left the study between age twelve and age fourteen had somewhat lower scores at age twelve for mathematics and the social and attitudinal competencies

FINDINGS AND EMERGING CONCLUSIONS

The age-fourteen analysis provides many interesting results, including the fact that social skills appear to be fixed very early in life, that there are strong interrelationships between social and attitudinal competencies and moderate relationships between these competencies and logical problem solving and with literacy (Wylie, Ferral, Hodgen, & Thompson, 2003). About 65 percent of the variance in literacy scores could be accounted for by scores in communication, curiosity, perseverance, and logical problem solving. The analysis also identified a strong association between math and literacy. These relationships would have to be taken into account in instruction.

The analyses at age sixteen reveal several additional interesting findings, including (Wylie & Hodgen, 2007):

> Literacy scores are reasonably consistent over time. From age 8 onwards the correlations between test scores at age 16 are at .6 or above. This implies, at a minimum, that initial education systems do little to attenuate early social inequality in literacy outcomes.
>
> Attitudinal competencies are less consistent, only reaching .6 between age 14 and 16 scores for the *"thinking & learning"* and *"focussed and responsible"* scores. This finding implies that these "skills" are much more malleable and may be thought of primarily as outcome rather than input measures.
>
> Social competencies are even less consistent, ranging from .32 for the correlation between social skills at age 14 and social difficulties at age 16 and to .46 for social skills at age 14 to social skills at age 16
>
> Around 80% of the variability of age 10–16 cognitive composite scores are accounted for by the cognitive composite scores 2 and 4 years before and the current attitudinal composite score. The stability of cognitive scores over time implies that early interventions would yield marked improvements in scores and reductions in social inequality in adult outcomes, including educational outcomes.

As illustrated in the following chart drawn from Wylie and Hodgen (2007), consistency of scores is strongest for students who were in the lowest and highest quartiles of literacy score in the previous period(s). This suggests that the underlying system is recursive and self-reinforcing in both positive and negative directions—a far more complex system than assumed in most policy. This implies that initial education systems amplify rather than attenuate literacy inequalities at school entry and that, unlike other outcomes that exhibit a considerable amount of interquartile mobility from period to period, literacy disadvantage (and advantage) are durable. On the one hand, this is unfair to those who enter the system with literacy scores in the lower part of the skill distribution. On the other hand, the findings suggest that interventions targeted on raising the skill level of children in this region might yield strong positive results.

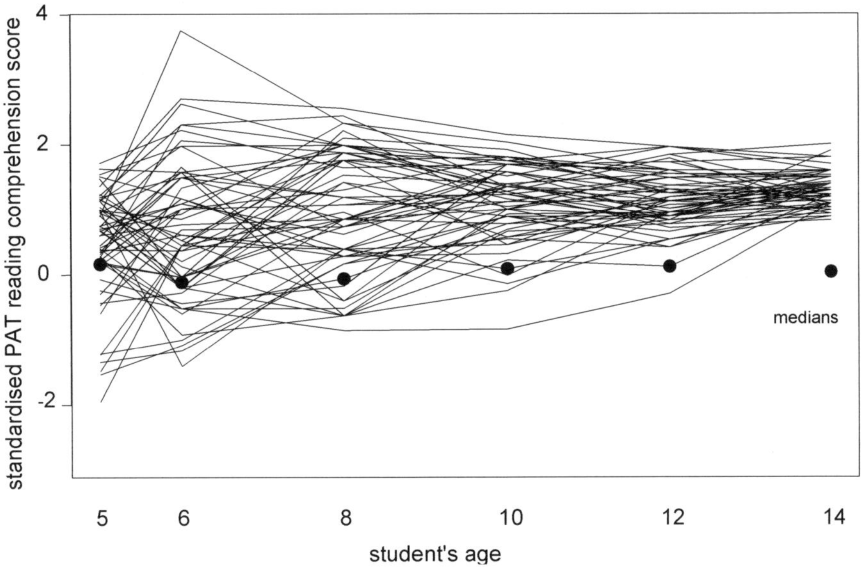

Figure 3.4 Standardized PAT reading comprehension scores, ages 5 to 14 for those in the highest quartile group at age 14, n = 59.

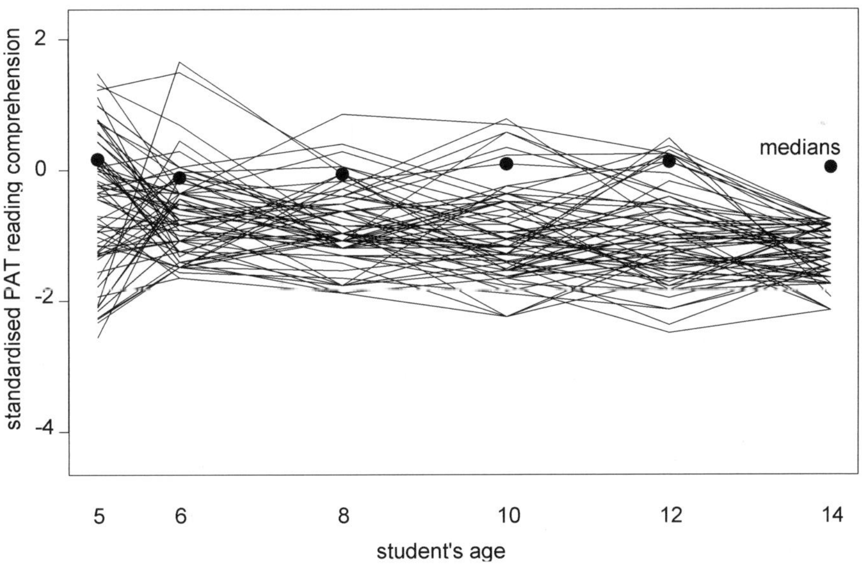

Figure 3.5 Standardized PAT reading comprehension scores, ages 5 to 14 for those in the lowest quartile group at age 14, n = 67.

This latter finding suggests that inequality becomes entrenched early in life. The authors conclude that where students become disengaged in learning they do so before the age of twelve. This suggests a need for early intervention.

The age-fourteen analyses revealed that literacy scores were far more consistent over time than other competencies, suggesting, at a minimum, that the initial education system does little to attenuate early social inequalities in literacy acquisition and that these inequalities persist and eventually have a profound impact on a range of adult outcomes. The central question for policy is what steps might be taken to reduce early literacy inequalities to the level observed in some Nordic countries.

The analyses of results at age fourteen also show quite clearly that there is little interquartile movement in key outcomes, particularly literacy skill. The age-sixteen analyses confirm this finding. This is an important finding for adult literacy programs because it suggests that there are significant barriers to be overcome to move adults in the bottom quartile. If the regular education system can't do it with a captive audience and all the tools, then adult educators need to be quite circumspect about what they can hope to achieve with adults.

Our hope is that we can talk the NZ researchers into retesting the literacy and numeracy skills of the same respondents towards the end of postsecondary. This would allow for an empirical analysis of skill gain associated with postsecondary participation and of skill loss, complete with controls for lots of other variables over the entire developmental trajectory.

CONCLUSION AND COMPARATIVE POSSIBILITIES

The three studies reviewed in this chapter attempt to explore the processes of skill gain and loss over the life course. The NZERC's longitudinal study traces literacy development trajectories from birth through age sixteen and allows for an exploration of their relationships to attitudinal competencies, other cognitive competencies, and social background variables. It confirms that inequalities in literacy skill emerge early and persist through to the age of sixteen. The initial education system tends to amplify, rather than attenuate, literacy inequalities over time, more so than for other competencies. Possible comparisons for further research could be made with the British birth cohorts and Canada's National Longitudinal Survey of Children and Youth (NLSCY) and with the Australian Longitudinal Survey of Australian Youth (LSAY) and Canada's 2000 YITS/PISA cohort (but involving fewer covariates/controls).

The Statistics Canada/HRSDC YITS/PISA longitudinal study traces the education, labor market, and social trajectories of a large cohort of fifteen-year-olds inducted in 2000 as part of the OECD PISA study. It confirms that differences in fifteen-year-olds' literacy have a marked impact on several key indicators of education success including high school completion and access to postsecondary education. The conditional probabilities are sufficiently large to expect that literacy will continue to exert its influence on postsecondary persistence and early labor market success. Similar synthetic cohort analyses can currently be undertaken for Norway, Switzerland, the United States, and Italy. Future cycles will add Hungary, Australia, New Zealand, and the Netherlands.

The IALS/ALL synthetic cohort analysis estimates the average skill gain (or loss) experienced by various population subgroups. It confirms that skill loss exists and that it is of a magnitude to have offset the entire skill gain from rising educational quality and quantity in the intervening nine-year period. It is possible to compare the YITS/PISA findings with those of the Australian Longitudinal Survey of Australian Youth (LSAY) study, the New Zealand Council for Educational Research's (NZCER) Competent Learner's study, the British Birth Cohorts, Canada's National Longitudinal Survey of Children and Youth (NLSCY), and with Swiss and German cohort studies based on PISA.

The fact that skill loss seems to be concentrated in adults with low socioeconomic backgrounds suggests that the social inequalities observed in the NZCER and YITS/PISA studies are further amplified in adulthood. To date, most policy frameworks have assumed a simple linear, sequential model of literacy development. The findings from these studies would suggest that the causal pathways are, in fact, much more complex. The systems would seem to be both recursive and self-reinforcing. Individuals in the bottom quartile of the skill distribution at a given age tend to accumulate further disadvantage, whereas individuals in the top quartile appear to accumulate advantage in both literacy and life outcomes. This finding implies that the returns to policies that seek to reduce social inequality in literacy outcomes might yield far larger returns than previously assumed.

The finding of skill loss also provides strong indirect evidence of inadequate social and economic demand for skill. This might be taken as evidence of general oversupply of skill and, hence, of a need for a reduction of educational investment. The fact that it is concentrated in occupations and industries that have been beneficiaries of government subsidies that serve to reduce the cost of financial and physical capital relative to human capital suggests that government policy may be partly responsible for dampening skill demand. This argues for a thorough review of all government policies that might knowingly, or inadvertently, reduce the demand for literacy skill. Governments have a huge array of policy instruments that might be put to use to raise the level of social and economic demand for literacy skill.

The fact that skill is concentrated in occupations and industries in which workers from lower socioeconomic backgrounds are concentrated suggests that economic policy serves to entrench skill-based social inequality in economic outcomes. While not unexpected, this is unfair and unnecessary in a modern "managed" economy.

The available macroeconomic evidence suggests that literacy has been the single most important determinant of GDP growth in the postwar period. Differences in the level of adult literacy skill explain 55 percent of differences in the long-term growth rates of GDP per capita. In addition to this level effect, the distribution of literacy skill seems to matter. The larger the percentage of adults with very low skills, the lower the overall growth rate (Coulombe, Tremblay, & Marchand, 2004). Thus, it is reasonable to believe that higher growth rates could be achieved by raising the average literacy level and that the highest return on investment would be achieved by reducing the proportions of adults with very low skills. Ironically, the evidence cited previously suggests that government education and economic policies may, in fact, be responsible for much of the social inequality in skills.

The microeconomic evidence suggests that employers can identify literacy skill and chose to reward it with higher wages, lower hours of work, and more stable employment across the entire skill distribution. This calls into question government affirmative-action programs that are predicated on an assumption of systemic discrimination and further buttresses the call for government investment in active education policies that seek to simultaneously raise the level of, and reduce social inequalities in, the literacy skill of students and adults.

The three studies reviewed in this paper build upon the seminal contributions of authors such as Bynner (Bynner & Parsons, this volume), who used data from the British Birth Cohorts to explore the relationship between cognitive skill and individual outcomes using longitudinal data.

The findings presented provide confirmation of several important facts, including that:

- Social inequalities in literacy skill emerge early in life.
- Primary and secondary education experiences tend to amplify, rather than attenuate, social inequalities in literacy skill.
- Social inequalities in literacy skill influence access and persistence in the postsecondary education system and eventually play a central role in generating social inequality in a range of individual health, social, educational, and labor market outcomes.
- Differences in the level and distribution of literacy skill appear to have been the single most important determinant of differences in GDP per capita over the long term.
- The existence of significant skill loss has reduced overall economic performance.
- The fact that skill loss is concentrated in workers from low socioeconomic backgrounds exacerbates social inequalities that are conditioned on skill.
- Ironically, some skill loss may be the product of government policies that unwittingly reduce the relative costs of physical and financial capital vs. human capital.

The policy prescription is simple. Governments need to invest over the life course to raise literacy levels and to reduce social inequalities in literacy.

This will necessarily involve interventions before the age of school entry to reduce social inequality in literacy to levels that can be managed by the school system. This will also involve reducing the growth of social inequality in literacy that appear to be one of the unintended consequences of primary and secondary education. Investment in these sectors would yield increases in the productivity of the initial educational process and would increase the efficiency of the selection processes into postsecondary education. Investment is also called for to improve adult literacy levels. This will require investment in raising the skill levels of Canadian-born and -educated adults who either left the educational system without the requisite literacy skills or whose skills have eroded since leaving formal education. It will also require investment in raising the language and literacy skills of a large numbers of immigrant adults. As noted earlier, the available evidence suggests that targeting these investments on individuals with the lowest skill levels would yield the largest economic returns.

Governments need to consider policies to raise the level of economic and social demand for literacy skill. This will help to reduce skill loss and will help to maximize the social and economic return on educational investments. Governments cannot assume that the economy will be able to absorb new skill supply.

This policy prescription challenges the prevailing assumptions that underpin public policy in Canada. Policymakers have tended to focus on the elite end of the skill distribution—on the postsecondary system and upon the social institutions that support research and innovation. They have also tended to pay much more attention to investments in financial capital and physical capital than to human capital. These policies have served large vested interests—the postsecondary institutions themselves and the firms that rely on the skills they create and upon technology companies and investment bankers. Evidence suggests that the skill of the average worker may become a much more important determinant of the relative economic success of OECD economies in the future. The insights afforded by the three studies that form the substance of this chapter have allowed Canadian public-policy makers to make much more balanced policy decisions than would otherwise be the case.

Some OECD governments have already begun to invest—the UK, Australia, and Ireland, to name a few. To date, Canada and the United States have been reluctant to do so. If the evidence presented in this chapter is correct, they will begin to pay an economic penalty for their reticence.

NOTES

1. As evidenced by the papyrus from 1210 BC Thebes reproduced in the foreword of *Reading the Future: A Portrait of Literacy in Canada.*
2. Subsequent development has yielded reliable measures of speaking and listening and of ICT skill.

REFERENCES

Canadian Council on Learning. (2007). *The state of learning in Canada: No time for complacency. Report on learning in Canada 2007.* Ottawa: Author.

Coulombe, S., & Tremblay, J. F. (2006a). *Human capital and Canadian provincial standards of Living.* Ottawa: Statistics Canada.

Coulombe, S., & Tremblay, J. F. (2006b). *Migration, human capital, and skills redistribution across the Canadian provinces.* Working Paper 2006 D-07. Industry Canada.

Coulombe, S., Tremblay, J. F., & Marchand, S. (2004). *Literacy scores, human capital and growth across fourteen OECD countries.* Ottawa: Statistics Canada. Cat. No. 89–552-MIE, no. 11.

Knighton T., & Bussière, P. (2006). *Educational outcomes at age 19 associated with reading ability at age 15.* Ottawa: Statistics Canada and HRSDC.

Macracken, M., & Murray, T. S. (2008). *The economic benefits of literacy: Theory, evidence and implications for public policy.* London and Ottawa: Canadian Language and Literacy Network and Canadian Council on Learning.

Maxwell, J., & Teplova, T. (2008). *The social cost of low literacy skills.* London and Ottawa: Canadian Language and Literacy Research Network and the Canadian Council on Learning.

Murray, T. S., Binkley, M., & Clermont, Y. (2005). *The Adult Literacy and Life Skills Survey: New frameworks for assessment.* Ottawa: Statistics Canada.

Rychen, D. S., & Salganik, L. H. (2001). *Defining and selecting key competencies.* Bern: Hogrefe and Huber.

Rychen, D. S., & Salagnik, L. H. (2003). *Key competencies for a successful life and a well-functioning society.* Bern: Hogrefe and Huber.

Statistics Canada & HRDC. (1995). *Reading the future: A portrait of literacy in Canada.* Ottawa: Author.

Statistics Canada and OECD. (2000). *Literacy skills for the information age: Final results of the International Adult Literacy Survey.* Ottawa and Paris: Author.

Statistics Canada, HRDC, & CMEC. (2001). *Measuring up: The performance of Canada's youth in reading, mathematics and science.* Ottawa: Author.

Statistics Canada, HRSDC, & CMEC. (2004). *Measuring up: Canadian results of the OECD PISA study.* Ottawa: Author.

Statistics Canada & OECD. (2005). *Learning a living: First results of the Adult Literacy and Life Skills Survey.* Ottawa and Paris: Author.

Willms, D., & Murray, T. S. (2007). *Gaining and losing literacy skills over the lifecourse.* Ottawa: Statistics Canada.

Wylie, C., Ferral, H., Hodgen, E., et al. (2003). *Competencies at age 14 and competency development for the Competent Learners study sample. First report from the Competent Learners study, age-14 phase.* Wellington, New Zealand: NZCER.

Wylie C., & Hodgen, E. (2007). *Competent Learners @16: Competency performance and development over time.* Wellington: NZCER.

Part II

Student, Teacher, and Classroom Studies

4 Developing Learners' Reading Skills in Adult Basic Education Programs

Judith A. Alamprese

INTRODUCTION

Since 2000, approximately 2.7 million adults have participated annually in adult basic education (ABE) programs funded by the U.S. Department of Education under the Adult Education and Family Literacy Act of 1998. Close to 40 percent had reading skills below the ninth-grade level upon enrolling in ABE programs, and half of these individuals were categorized as low-level learners—scoring at the zero to sixth-grade equivalence on a standardized literacy test or placed in one of the first three levels of the National Reporting System (Beginning Literacy, Beginning Basic Education, and Low-Intermediate Basic Education) (U.S. Department of Education, 2002, 2006). While most adults with low literacy and numeracy enroll in ABE programs with the intent of earning a General Educational Development (GED) credential, obtaining employment, or improving their status in their current job, many are likely to have reading skills below the level needed to achieve their goals. As the number of adults in ABE programs with low literacy and numeracy has remained steady, there has been an increased interest among federal agencies in understanding the types of reading instruction that are provided to adults in ABE programs and the extent to which adults improve their reading skills as a result of their participation in these services.

One of the first longitudinal studies to investigate the association between reading instruction and reading skill development in ABE programs was conducted by Abt Associates Inc. during 1996–2003. The purpose of the *Study of Reading Instruction for Low-Level Learners in Adult Basic Education Programs* was to assess whether adults with low literacy skills (i.e., skills below a seventh-grade equivalence in reading comprehension) who participate in class-based instruction in naturalistic ABE settings learn to read and, if so, what factors account for their success—their personal characteristics, the type of instruction that they receive, or the services that are offered by the ABE programs in which they participate. The study analyzed the background characteristics of low-literacy adults who participated in reading classes in the study's 35 ABE programs, their hours and weeks of attendance in these classes, their use of literacy at home, at work, and in the community, and the operational characteristics of the ABE programs in which they participated. A key feature of the study was the examination of the reading instruction provided in the study's classes, which represented a range of approaches to teaching reading. These approaches included reading methods that emphasized phonics instruction and used a published scope and sequence (e.g., Wilson Reading System[1]), instruction that included phonics but was not based on a published curriculum, and instruction that emphasized life skills' content and did not follow a prescribed scope and sequence. The reading instruction provided in the classes was analyzed in terms of the reading content (i.e., types of reading component skills [Strucker, 1997]) that was taught, the amount of time that was focused on teaching reading component skills, and the instructional strategies that were used in teaching.

The U.S. Department of Education's Office of Vocational and Adult Education and Policy and Program Studies Services funded this research. The Office of Vocational and Adult Education is the entity that disperses federal funds for the support of local ABE programs.

This chapter discusses key findings from Abt Associates' study concerning the characteristics of the adults participating in the study, the types of class-based reading instruction that was provided to participants, their reading outcomes, and the relationship of learners' reading outcomes to the type of reading instruction that they received.[2] The study's results indicated that learners made significant gains on the six standardized reading tests that were used to assess their word recognition, decoding, vocabulary, comprehension, and spelling skills from pre- to posttest (nine months) and from pre- to follow-up test (eighteen months). Furthermore, learners in classes that emphasized phonics instruction and used a published scope and sequence had larger gains on decoding than did learners in the study's other classes.

These results provide an initial understanding of the types of reading instruction that is provided to low-literacy ABE learners and the factors that are associated with learners' development of their decoding skills. Abt Associates' study also was notable for the large samples of adult learners that participated: 643 learners from 130 ABE classes in thirty-five programs across sixteen states.

As interest in adult reading instruction has increased, investments in research have continued and experimental studies are underway to assess the impact of varied reading instructional approaches on low literacy adults' development of their reading skills.[3] The current adult reading research is a rigorous test of the some of the instructional approaches investigated in Abt Associates' study and should provide critical information for ABE instructors.

QUESTIONS GUIDING STUDY OF READING INSTRUCTION FOR LOW-LEVEL LEARNERS IN ABE

Abt Associates' study investigated three questions:

- How much do low-level adults who participate in ABE programs improve their reading skills?
- What characteristics of low-level adult learners (demographic, personal, and amount of exposure to ABE) affect the amount of improvement that they make in their reading skills?
- How are the instructional and operational characteristics of ABE programs related to the amount of low-level learners' improvement in their reading skills?

These questions were intended to provide in-depth descriptive information about low-literacy adults enrolled in ABE reading classes in order to assist policymakers and literacy service providers in understanding the variety of adults who seek reading instruction so that policies and services could be refined to address the needs of low-literacy adults. Furthermore, the study was to explore through correlational analyses various factors that are associated with learners' development of their reading skills as an initial step in determining effective reading instruction.

STUDY DESIGN

The study employed a pre-post-follow-up, nested-sample design, and correlational analytic methods to assess the amount of progress that low-level learners make in ABE reading classes. The design was nested because learners were enrolled in 130 classes that were selected from a sample of thirty-five programs. Study learners were assessed on five norm-referenced reading tests and a spelling test at the beginning of their participation in ABE classes, at the end of the class at nine months, and approximately twelve months later. Hierarchical linear regression models were used to estimate the average gains made by learners across the multiple ABE classes and programs. The assumption guiding the research was that any significant gains made by learners over the period of the study's classes represent the effects of ABE instruction. This assumption was based on analyses of existing data from the Wide Range Achievement Test (WRAT3) Word Reading measure (Wilkinson, 1993) and the Woodcock-Johnson Revised Word Attack and Letter-Word Identification measures (Woodcock & Johnson, 1989, 1990). The analyses on growth trends for adults (age sixteen years and older) indicated that the normal developmental trajectory for adult learners on standardized reading assessments is essentially flat in the posthigh-school-age period (Alamprese, Goodson, & Tao, 1999).

In addition to assessing learners' gains in reading, the study measured demographic and attitudinal characteristics of learners at baseline, the amount of instruction that learners received, characteristics of the classroom reading instruction, and characteristics of the ABE programs and their operations. These data were used to extend the statistical models describing learner gains in reading skills to estimate the relationship between these gains and the characteristics of learners, the characteristics of reading instruction, and the characteristics of the ABE programs.

STUDY SAMPLE

Thirty-five ABE programs formed the sample for the study. The sample was a purposive sample of programs that met three criteria: a) provided class-based reading instruction to English-speaking adults whose reading skill level in comprehension was lower than a seventh-grade equivalence level, b) had a basic level of operations in learner recruitment, learner assessment, program management, program improvement, and support services (Alamprese, 1993), and c) had instructors who were trained or experienced in teaching reading and whose instruction followed a discernable scope and sequence. Twenty-six of the programs had more than one reading class in the study.

In each class in the study, all learners were recruited to participate in the data collection. Participation in the study was voluntary for learners, and all but two learners from the 130 classes agreed to participate. While the study's intent was to examine the outcomes of native-born, English-speaking adults in low-level reading classes, almost all of the reading classes selected for the study had some learners who were born and educated outside of the United States but who had sufficient oral fluency in English to benefit from an ABE reading class. Since the study was designed to examine the instruction and learners in operating ABE reading classes, both native-born and nonnative-born adult learners were included in the study. Approximately 70 percent of the sample was native-born learners who were educated within the United States (including a small number of learners who were born outside of the United States but who had been educated in the United States and were fluent in English).

Programs entered the study at three different time points (or waves), and this affected how many cohorts of learners participated in the study from each class. The classes for Wave 1 and Wave 2 programs provided a new group of study participants in the subsequent years in which they participated in the study, while Wave 3 programs provided study participants for one year. Participants from Waves 1 and 2 were eligible to participate in the pre-post and 12-month follow-up data collection, and Wave 3 learners were eligible only for pre-post data collection.

The data collection schedule resulted in three groups of learners from the thirty-five programs:

- Learners who were pretested at the beginning of the program year but who were not available for posttesting at the end of the class;
- Learners who were pretested and posttested at the end of the program year but who were not available at follow-up testing, either because they could not be located or the date of their baseline testing did not allow time for the follow-up period to elapse before the end of the study; and
- Learners who were pretested, posttested at the end of the program year, and who participated in a follow-up data collection approximately twelve months after the pretest.

The total pretest sample was 920 learners. Of these learners, 643 (70 percent) also were posttested at the end of the class session. The sample of 643 learners was more than sufficient for the study to detect even a small pre-post gain (three points) on standardized reading measures and a pre-post difference of 5 percentage points on the nonstandardized measures with high power (95 percent and 80 percent, respectively, for the standardized and nonstandardized measures). Of the 643 learners with pre-post data, 332 were eligible for follow-up testing and 245 (74 percent) were assessed. The sample of learners with pre- and posttests was 66 percent native-born adults, and the twelve-month follow-up sample was 63 percent native-born adults.[4]

DATA COLLECTION

Three types of data were collected in the study: learner, class, and program. The data collection for learners consisted of individual and group administration of five reading tests, a spelling test, and an in-person interview. These data were collected at baseline for each cohort of learners and at the end of the class session (spring of the next year). Follow-up testing and interviewing were conducted one year later for eligible learners. Data on classroom practices were obtained through class observations and individual interviews with instructors. Data on program operations came from interviews with program directors and staff as well as from reviews of program documents. The majority of classes (86 percent) were observed twice over the data collection period and the other classes were observed once. The interviews with staff were conducted during site visits to the programs—once during the first year of a program's participation and again during the final year of the study.

Two types of data collection staff were involved in the study. Staff who administered the reading assessments and interviews with learners were selected or referred from each program. These data collectors had experience in working with adult learners as well as in conducting assessments and interviews. Abt Associates' senior staff trained all data collectors in procedures for test administration and data collection prior to the initial data collection period to establish reliability, and conducted a follow-up training session prior to the second year of data collection. Abt's staff also monitored the data collection and conducted weekly quality control during the periods of data collection.

Abt's senior researchers conducted all of the study's class observations and interviews with instructors and program staff. These data collectors established interobserver reliability in recording instructors' and learners' behaviors prior to conducting any class observations. Reliability in interviewing instructors and program staff also was established.

CONSTRUCTS AND MEASURES

The constructs measured in the study and the instruments used to collect the study's three types of data—learner, class, and program—are summarized in Table 4.1.

Described following are the constructs and measures that relate to the learner analyses discussed in this chapter.

Learner Reading Measures

Since the study was focused on assessing low-literacy adults' development of reading skills and no one reading instrument measures the key reading skills of interest, five reading assessments and a spelling assessment were used to assess learners' skills. These were:

- Woodcock-Johnson-Revised: Word Attack (Woodcock & Johnson, 1989, 1990) (decoding of nonsense words);
- Woodcock-Johnson-Revised: Letter-Word Identification (word recognition);
- WRAT3-Word Reading (Wilkinson, 1993) (word recognition);
- WRAT3-Word Spelling;
- Nelson Word Meaning (Hanna, Schell, & Schreiner, 1977) (vocabulary); and
- Nelson Reading Comprehension.

The reading instruments were selected based on their: 1) appropriateness in assessing the specific constructs and reading skills examined in the study; 2) psychometric properties (availability of national norms, adequate reliability, and validity); 3) appropriateness for use with low-level adult readers; 4) availability of alternate, parallel forms to be used for pre- and posttests; and 5) relatively short length to minimize respondent errors associated with fatigue. In addition to the tests of reading skills, the Social Studies subscale of the Woodcock-Johnson-Revised was administered to measure adult

Table 4.1 Study Constructs, Instruments, and Data Collection Methods

Types of Data Collection/ Constructs	*Instrument*	*Data Collection Method*
Learner Data 1. Reading skills		I = Individual testing G = Group testing
• Word recognition	Woodcock-Johnson-R: Letter-Word Identification	I
• Word analysis	Woodcock-Johnson-R: Word Attack	I
• Word recognition	WRAT-3: Word Reading	I
• Vocabulary	Nelson Reading: Word Meaning	
• Reading comprehension	Nelson Reading: Reading Comprehension	G
• Background knowledge	Woodcock-Johnson-R: Social Studies	G
2. Spelling	WRAT-3: Word Spelling	I
3. Use of literacy skills outside of class	Learner Background Interview Form	Learner self-report
4. Self-empowerment, self-esteem, academic self-efficacy	Generalized Perceived Self-Efficacy Scale Rosenberg Self-Esteem Measure Adult and ESL Literacy Learning Self-Efficacy Scale	Learner self-report
5. Social Support Networks	Learner's Social Support Network	Learner self-report
6. Learner's background (e.g., education, employment, health); demographics; perceptions of instructional content and methods, and skills learned	Learner Background Interview Form	Learner self-report
7. Learner class attendance	Learner Attendance Form	Program records
8. Learner's overall class performance	Instructor's Review Form	Instructor rating
B. Class Data Reading methods and instructional strategies	Class Observation Form and Instructor Interview Form	Direct observation, instructor interview

(continued)

Table 4.1 Study Constructs, Instruments, and Data Collection Methods

Types of Data Collection/ Constructs	*Instrument*	*Data Collection Method*
2. Instructor's perception of teaching of reading methods and instructional strategies; instructor's background characteristics	Instructor Interview Form	Instructor interview
C. Program Operations Data • Program management • Program improvement • Learner recruitment and intake • Learner assessment • Support services • Learner transition	Site Visit Topic Guide	Interviews with program staff

learners' general background knowledge, which is hypothesized to be a factor in literacy learning among adults.

Learners' Personal Characteristics

In order to understand learners' personal characteristics at the time of the pretests and how these changed over the period of their participation in the study, an interview protocol was developed for collecting information about learners' demographic characteristics, education and work status, and health status. The protocol also included questions about learners' perceptions of health and learning problems or disabilities during childhood and at the time of the interview (Israel, Checkoway & Schultz, 1994). Another topic addressed was learners' perceptions about the instruction that they received. Questions were included in the postlearner interview regarding the skills that adults learned in the reading class, and the characteristics of the instructor, teaching method, and materials that assisted their learning.

Learners' Attendance

The study programs provided the hours of learners' attendance for the period in which learners participated in the study. The programs also submitted information about learners' participation in ABE classes in addition to the reading classes in the study.

DATA ANALYSIS

The study's findings concerning the characteristics of learners who participated in the low-level reading classes in the study, the types of reading gains that these learners made, and the relationship of the type of instruction that learners received to their reading gains are discussed below. The data analyses procedures that were conducted to determine these outcomes are presented in a brief summary.

Learner Characteristics

The description of learners was performed for the total sample as well as for subgroups of learners (e.g., native-born versus nonnative-born, low versus higher pretest skill levels). These analyses provided information about the learners who participated in the study and identified key variables to use in the analyses estimating the outcomes for learners.

Changes in Learners' Reading Component Skills

For each reading test, learners' raw scores were converted into 1) grade equivalent scores, used for descriptive purposes, and 2) scale scores, appropriate for more complex statistical analyses. Change scores that represent the amount of a learner's skill change from pre- to post- to follow-up tests were constructed by subtracting each learner's pretest scale score from his/her posttest and follow-up scale scores. The statistical significance of the mean gain was assessed using the paired t-test.

The mean test score change also was expressed as an effect size—that is, the mean gain as a number of standard deviation (SD) units. Expressing mean gain as an effect size allowed Abt's researchers to compare mean gains obtained across six reading tests. Also, there is a convention for evaluating the magnitude of effect sizes (Cohen, 1988). For effect sizes associated with changes in mean test score gains, effect sizes of .20 SD units or lower are generally considered to be "small," a "medium" effect size is approximately .50 SD units, and a "large" effect size is .80 SD units or greater.

Relationship Between Reading Skill Change and Characteristics of Reading Classes

Regression analyses were conducted to identify which learner characteristics were related to gains in reading skills. Once the key learner characteristics that predict learner skill gains were identified, they were entered, along with key measures of class instruction, in hierarchical linear modeling (HLM) analyses to determine the relative contribution of each measure in predicting learners' skill change (Bryk & Raudenbush, 1992).

DESCRIPTION OF ABT'S STUDY'S LEARNERS

The adult learners who participated in the study's low-level reading classes varied in their background characteristics and the amount of formal education they had received. While data on the universe of learners enrolled in ABE reading classes at the time of Abt's study are not available as a comparison for the study's learners, there is limited information on the characteristics of all learners who participated in programs funded under the Adult Education and Family Literacy Act (AEFLA) in program year 2001–2002 (the last year of the study's data collection). Relative to all adults in AEFLA-funded programs at the time of Abt's study, the study's participants tended to be older—81 percent were twenty-five years of age or older compared with 59 percent of adults enrolled in AEFLA programs. (The mean age of the study's learners was forty years.) The distribution of ethnic/cultural groups in the study's sample also differed from ABE learners nationally. The study included higher percentages of African Americans, white (non-Hispanic) adults, and American Indian/Alaskan Natives. (Some of these differences may have been caused by the location of a quarter of the study programs in the Pacific Northwest, where there are large concentrations of American Indian/Alaska Natives.) The study's focus on ABE reading classes serving native-born learners is reflected in the 14 percent of adults in the study who were Hispanic or Latino, compared with 40 percent of adults nationally who participated in AEFLA-funded programs in 2001–2002. Study participants' distribution by gender (53 percent female) was similar to that of all adult education participants (fifty-seven participants) (U.S. Department of Education, 2002).

Study participants' level of formal education, employment status, and income are of particular interest. While the classes in the study were aimed at serving adults below a seventh-grade equivalence reading level, participants' level of education at pretest indicated that 73 percent had completed the seventh grade or higher of formal education. Seventeen percent of participants had received a high school diploma or a GED credential, and 7 percent had college experience.[5]

Data on employment status and income provide another lens for understanding the adults who enrolled in the study's classes. Although 41 percent of participants were employed at the time of the study's pretest, only half of these individuals had permanent, full-time jobs. Half of participants (52 percent) reported having jobs in one of three occupational areas: food preparation and serving (e.g., cook, dishwasher, waitress), building and grounds cleaning and maintenance (e.g., custodian, housekeeper, and janitor), and production (e.g., factory assembly, machinist, and seamstress). Overall, participants' incomes were low, with 65 percent having an annual personal income of $12,000 or less.

While Abt's study was intended to have a sample of only native-born, ABE learners, all of the ABE reading classes in the study included adults who were not born in this country. ABE staff were consistent in reporting the criterion for placing nonnative-born learners in ABE reading classes—learners had to be sufficiently fluent in oral English to enroll in the class. Across the 130 classes in the study, 59 percent of learners were born and educated in the United States and another 7 percent were born outside the United States but educated in the U.S. (These individuals were included in the native-born group.) Thirty-four percent of learners were born and educated outside the United States and were judged by ABE program staff to be sufficiently fluent in oral English to be enrolled in the same reading classes with native English speakers.

The nonnative-born learners differed from native-born learners on all characteristics except type of employment. They were older, more likely to be female, more likely to be married, and more likely to have children living in their households. The nonnative-born learners also had less formal education, were more likely to be employed, and had higher personal incomes. These characteristics highlighted that these were two different groups of learners and the importance of accounting for this difference in the study's data analysis.

LEARNERS' ATTENDANCE

A key challenge for ABE programs is encouraging learners to access as much instruction as they can given the demands on their daily lives. In Abt's study, learners' attendance was measured in three ways: the average hours they attended the study's reading class, the number of weeks they attended the class, and the percent of the total hours of the class they accessed. Participants' average hours of attendance during the study period was 124 hours, which exceeds the national yearly average of sixty-six hours reported by the U.S. Department of Education (2003) and learners' mean attendance of between 80–100 hours cited in the report on the results from the Adult Education Program Survey that was undertaken during the same time period as the data collection for Abt's study (Tamassia, Lennon, Yamamoto, & Kirsch, 2007). About one-third of the learners received from 100 to 200 hours of instruction, and the majority (75 percent) received these hours over a period of fifteen to thirty-seven weeks. Study learners had an average participation of twenty-two weeks, which is the equivalent of about two instructional sessions (e.g., fall and winter). About one-third of the learners attended the study's classes for more than two sessions. On average, study participants accessed 63 percent of the reading class hours that were available to them.[6]

LEARNERS' DEVELOPMENT OF READING SKILLS

This section discusses the amount of improvement that learners made in their reading skills after participating in their reading classes. Outcomes are examined for learners who participated in pre- and posttests and for those who also were in the twelve-month follow-up.

Learners' Reading Skills at Pretest

As previously noted, six standardized tests were used to assess learners' reading and spelling skills. On each of the six tests at pretest, at least half of the learners scored at the equivalent of grade 3 or lower, with few scoring at an equivalent of seventh grade or higher. These results corroborated that the study's learners were the intended target population for the study—first-level learners in the range of 0–6th grade equivalence.

Compared with the native-born learners, the nonnative-born learners performed better at pretest on decoding and word recognition (as measured by Woodcock-Johnson Revised Word Attack and WRAT3 Word Reading), but less well on vocabulary and reading comprehension (as measured by Nelson Word Meaning and Nelson Comprehension).

Definition of Learner Skill-Level Groups

To prepare for the analyses of gains on the six tests, Abt's researchers examined the correlations of pretest scores across the five reading measures to determine if there were really discrete reading skills being measured. Three of the reading measures—Woodcock-Johnson Revised Word Attack and Letter-Word Identification and WRAT3 Word Reading—are defined as assessing the related skills of decoding and word recognition. The scores on these tests were highly correlated. A high correlation (.81) also was found between the instruments measuring vocabulary and comprehension—Nelson Word Meaning and Nelson Reading Comprehension. The lowest correlation was between Word Attack and Reading Comprehension (r = .55), which was expected given some low-level learners' discrepancy between their decoding and comprehension skills.

Abt's researchers examined the correlations among the study's reading tests, which indicated that decoding and comprehension scores were not necessarily related and the emerging evidence from research (e.g., Kruidenier, 2002; Strucker, 1995) that reading component skills are not necessarily at the same levels of proficiency (i.e., adults who have low scores on decoding may have higher comprehension scores and vice versa). Based on this information, they conducted the analyses of gains in reading separately for subgroups of adults who differed on their profile of reading skills at pretest.

Study participants were divided into four groups based on the pattern of their scores on decoding and comprehension skills. In constructing the groups, one test was used to represent decoding skills and a second test was used to represent comprehension skills. Since the intercorrelations among the reading tests indicated that the three decoding measures (Woodcock-Johnson Revised Word Attack and Letter-Word identification and WRAT3 Word Reading) were highly related, as were the vocabulary (Nelson Word Meaning) and comprehension (Nelson Reading Comprehension) tests, the Word Attack test was selected to represent decoding

Table 4.2 Reading Skill Group at Pretest

Comprehension (Nelson Reading Comprehension Measure)	*Decoding* (*WJ-R Word Attack Measure)*	
	Low (0–3rd grade equivalent)	*Medium/High* (4th–6th grade equivalent)
	A	B
Low (0–3rd grade equivalent)	Low Decoding/ Low Comprehension N = 347	Med/High Decoding/ Low Comprehension N = 27
Medium/High (4th–6th grade equivalent)	C Low Decoding/Med/ High Comprehension N = 161	D Med/High Decoding/ Med/High Comprehension N = 108

Having a lower pre-test score was a significant predictor of higher gains on the WJ-R Word Attack. *Source*: Alamprese, J., Tao, F., & Price, C. (2003). *Study of Reading Instruction for Low-Level Learners in Adult Basic Education Programs, Volume I: Study Findings*. Bethesda, MA: Abt Associates Inc.

and the Reading Comprehension was selected to represent comprehension in constructing the groups. Four skill groups were formed based on learners' pretest scores on two tests.

Over half of the learners (54 percent) were the group A with low decoding and low comprehension skills. Another quarter of the learners (group C) had low decoding skills but medium-high comprehension. The distribution of the sample by learners' place of birth and education revealed only slight differences. When compared to native-born learners, larger percentages of nonnative-born learners were in groups A and C (low in both decoding and comprehension or only low in decoding).

Learners' Pre-Post Reading Gains

The average pre-post gains for learners were statistically significant for at least one skill group of learners on all six tests. The gains for the decoding and word recognition measures—Woodcock-Johnson Revised Word Attack and WRAT3 Word Reading—had an effect size of .20 SD units or greater (Word Attack: .33 SD units and Word Reading: .20 SD units), while the gains for the other tests were very small, ranging from effect sizes of .05 SD units to .13 SD units. Learners with low decoding skills achieved the greatest gains. On the Word Attack test, learners in group A (low decoding and comprehension skills) gained seven points from pre- to posttest, on average, which is an effect size of

.42 SD units (near the conventional criterion for a "moderate" effect). On the WRAT3 reading test, these same learners made smaller but still meaningful gains (effect size of .20 SD units). Learners with low decoding skills and medium-high comprehension skills also increased their vocabulary skills by a meaningful amount (effect size of .24 SD units). Learners with higher comprehension skills or higher skills in both areas made smaller gains.

Learners who were neither born nor educated in the U.S. had greater reading gains than those who were born and educated in this country. Nonnative-born learners made gains that were related to effect sizes of greater than .20 SD units on the two decoding and the word recognition measures, with their strongest gains on Word Attack (effect size of .49 SD units).

Of critical interest to the field of adult reading is whether learners with low decoding skills are able to increase their vocabulary and comprehension through their participation in ABE. Given that over half of Abt's study participants (54 percent) had low decoding skills at pretest, it was not expected that study participants would necessarily increase their comprehension test scores during their period of participation in the study (a maximum of nine months of instruction). However, adults with low decoding and medium-high comprehension scores at pretest were able to make small but potentially meaningful gains in vocabulary (effect size of .24 SD units), which represented a step in developing comprehension.

Reading Gains and the Follow-up Sample

In order to understand the effects of ABE on longer-term outcomes for low literacy learners, Abt's researchers examined reading gains through the twelve-month follow-up testing for the five reading and the spelling tests. These analyses were conducted separately for learners who participated in ABE reading classes during the follow-up period and for those who did not continue their education. Compared with learners who did not remain in ABE during the follow-up period, learners who continued their participation in ABE obtained statistically significant gains on all of the tests from the pretest to the follow-up testing. Of particular note was the pattern of gain for vocabulary (Nelson Word Meaning) and comprehension (Nelson Reading Comprehension). On these two tests, learners who were in ABE at follow-up made progressively larger gains from pretesting to posttest to follow-up; the effect size for the overall gain from pretest to follow-up was .25 SD units on both tests. Learners who remained in ABE also made gains on Word Attack (effect size of .33 SD units) and WRAT3 Word Reading (effect size of .27 SD units) from pretest to follow-up. These data suggest that on average, learners' continued participation in ABE was beneficial to their reading development.[7]

Factors Predicting Amount of Learners' Reading Gains

HLM analyses were conducted to identify the factors that predicted learners' reading gains. The HLM tested the relationship between different learner characteristics and the amount of pre-post gains on the six tests. The first factor tested was learners' reading level at pretest. As expected, pretest scores were strong, significant predictors of pre-post change for all reading assessment measures.

Table 4.3 Key Demographic Predictors of Pre-Post Skill Gains

Key Demographic Variables	*Decoding and Word Recognition*			*Vocabulary and Comprehension*	
	WJ-R Word Attack	*WRAT3 Reading*	*WJ-R Letter Word ID*	*Nelson Word Meaning*	*Nelson Comp-rehension*
Lower pretest score	Higher gains****	Higher gains****	Higher gains****	Higher gains****	Higher gains****
Nonnative-born learner	Higher gains****	Higher gains****	Higher gains****	Lower gains***	
Attended ABE prior to current class	Lower gains****	Lower gains**		Lower gains***	
Has current learning prob-lem	Lower gains***				
Is currently employed		Lower gains **			
Had poor school experience as child			Lower gains ***		
Demographic Variables That Are Related with Vocabulary and Reading Comprehension Outcomes Only					
Has higher background knowledge				Higher gains***	Higher gains****
Is concurrently enrollment in another ABE class				Higher gains***	
Older adults					Lower gains**

*The relationships between learner demographic variables and pre-post gain scores on each reading test are significant at the following levels: $^{**}p < .05$, $^{***}p < .01$, $^{****}p < .001$ (based on beta coefficients resulting from HLM modeling with all learner predictor variables entered simultaneously in which the significant predictors were identified through a backwards and forwards elimination process).

Lower average pretest scores predicted higher average gains on the reading tests. One interpretation of this finding is that it could be due to regression-to-the-mean, which is a possibility. However, it is not likely that regression-to-the-mean would be responsible for the statistically significant pre-post gains across the five reading assessments. The results of the paired tests of the null hypothesis that there would be zero gain demonstrate unequivocally that, relative to pretest scores, the distribution of the posttest scores had shifted to higher mean levels. Another possibility is that ceiling effects could be involved in the correlation between pretest and gain; however, no ceiling effects were found.

Another significant predictor of amount of gain was learners' place of birth and place of education. Learners who were neither born nor educated in the United States made larger average gains on the decoding and word recognition measures, compared with learners who were born in the United States. On the vocabulary test, native-born learners had higher pre-post gains.

Variables with significant but less strong relationships to pre-post gains on reading scores were learners' self-report of having a learning problem or disability, baseline employment status, and their school experience as a child. Learners who said that they had current learning problems tended to have smaller pre-post gains on the decoding measure. These learners also were more likely to have reported learning problems as a child, current or childhood health problems, stability problems in their homes while growing up, current and childhood functional disabilities, and poorer overall health.

Three other learner characteristics were significantly related to gains on the vocabulary and comprehension measures but not on the decoding and word recognition tests. These were learner background knowledge (as measured by the Woodcock-Johnson Revised Social Studies score at pretest), concurrent enrollment in other ABE classes, and gender. For both the vocabulary and reading comprehension measures, higher social studies pretest scores were associated with larger pre-post gains. Learners who were enrolled concurrently in a study class and a nonstudy ABE class also had greater gains on the vocabulary measure. The final demographic characteristic that was a predictor was gender, with males having slightly smaller gains than females.

Based on these HLM analyses, pretest scores and learners' place of birth and education were used as grouping variables in further examinations of gains.

Attendance and Reading Gains

An important policy question in ABE is the extent to which the amount of learners' participation in instruction affects the development of their skills. In Abt's study, analyses were conducted to examine whether hours or weeks of attendance predicted reading gains. (The two reading measures used in these analyses were the Word Attack and the Word Reading tests, since these tests showed the most consistent pre-post gains.) On average, across all learners, there was not a significant relationship between hours of attendance and reading gains. There was a relationship between weeks of attendance and gains on decoding skills. In analyses that controlled for other learner characteristics, differences were shown between learners who had attended class for more than seven weeks and those who had attended for seven weeks or less.

Further, these patterns differed for native-born and nonnative-born learners. For native-born learners who attended class for periods from eight to fourteen weeks, there were significant gains on both Word Attack (effect size of .36 SD units) and Word Reading (effect size of .27 SD units). Native-born learners who attended longer, in the range of fifteen to thirty-six weeks, also increased their Word Attack scores, but the effect sizes were smaller.

There was a different relationship between attendance and skill gain for nonnative-born learners—one that is somewhat more linear. These learners who had fifteen to twenty-four weeks of attendance made greater gains than those with eight to fourteen weeks, on both Word Attack (effect sizes of .53 SD units vs. .40 SD units) and Word Reading (effect sizes of .23 SD units vs. .13 SD units). Thus, for nonnative-born learners, attendance was positively related to growth in decoding skills.

ANALYSIS OF READING CLASSES

A key area of interest to ABE instructors and policymakers concerns the types of instruction that can best assist ABE learners in developing their skills. Given the limited research on ABE reading instruction, this study developed a base from which to examine instructional practices in adult reading classes. Using the data from the class observations and instructor interviews that Abt's senior researchers conducted, a typology of the types of reading classes in the study was developed by combining three sources of data: a) the types of reading content taught in the study's reading classes, b) the ways in which the reading classes were organized, and c) instructors' descriptions of their approach to teaching reading.

Based on these data, the 130 reading classes were grouped into the following four types:

- *Structured Reading Using Phonics*[8] (n = 41 classes): In these classes, activities were highly structured according to an established reading curriculum that emphasized the use of phonics in developing learners' alphabetic skills and included activities aimed at increasing word fluency. There was limited instruction on developing learners' passage comprehension skills.

- *Structured Reading with Phonics and Comprehension* (n = 48 classes): The class activities were highly structured according to an established reading curriculum that used phonics in developing learners' alphabetic skills and included activities aimed at increasing learners' fluency and passage comprehension.
- *Combination of Various Reading Components* (n = 31 classes): The class activities involved teaching the components of reading; activities were organized by the instructor and were not based on one established reading curriculum (although they may have included materials from established curricula). These classes varied in the types of reading component skills that were emphasized. Also, the extent to which decoding, fluency, vocabulary, and reading comprehension were taught varied according to the philosophy, training, and experience of individual instructors.
- *Life Skills Method* (n = 10 classes): The class activities were designed and organized with the purpose of assisting learners in applying reading skills to real-life situations and tasks. Learners received limited specific instruction of reading component skills. Reading skills were taught in the context of real-life reading materials and situations.

The validity of the categorization of the types of classes into four groups was predicated on the assumption that there were distinct differences in the instructional content among the four groups of classes. To test this assumption, a series of discriminant analyses was performed. These analyses indicated that four class types were differentiated in terms of: a) the use of phonics, b) use of word and structural analyses, c) the conduct of spelling activities, d) the provision of opportunities for silent reading, e) the conduct of nonreading activities, and f) the provision of opportunities for reading comprehension.

A further analysis was performed on the instructional content data to identify a set of instructional content codes that represented each of the key reading components, independent of class type, with a high degree of internal consistency. Eight instructional content scales were derived through this process. The calculation of the Cronbach's Alpha coefficients revealed that these eight scales were internally consistent and the coefficients ranged from .72 to .88.

The four types of classes varied in the mean prevalence of the reading instructional methods that were used. The structured phonics classes (with and without an emphasis on comprehension) used word analysis and phonics more frequently than did the two other types of classes, although both the combination of reading components classes and the life skills classes also used both methods. In terms of comprehension development, the phonics classes with comprehension and those teaching a combination of reading components spent significantly more time in developing learners' comprehension skills than did classes focusing on phonics or life skills. Across all types of classes, instructors used a variety of activities to teach comprehension, with a particular emphasis on having learners answer literal as well as inferential questions, responding to background information, and synthesizing information. Classes were the least varied in their teaching of vocabulary, and used a range of activities including reviewing the definitions of words and highlighting common and uncommon words.

Analysis of the amount of time that the reading component skills were taught in each types of class provided another way of understanding the differences among class types. Almost half (45 percent) of the time in structured reading classes using phonics was spent on decoding, word recognition, and the use of phonics. These classes differed significantly from the other types of classes on this dimension. Also of interest was the use of nonreading activities, on which significantly more time was spent in life skills classes (24 percent) than in the other classes.

To understand the relationship between learners' reading gains and the types of reading instruction that they received, learners' gains on the six tests were analyzed according to class type. Learners in classes using structured phonics with and without comprehension had larger gains on decoding (Word Attack and WRAT3 Word Reading) than did learners in classes taught with a combination of reading components or life skills classes. On Word Attack, the effect sizes were .37 SD units for structured phonics and .42 SD units for structured phonics with comprehension compared with effect sizes of .25 SD units for a combination of components and .18 SD units for life skills classes.

An examination of gains by class types and learners' skills at pretest indicated that within the structured phonics classes, learners who had low decoding and medium-high comprehension skills at pretest made moderate gains (effect size of .58 SD units) on Word Attack and small-moderate gains (.32 SD units) on WRAT3 Word Reading. Of particular note were the gains for learners with medium-high decoding and medium-high comprehension at pretest. Those groups of learners who attended classes using structured phonics achieved gains with an effect size of .61SD units on comprehension. The low-decoding and low-comprehension learners in life skills classes also had a small (effect size of .24 SD units) increase in their comprehension skills. Overall, the analysis of the relationship between reading instructional methods and learners' skill gains indicated that the type of reading instruction that study participants received was related to their development of decoding skills as well as their vocabulary and comprehension skills.

SUMMARY AND DISCUSSION

The *Study of Reading Instruction for Low-Level Learners in ABE Programs* provided new information about the range of adults who seek reading assistance in ABE settings and factors that are related to these adults' development of their reading skills. The majority of the learners who were enrolled in the study's 130 reading classes had a formal education at the seventh grade or higher, yet their reading decoding skills were in the zero to third-grade equivalence level as measured by standardized reading tests. Approximately one-third of the learners were neither born nor educated in this country but had sufficient oral fluency in English to benefit from ABE reading classes. Learners in the study had participation rates that exceeded the average hours of attendance reported for adults participating in federally funded ABE programs.

The analysis of the relationship between learners' characteristics and reading outcomes indicated that learners' pretest score in reading was the strongest predictor of their reading gains. Learners' place of birth and place of education also were important factors in determining their reading progress. Overall, study participants made small (in terms of effect sizes) but significant gains in their decoding skills from pre- to posttest, with learners who were born and educated outside of this country making larger gains than native-born learners. Learners who were assessed twelve months after the posttest also made small but significant additional increases in their vocabulary and reading comprehension. Learners who continued to participate in ABE after the posttest made greater gains than those who did not continue their education.

Learners' attendance in terms of the number of weeks (rather than the total hours) that they participated in ABE reading classes was related to their development of decoding skills but not to the development of their reading comprehension. This finding was stronger for nonnative-born learners than for learners born in this country.

The study also had important results concerning the delivery of class-based reading instruction in ABE programs. Data from the study's thirty-five programs and 130 classes provided a preliminary understanding of the types of reading methods and general instructional strategies that are being used in ABE classes for first-level learners. The ABE classes in the study taught reading in one of the following ways: a) adapting K–12-based reading methods that incorporate phonics and that may or may not include the teaching of comprehension; b) teaching a variety of reading component skills, but not according to a prescribed sequence; or c) using context-based life skills as a focus of instruction in teaching reading. This study found that the use of structured reading instruction where phonics is a dominant method was associated with greater decoding gains in low-level learners. There also was a preliminary finding for learners with medium-high skills at pretest in both decoding and comprehension, where their participation in structured phonics classes was associated with a moderate increase in their comprehension skills.

IMPLICATIONS FOR POLICY, RESEARCH, AND PRACTICE

The study's results have implications for adult education policy, practice, and research. One important policy issue concerns the measurement of adults' reading skills and the types of professional development that can best prepare instructors to teach reading effectively. Currently, adult learners' progress is reported to the U.S. Department of Education's National Reporting System (NRS) for Adult Education in terms of their results on general literacy measures. Given this study's findings that suggest that low-level adults make progress in decoding before increasing their skills in other reading components, the use of general literacy measures may not be adequate to document the gains of these learners. Rather, instruments that measure decoding as a discrete skill may need to be included in the portfolio of acceptable instruments for NRS reporting.

The study's results on learner outcomes, instruction, and program operations have implications for the delivery of services to ABE learners. These implications are the following.

Learner Intake and Assessment

The study's finding that learners' pretest is the strongest predictor of learner gain suggests the importance of conducting an in-depth assessment at intake to obtain information about learners' reading skills. Having this information will facilitate the placement of learners into instruction that is aligned with their needs. The use of pre- and posttest measures of reading also will enable programs to assess and report learners' reading gains more accurately.

Learner Instruction

In this study, both native-born and nonnative-born learners participated in low-level reading classes and were able to develop their decoding skills. This finding is important for ABE programs that want to transition ESL learners into ABE and need information about how to organize reading instruction. The study's results on reading methods also suggest that organized, structured teaching with a scope and sequence that includes phonics can assist low-level learners' development of decoding skills, regardless of their place of birth and education.

Program Management

The use of organized, structured teaching, as described earlier, may require new program policies or scheduling to assist learners in obtaining the maximum benefits from instruction. An example of a policy is managed enrollment, where learners are permitted to enter a class only within a limited period of time after an instructional cycle begins. Learners who are waiting to enter class can be placed in a computer lab or with a tutor until the class cycle begins. A scheduling change may involve offering classes in eight- to ten-week cycles to reduce the amount of time learners have to wait to enter a class cycle.

Program Improvement and Professional Development

The study's findings about the use of organized, structured reading methods with phonics in teaching low-level learners suggests that professional development may be necessary to assist teachers in learning about and incorporating new instructional content and methods in their practice. The experience of the study's programs in supporting instructor change suggests that administrative support is a facilitating factor in instructors' use of the information acquired through professional development, and that having teams of instructors work together on new teaching methods helps to ensure their successful adoption or adaptation of these methods.

The final implication from this study concerns future research on adults' development of reading skills. This correlational study provided rich descriptive data about the types of learners in ABE reading classes and the instruction that they received. The study also found that learners' characteristics and reading instructional methods were associated with learners' development of decoding skills. Furthermore, learners who continued their participation in ABE for a second year were able to increase further their vocabulary and comprehension skills.

These findings suggest that rigorous research using an experimental design is needed to understand the types of specific methods for teaching reading component skills that instructors can use to facilitate learners' development of reading skills. Since adults are available for instruction for limited time periods, evidence-based information on the use of targeted strategies would assist instructors in using the time that is available for instruction more effectively. Furthermore, given that the duration of learners' participation appears to be associated with the development of their reading skills, further investigation is needed to identify the types of programmatic and other arrangements that can support adult learners in their reading development both in and outside of the classroom.

NOTES

1. Wilson, B. (1998). Matching student needs to instruction. In S. Vogel and S. Reder (eds.) *Adult literacy, education and learning disabilities.* Baltimore: Paul Brookes, Inc.
2. For more details on the study's participants, findings, methodology, and statistical analyses, see the final report for the study (Alamprese, Tao, & Price, 2003).
3. For descriptions of the six adult reading studies that comprise the National Reading Program supported by the National Institute of Child Health and Human Development, the National Institute for Literacy, and the U.S. Department of Education, see http://www.nifl.gov/nifl/nat_research.html.
4. Analyses were conducted to examine differences in the characteristics of the samples of learners who had: a) only a pre-test, b) a pre- and posttest, and c) a pre-, post-, and follow-up tests. The analyses indicated that there were no significant differences in the characteristics of the sample except on age and place of birth. While the learners in the pretest group tended to be younger and more likely to have been born in the United States, learners with these characteristics still were represented in the posttest and follow-up groups. Differences on demographic and health characteristics were examined between learners with a pretest and those with a pre- and posttest. There were no significant differences between these groups except that learners with only a pretest had, on average, higher pretest reading scores, were younger, were more likely to be native born, and more likely to report having a current learning problem. The comparison of the groups of learners who had pre-post tests and were eligible to participate in the follow-up test but did not with those who had follow-up data revealed no significant differences except on two characteristics. The group with only a posttest had higher mean baseline reading scores and was younger.
5. In order to earn a GED credential, an individual must pass a battery of five tests that are designed to measure the academic skills normally required by a typical high school program of study. The five tests (Language Arts-Reading, Social Studies, Science, Mathematics, and Language Arts-Writing) use a multiple-choice format and include a timed essay on an assigned topic. Less than half of the 17 percent of Abt's study learners with a high school diploma or GED at pretest had earned the GED certificate. While the GED test measures primarily reading skills above the ninth-grade equivalence (GE) level, it is possible but unusual that adults with reading skills below the seventh GE level (the level of Abt's study learners) are able to earn the GED credential.
6. Two factors may explain why the study sample's average hours of attendance were greater than the national average of ABE program participants. The study's learners were adults with low reading skills—a subpopulation of national ABE participants who typically attend ABE programs for more hours than learners who enter ABE with higher skill levels. The sample of thirty-five ABE programs in the sample also were programs that met a basic level of operationally functioning and, as such, were programs that had a number of characteristics that should support learners' persistence.
7. Of the learners who participated in the follow-up testing but who did not continue their participation in an ABE reading class, only 5 percent indicated that they did not continue education because they were not learning fast

enough or the class was too difficult. The main reasons that learners gave for not continuing in ABE were that they had a conflict with their work schedule (39 percent) or that they had a conflict with family responsibilities or did not have child care (20 percent).

8. The reading curricula that instructors used were: Wilson Reading System, The Slingerland Multisensory Approach, Lindamood-Bell Learning Process, WORDS, Neuhaus Multisensory Reading and Spelling (MRS), SRA Corrective Reading, and the Texas Scottish Rite Hospital Literacy Program.

REFERENCES

Alamprese, J. A. (1993). Key components of workplace literacy projects and definitions of project "models." In Research Triangle Institute, *Alternative designs for evaluating workplace literacy programs*. Research Triangle Park, NC: Research Triangle Institute.

Alamprese, J. A., Goodson, B., & Tao, F. (1999). *Request for OMB review: Evaluation of effective adult basic education programs and practices*. Bethesda, MD: Abt Associates Inc.

Alamprese, J. A., Tao, F., & Price, C. (2003). *Study of reading instruction for low-level learners in adult basic education programs, Volume I: Study findings*. Bethesda, MD: Abt Associates Inc.

Bryk, A., & Raudenbush, S. (1992). *Hierarchical linear models*. Newbury Park, CA: Sage.

Cohen, J. (1988). *Statistical power analysis for the behavioral sciences* (2nd ed.). Hillsdale, NJ: Lawrence Erlbaum Associates.

Hanna, G., Schell, L. M., & Schreiner, R. (1977). *The Nelson Reading Skills Test*. Itasaca, IL: Riverside.

Israel, B. A., Checkoway, B., Schultz, A., et al. (1994). Health education and community empowerment: Conceptualizing and measuring perceptions of individual, organizational, and community control. *Health Education Quarterly*, 21(2), 149–70.

Kruidenier, J. (2002). *Research-based principles for adult basic education: Reading instruction*. Washington, DC: National Institute for Literacy.

Strucker, J. (1995). *Patterns of reading in adult basic education*. Unpublished doctoral dissertation, Cambridge, MA, Harvard University Graduate School of Education, Gutman Library.

Strucker, J. (1997). *The Reading Components Approach*. Cambridge, MA: National Center for the Study of Adult Learning and Literacy.

Tamassia, C., Lennon, M., Yamamoto, et al. (2007). *Adult education in America: A first look at results from the Adult Education Program and Learner Surveys*. Princeton, NJ: Educational Testing Service.

Tao, F., Price, C., & Alamprese, J. (2003). *Study of reading instruction for low-level learners in adult basic education programs, Volume II: Study methodology*. Bethesda, MD: Abt Associates Inc.

U.S. Department of Education, Division of Adult Education and Literacy. (2002). *FY2000 state-administered adult education program: Data and statistics*. Washington, DC.

U.S. Department of Education, Division of Adult Education and Literacy. (2003). *Data on contact hours*. Washington, DC.

U.S. Department of Education, Division of Adult Education and Literacy. (2006). *State administered adult education program: Program year 2004–2005 enrollment*. Washington, DC.

Wilkinson, G. S. (1993). *Wide Range Achievement Test–Revision 3*. Wilmington, DE: Jastak Associates, Inc.

Wilson, B. (1998). Matching student needs to instruction. In S. Vogel and S. Reder (eds.) *Adult literacy, education and learning diabilities*. Baltimore: Paul Brooks, Inc.

Woodcock, R. W., & Johnson, M.B. (1989, 1990). *Woodcock-Johnson Psycho Educational Batter–Revised*. Itasca, IL: Riverside.

5 "What Works" for Adult Literacy Students of English as a Second Language

Larry Condelli, Heide Spruck Wrigley, and Kwang Suk Yoon

The focus of the *What Works Study* is on adult English as a Second Language (ESL) *literacy* students—ESL learners who lack basic literacy skills and have minimal proficiency in English. These learners face the dual challenge of developing basic skills, decoding, comprehending, and producing print. Until the last quarter century, schools and resettlement agencies designed ESL classes on the assumption that adult students had the basic education and literacy skills to learn another language (Van de Craats, Kurvers, & Young-Scholten, 2006). A wave of immigrants that arrived beginning in the late 1970s challenged this assumption, as these new immigrants did not have the strong educational experiences upon which literacy is built. Without the basic text-processing skills that allow them to follow text that appeared in class and in textbooks, these students became frustrated, overwhelmed, and had a high dropout rate due to their inability to catch up the missing literacy skills and keep up with the more literate students in the class (Wrigley & Guth, 1992). Recent trends indicate that the number of new immigrants to the United States who have very low levels of literacy is continuing to increase (Fix, Passel, & Sucher, 2003).

The *What Works Study* is the first large, national study to examine these students: who they are, their instructional experiences, and what they learn. However, the goal of this study was not merely descriptive: it also sought to identify "what works"—the instructional activities that help to develop and improve ESL literacy students' English literacy skills and their ability to communicate in English, by addressing the following questions:

- Who are adult ESL literacy students? What are their English literacy and language abilities?
- What types of instructional approaches do teachers of adult ESL literacy students use?
- What classroom and instructional variables are correlated with improving adult ESL literacy students' literacy and language development and

how do these relationships vary by students' initial literacy level, native language, age, or other characteristics?

- What are the attendance and persistence patterns of adult ESL literacy students and what instructional factors affect persistence?

The study was to suggest changes in program design, resources, and instruction that are needed to improve instruction to, and literacy and language development of, adult ESL literacy students.[1]

CONCEPTUAL FRAMEWORK

Teaching ESL literacy requires instruction in (1) the *language skills* necessary to communicate in English, including subskills related to sentence structure, pronunciation, word endings, tenses; and (2) the *literacy* or reading and writing skills necessary to process print and gain meaning from the written word. Cognizant of the importance of these key components in teaching and learning, we based our approach to the study on the literature in second-language learning, theories of adult learning, studies in cognition, and research in reading. When we began the study, very few studies existed that systematically explored the intersection of literacy development and language learning in students who were nonliterate or only marginally literate in their native language. Since there was no direct research upon which to build our study, the conceptual framework we created had to be based on implications from related research.

To guide our efforts, we built on a set of principles, originally developed as part of the national Cyberstep Project (Wrigley,. 2001), outlining what it takes for adults to learn. We wed these principles to research in basic reading and created a framework that looked at key components needed to enable students to learn to read and write for the first time in English. These components included reading as meaning-making, reading for practice, and activities related to oral fluency, grammatical accuracy, and vocabulary development. We also included strategies likely to improve literacy development, such as connecting prior information with new information and focusing on both global meaning and details, and subskills related to phonemic awareness, decoding, and pronunciation.

There is very little research on outcomes related to adult literacy students, and almost no such research on adult ESL literacy students, to inform development of a growth model.

However, the broader research on adult second-language acquisition suggests that the following could affect English-language learning and literacy development of our students:

- **English oral language** ability could influence second-language reading development (Alderson,1984; Bernhard & Kamil, 1995; Carrell 1991).

- **Literacy in context**—use of language in rich contexts in instruction (such as connection to real-world tasks and materials) promotes second-language learning (Bransford, Brown, & Cocking, 2000; Brown, Pressley, Van Meter, & Schuder, 1996; Greeno, Collins, & Resnick, 1996; Purcell-Gates, Degener, Jacobson, & Soler, 2002).
- **Native language support**—literacy in the native language promotes second-language development (Ramirez, Yuen, & Ramey, 1991; Slavin & Cheung, 2003; Thomas & Collier, 2002).[2]

METHODOLOGY OF THE STUDY

We collected data over a two-year period in thirty-eight classes from thirteen adult ESL programs in seven states (Arizona, California, Illinois, Minnesota, New York, Texas, and Washington), and we had a final sample size of 495 students. The sample included two cohorts of students who were followed from the time of entry into class for nine months. Data collectors assessed students at entry (initial assessment), approximately three months after enrollment (second assessment), and about nine months after enrollment (final assessment), regardless of how long the student remained enrolled. Data collectors also observed each class an average of nine times over the data collection period and coded instructional activities.

SITE AND STUDENT SELECTION

We identified ESL literacy students as those who had three years of schooling or fewer and selected a sample of candidate sites (over twenty-five in all) chosen from a mail survey. We recruited 558 students, of which fifty-eight left classes less than four weeks after recruitment, the minimum time we required for inclusion in the study, leaving five hundred eligible students in the sample. Data for five students were incomplete or inaccurate, giving us a final study sample of 495.

LONGITUDINAL ISSUES: MAINTAINING CONTACT WITH STUDENTS

Although the study had a nine-month follow-up period, we did not expect learners to remain in class more than three or four months. The nine-month follow-up assessment, however, allowed us to take a longitudinal look at student learning after instruction ended. The extra time also provided an opportunity to look at student attendance and retention in adult ESL classes.

We knew that keeping track of adult ESL students for even a relatively short time would be a challenge. Our main strategy for tracking students was to use individuals from the community as our local field staff. These "study liaisons" lived in the communities of each local adult school in the study and most were former adult ESL teachers. They knew the issues and concerns facing students in the program and were experienced in working with the study population. In Latino communities in the study, all liaisons were English-Spanish bilingual and one had been a student in the school where the study was being conducted.

The study liaisons recruited students into the study, observed classes, and had ongoing, direct contact with each student in the study. They collected extensive contact information on each student, stressed the importance of remaining in touch for the study, and encouraged students to inform them of planned leaves or absences. If students failed to attend class, liaisons would try to locate them, either directly or by asking other students in the class or the community to assist in locating them. Through this personal relationship between student and liaison, we hoped to minimize study attrition.

MEASURING INSTRUCTION: CLASSROOM OBSERVATIONS

Guided by our conceptual framework and our observations of over seventy-five classes, we outlined the learning tasks and teaching strategies associated with both literacy development and second-language development and created codes that described the components of learning and instruction associated with each.

We quantified instructional activities (using percent of observed time on the activity) as well as observer ratings of teachers' use of instructional strategies, with four-point Likert scales. We created two categories of measures: *instructional emphasis* measures, which describe the content of the instruction in terms of the language or literacy focus, and *instructional strategies*, the activities teachers used to organize and teach the lesson.

While these strategies and emphases characterize how instruction was provided, they were not mutually exclusive or independent of each other. In fact, teachers that used one set of strategies often used combinations of them over time or within a single class session. The following instructional variables were used in the analyses.

Instructional Emphasis Variables

- *Literacy development emphasis*—Main focus on reading and writing development.
- *ESL acquisition emphasis*—Main focus on fundamentals of English (e.g., how English works).

- *Functional skills emphasis*—Main focus on functional literacy (e.g., interpreting forms, labels, using money, maps).
- *Basic literacy skills emphasis*—Main focus on print awareness, fluency, and basic reading skills.
- *Reading comprehension emphasis*—Main focus on comprehension strategies.
- *Writing emphasis*—Main focus on writing fluency, writing practice.
- *Oral communication emphasis*—Main focus on speaking and listening practice.

Instructional Strategies Variables

- *Varied practice and interaction*—Teachers provide students with opportunities to learn in a variety of ways and modalities (e.g., speaking, reading, writing) and by having students interact with each other.
- *Open communication*—Teachers are flexible and respond to students' concerns as they arise; ask for open-ended responses; support authentic communication.
- *Connection to the "outside"*—Teachers link what is being learned to life outside classroom and bring the "outside" into the class through use of field trips, speakers, and real-life materials.

Another instructional strategy we coded was the teacher's use of students' native language in instruction. We constructed a scale of the use of this instructional strategy by first conducting a factor analysis of the four measures we used of how native-language use was incorporated into classes: to explain concepts, give directions, for students to ask questions, and to do written assignments. The analysis identified only one factor, which incorporated all of the measures. We combined these four items into a single index representing the average proportion of use of the four native-language instructional activities in each class. The scale ranged from zero (use of no activities) to one (use of all four activities). We then averaged the scores across observations.

To ensure reliability, we trained observers extensively and reviewed coding of all observations. The framework and the resulting observation instrument not only helped to guide the study, but also served as a blueprint of sorts to the kind of teaching and learning that make up adult ESL literacy.

MEASURING STUDENT LEARNING: OUTCOME MEASURES

Assessment in adult ESL requires measurement of skills in two domains: English language proficiency and literacy ability. Knowledge of English is interwoven with the ability to process print. To assess students' knowledge of English, regardless of their ability to read and write, we needed assessments that measured speaking and listening and did not require reading instructions or other literacy-based skills, to respond. Conversely, to find out if students could complete literacy tasks, we had to make sure that they understood the instructions. To that end, we gave instructions to each task in English and the native language.

To capture these complexities, we used a multidimensional, multimethod approach to assessment with a battery of standardized and nonstandardized tests that included:

- The Woodcock-Johnson Basic Reading Skills Cluster (WJBRSC) and Reading Comprehension Cluster (WJRCC), which measured basic reading and comprehension abilities;
- The oral Basic English Skills Test (BEST), measured English speaking and listening; and
- Adult Language Assessment Scales Writing Assessment (A-LASW) measured writing ability.

The study also included an interview about student literacy practices in both English and the native language and a reading demonstration task, which measured student English fluency and comprehension through reading of authentic materials. Each assessment was conducted individually. Instructions were translated and the literacy practices interview was conducted entirely in the native language.

STUDENT ATTENDANCE MEASURES

Students enrolled in adult basic literacy and ESL classes typically attend for relatively brief periods of time. These students face many barriers to attending regularly or completing a course, including the lack of available childcare, family problems, transportation difficulties, and the trade-off between work and going to class (Chisman, Wrigley, & Ewen, 1993), which may cause them to drop out of class. However, students may also leave for positive reasons, such as meeting their educational goals and some may "stop out" and return to the class at a later time (Comings, Cuban, Bos, & Porter, 2003; Quigley, 1996). Many adult educators fear this brief attendance prevents students from achieving significant improvements in their literacy skills.

We studied student attendance patterns through the following measures:

- *Total hours*—total number of instructional hours attended;
- *Total weeks*—total number of weeks attended;
- *Rate of attendance*—proportion of total hours attended out of hours possible to attend; and

- *Intensity*—average number of hours attended per week.

Each measure of attendance provides us with different information about student attendance patterns.

DESCRIPTIVE FINDINGS

Of the 495 participants in the study, 258 provided assessment data for all of the three assessment periods, 135 provided initial assessment data only, and 102 provided data at other times. Of those 102 students, ninety-seven completed the initial and second assessments but were inaccessible for the final assessment. Consequently, of 495 students that took the initial assessments, 356 students (72 percent) took the second assessments and 263 students (53 percent) took the third set of assessments.

The study attrition for the third assessment was quite high, due primarily to the loss of students during the summer months in the first year of the study. While we tried to begin the study in the fall each year, during the first year some sites did not begin data collection until the winter; and in those sites, the third assessment period was in the fall. Over the summer, many students left the area and liaisons lost contact with students. During the second year of the study, when the final assessments were completed by June, the retention was over 70 percent.[3]

The average age of students in the study was forty; they were 72 percent female and had an average of 3.1 years of schooling in their home country. Table 5.1 summarizes the characteristics of students in the study.

Table 5.1 Education in Home Country, by Language Background

Student Language Background	*Number of Students*	*Mean Years of Education in Home Country*	*SD of Mean Years*	*Percent of Students with No Formal Education*
All What Works Participants	490	3.1	2.8	33.1
Spanish—Mexican	285	4.0	2.7	17.9
Spanish—non-Mexican	43	3.8	2.2	11.6
Hmong	38	0.3	0.9	81.6
Somali	47	1.7	2.9	66.0
All others*	77	1.8	2.5	57.1

Note: Prior education data were missing from five students in the sample of 495. *More than 30 other languages are included in this group.

Students in the study attended an average of about sixteen weeks and 128 total hours and attended about two-thirds of possible time (rate of 0.64). This rate had very little variability, and further analysis revealed that type of instruction and scheduled class time had no significant effect on rate of attendance. These findings suggest that students devote a set proportion of their time to attend class, which appears to be difficult to change through schedules, class content, or attendance requirements. The lack of effect of these variables on rate of attendance may be due to life circumstances of ESL literacy students, which make it hard for them to attend more frequently, or it may be an implicit decision by students to devote only a certain amount of time to class attendance.[4]

LITERACY AND LANGUAGE SKILLS

The students in the study were indeed very low literate and had minimal oral language skills in English, as measured by the standardized assessments. According to the WJRBRSC subtests, students started at an average of about a 1.5 grade level and improved to second-grade level at the final assessment. Students' reading comprehension scores as measured by the subtests were also very low and changed little over the study period, remaining at just about the first grade level. According to the BEST Oral English assessment, students had very little oral English ability, with almost 70 percent of them initially at SPL II or lower. In contrast to literacy skills, however, BEST scores improved dramatically over the study period, with only about 40 percent of students at these levels and increased percentages at higher SPLs at the final assessment.

The Adult Language Assessment Scales Writing Test (ALAS-W) presented a challenge for students. Most students were able to write few, if any, English words and most words were simple nouns and pronouns. It was not uncommon for students to write partially or even exclusively in their native languages or to return blank test forms. By the final assessment, the average scores on both subtests increased little and the increase was not statistically significant.

STATISTICAL MODELING OF GROWTH OF LITERACY AND LANGUAGE SKILLS

The study examined the relationship of instructional content, instructional strategies, attendance, student characteristics, teacher characteristics, and class variables on student outcomes using a latent growth modeling within a hierarchical linear model (HLM) framework (Bryk and Raudenbush, 1992). The latent growth modeling technique is designed to capture the underlying growth that takes places over time by using each individual student's data to draw a single, underlying growth trajectory that fits a straight line or smooth curve. The statistical parameters underlying the line or curve can then be used to describe students' literacy growth in terms of their *initial status*, or where they started, and the *linear* and *quadratic* (or nonlinear) rates and direction of change. The parameters allow prediction of the effect of variables in the model that relate to growth. In other words, using this technique, we can estimate where students were on the measures when they enrolled, how fast they grew on the measures over the course of their class participation, and estimate continued growth. We can also relate this growth to specific variables in the model to predict which ones correlate with faster (or slower) growth.

VARIABLES USED IN THE MODEL

The conduct of latent growth modeling requires identification of predictor variables and a model of how the predictors relate to an outcome measure within the different levels of analysis. The researcher must identify the predictor variables and model through theory, prior research, and the statistical nature of the data collected.

We first conducted preliminary analyses on student, class, teacher, and instructional variables that we believed would influence language learning and, based on these analyses, dropped variables due to lack of variance on measures, to avoid redundant predictors and to achieve a parsimonious model.[5]

Figure 5.1 shows the variables used in the model, which include (1) student background variables, (2) teacher characteristics, (3) class types, (4) instructional variables, (5) attendance measures and the English literacy

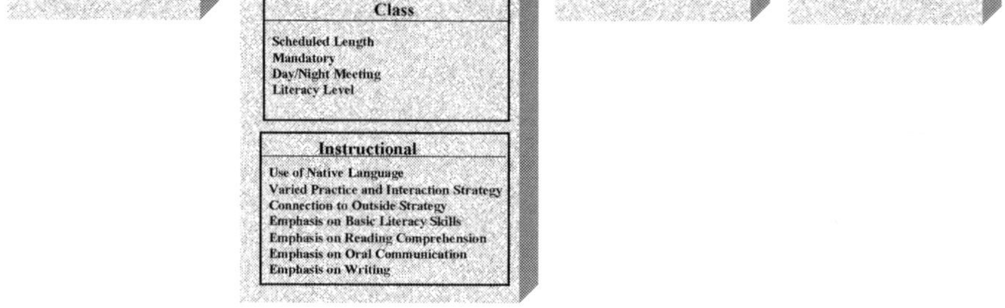

Figure 5.1 Variables used in the HLM growth modeling of literacy and language development.

and language test scores we used as outcome measures. The student background, attendance, and outcomes variables are student-level measures, and the remaining variables are class-level measures. Not all the measures in Figure 5.1 were used in all analyses. Instead, we had a core set of measures, which served as the basic model, and then used additional measures that were appropriate to the outcome under investigation.[6]

VARIABLES RELATED TO GROWTH IN BASIC READING SKILLS (WJBRSC)

Table 5.2 shows the results of the HLM modeling on the WJBRSC, which assessed students' basic reading skills, including letter-word identification and knowledge of phonics. Two student variables, age and years of formal schooling, were significantly related to growth in basic reading skills. Age was negatively related to linear growth rate, meaning that older students acquired these skills more slowly. Even though younger students started lower on this measure (as revealed by preliminary analysis), they made up for their initial disadvantage in basic reading skills by learning faster.

Students' years of formal schooling in the home country was also positively associated with linear growth rate. Students with more education both started at a higher level and learned faster than their less educated peers. However, students' years of formal schooling in the home country became less important over time, as indicated by the negative quadratic growth coefficient. This means that the initial positive effect of formal schooling in the native country on linear growth fades over time. While prior education initially helps ESL literacy students acquire basic reading skills, this initial advantage does not help later.

Students' oral English skills, as measured by the BEST, were also positively related with the linear growth in basic reading skills (with marginal statistical significance). This finding may indicate that some proficiency in oral English-language skills may work to assist learning of basic reading skills.

The only class-level variable related to growth in basic reading skills was the length of the scheduled hours per week of class meeting time. Students in classes with longer scheduled hours showed less growth than students in classes with fewer scheduled hours. Other things being equal, including students' attendance and persistence, the longer the classes' weekly scheduled meeting hours, the slower the rate of students' learning in basic reading skills.

The use of the instruction strategy we called "connection to the outside," where teachers brought real-world materials and examples into their instruction, had a positive effect on the linear growth in basic reading skills. The use of this strategy was effective in raising the level of students' mastery in basic reading skills. The analyses suggest that that all else being equal, the use of the connection to the outside strategy results in an increase in basic skills development over time.

VARIABLES RELATED TO GROWTH IN BASIC READING COMPREHENSION (WJRCC)

The latent growth modeling analysis for the reading comprehension measure (the WJRCC), showed an average steady linear growth over time of about 1.2 points per month. We examined within the model the relationship of students' basic reading skills at entry in class on growth in reading comprehension. The results of the analysis, shown in Table 5.3, revealed both a significant negative linear growth and a positive quadratic growth curve. We interpret this finding to mean that the reading comprehension of students with higher BRSC scores at class entry grew very little at first, but over time this growth accelerated more dramatically. In contrast, students with little or no basic reading skills at entry showed a small amount of growth initially, but then failed to improve their reading comprehension skills over time. In other words, adult ESL literacy students who entered class with some basic reading skills showed significant growth in reading comprehension compared to students who had little or no basic reading skills, but this took time to appear. Initially, students with low basic reading skills improved slightly, but then later showed no growth in their reading comprehension skills.

The model identified the rate of attendance (proportion of hours attended to scheduled hours) as positively related to linear growth in reading comprehension. The coefficient of 0.02 for the attendance rate means that there was a 0.2-point increase per month with each 10 percent increase in attendance rate, which was significant even after controlling for the total attendance time in hours. Thus, students who attended more regularly improved their reading comprehension skills, no matter how many hours they attended. The scheduled length of class in hours per week was also related to positive growth in reading comprehension. Students in class with more scheduled hours per week had more growth in reading comprehension.

The use of a native language in class, a measure of how teachers used the students' native language for clarification during instruction, had a positive effect on the linear growth in reading comprehension. In other words, the more teachers used students' native language to do such things as give directions about class activities or to clarify concepts, the faster students' reading comprehension grew. This relationship was quite strong: the coefficient of 3.44 for the variable can be translated to a gain of 8.2 points over a year with a 20 percent more use of native language.

Table 5.2 Growth in Basic Reading Skills (WJBRSC): Predictors of Initial Status, Linear and Quadratic Growth Rates

Parameter	*Predictor*	*Co-efficient*	*se*	*df*	*t-ratio*	*p-value*
Initial Status	Intercept (Base Level)	428.78	6.94	27	61.77	<.001***
	Student Variables					
	Age	0.15	0.07	855	2.01	.04*
	Formal Schooling at Home Country	2.76	0.39	855	7.14	<.001***
	Employed	0.71	2.17	855	0.33	.74 ns
	Hispanic Student	13.97	3.94	855	3.54	.00***
	Hmong Student	–2.12	5.00	855	–0.42	.67 ns
	Somali Student	7.16	3.56	855	2.01	.04*
	Basic Oral English Skills (BEST)	0.49	0.06	855	8.68	<.001***
	Attendance Variables					
	Attendance Rate	0.01	0.07	855	0.15	.88 ns
	Total Attendance Time (in hours)	0.01	0.02	855	0.92	.36 ns
	Teacher Variables					
	Hispanic Teacher	15.41	7.24	27	2.13	.04*
	Class Variables					
	Length of Class (in hours per week)	0.22	0.75	27	0.29	.78 ns
	Mandatory Class	–1.30	5.29	27	–0.25	.81 ns
	Day Class	4.67	5.78	27	0.81	.43 ns
	Mixed Class	6.56	4.60	27	1.42	.17 ns
	Instructional Variables					
	Use of Native Language	11.13	14.62	27	0.76	.45 ns
	Practice Strategy	12.72	6.46	27	1.97	.06 $
	Connection Strategy	–0.40	4.79	27	–0.08	.93 ns
	Emphasis on Basic Literacy Skills	2.57	20.69	27	0.12	.90 ns

(continued)

Table 5.2 Growth in Basic Reading Skills (WJBRSC): Predictors of Initial Status, Linear and Quadratic Growth Rates

Parameter	*Predictor*	*Co-efficient*	*se*	*df*	*t-ratio*	*p-value*
Linear Growth Rates	Intercept (Base Level)	−0.29	0.51	855	−0.58	.56 ns
	Student Variables					
	Age	−0.03	0.01	855	−2.52	.01**
	Formal Schooling in Home Country	0.32	0.15	855	2.17	.03*
	Employed	0.39	0.30	855	1.32	.19 ns
	Basic Oral English Skills (BEST)	0.04	0.02	855	1.78	.08 $
	Attendance Variables					
	Attendance Rate	0.00	0.01	855	−0.37	.71 ns
	Total Attendance Time (in hours)	0.00	0.00	855	0.63	.53 ns
	Teacher Variable					
	Hispanic Teacher	0.09	0.46	855	0.20	.84 ns
	Class Variable					
	Length of Class (in hours per week)	−0.14	0.05	855	−2.58	.01**
	Instructional Variables					
	Use of Native Language	−0.83	0.99	855	−0.84	.40 ns
	Varied Practice Strategy	−0.42	0.41	855	−1.03	.30 ns
	Connection to Outside Strategy	0.62	0.32	855	1.95	.05*
	Emphasis on Basic Literacy Skills	−1.92	1.45	855	−1.32	.19 ns
Quadratic Growth Rates	Intercept (Base Level)	0.07	0.05	855	1.60	.11 ns
	Formal Schooling at Home Country	−0.05	0.02	855	−2.98	.01**
	Basic Oral English Skills (BEST)	0.00	0.00	855	−1.50	.13 ns

Note 1: se = standard error; df = degrees of freedom; ns = nonsignificant finding.

$ p < .10; *p < .05; **p < .01; ***p < .001

Note 2: The intercept is the estimate of the outcome measure when all the variables in the model take on a base or reference value. In other words, age = mean of 40.5, formal schooling = mean of 3.13 years, and so forth.

Variance Decomposition of Growth in Basic Reading Skills (WJBRSC).

Variance Component	*Estimate*	*se*	*Z-value*	*p*
Between-class variance in initial status	386.55	104.14	3.71	0.00***
Covariation in class-level initial status & linear growth	−6.48	4.48	−1.45	0.15 ns
Between-class variance in linear change	0.56	0.31	1.82	0.03*
Between-student variance in initial status	350.31	35.02	10.00	<.0001 ***
Covariation in student-level initial status & linear growth	−2.44	3.73	−0.65	0.51 ns
Between-student variance in linear change	0.71	0.58	1.24	0.11 ns
Residual	169.66	13.56	12.51	<.0001 ***
Total	907.79			

Table 5.3 Growth in Reading Comprehension (WJRCC): Predictors of Initial Status, Linear and Quadratic Growth Rates

Parameter	*Predictor*	*Co-efficient*	*se*	*df*	*t-ratio*	*p-value*
Initial Status	Intercept (Base Level)	430.83	3.54	27	121.62	<.001***
	Student Variables					
	Age	−0.02	0.05	951	−0.34	.73ns
	Formal Schooling in Home Country	0.59	0.24	951	2.49	.01**
	Employed	−0.62	1.32	951	−0.47	.64 ns
	Hispanic Student	−5.75	2.39	951	−2.40	.02 *
	Hmong Student	−9.25	3.04	951	−3.05	.01 **
	Somali Student	−1.45	2.29	951	−0.63	.53 ns
	Basic Reading Skills (BRSC)	0.35	0.02	951	15.71	<.001***
	Attendance Variables					
	Attendance Rate	−0.02	0.04	951	−0.49	.62 ns
	Total Attendance Time (in hours)	−0.02	0.01	951	−2.19	.03 *
	Teacher Variable					
	Hispanic Teacher	−2.02	3.51	27	−0.58	.57 ns
	Class Variables					
	Length of Class (in hours per week)	−0.57	0.41	27	−1.39	.17 ns
	Mandatory Class	−2.36	2.73	27	−0.87	.39 ns
	Day Class	2.63	2.89	27	0.91	.37 ns

(continued)

Table 5.3 Growth in Reading Comprehension (WJRCC): Predictors of Initial Status, Linear and Quadratic Growth Rates (continued)

Parameter	*Predictor*	*Co-efficient*	*se*	*df*	*t-ratio*	*p-value*
	Mixed Class	2.83	2.36	27	1.20	.24 ns
	Instructional Variables					
	Use of Native Language	−1.71	6.90	27	−0.25	.81 ns
	Varied Practice Strategy	−3.59	3.15	27	−1.14	.26 ns
	Connection to Outside Strategy	−1.76	2.25	27	−0.78	.44 ns
	Emphasis on Comprehension	35.46	11.62	27	3.05	.01 **
Linear Growth Rates	Intercept (Base Rate)	0.80	0.31	951	2.57	.01 **
	Student Variables					
	Age	−0.01	0.01	951	−1.54	.12 ns
	Formal Schooling in Home Country	0.03	0.04	951	0.88	.38 ns
	Employed	0.08	0.19	951	0.44	.66 ns
	Basic Reading Skills (BRSC)	−0.02	0.01	951	−2.00	.05 *
	Attendance Variables					
	Attendance Rate	0.02	0.01	951	2.85	.01 **
	Total Attendance Time (in Hours)	0.00	0.00	951	−0.62	.53 ns
	Teacher Variable					
	Hispanic Teacher	−0.41	0.33	951	−1.24	.21 ns
	Class Variable					
	Length of Class (in hours per week)	0.07	0.04	951	1.97	.05 *
	Instructional Variables					
	Use of Native Language	3.44	1.07	951	3.22	.001 **
	Varied Practice Strategy	0.16	0.30	951	0.52	.60 ns
	Connection to Outside Strategy	−0.01	0.22	951	−0.05	.96 ns
	Emphasis on Comprehension	−1.27	1.15	951	−1.10	.27 ns
Quadratic Growth Rates	Intercept (Base Rate)	−0.02	0.03	951	−0.86	.39 ns
	Basic Reading Skills (BRSC)	0.002	0.00	951	2.71	.01 **
	Use of Native Language	−0.13	0.10	951	−1.34	.18 ns

Note 1: se = standard error; df = degrees of freedom; ns = nonsignificant finding.
$ p < .10; *p < .05; **p < .01; ***p < .001
Note 2: The intercept is the estimate of the outcome measure when all the variables in the model take on a base or reference value. In other words, age = mean of 40.5, formal schooling = mean of 3.13 years, and so forth.

Variance Decomposition of Growth in Reading Comprehension (WJRCC).

Variance Component	*Estimate*	*se*	*Z-value*	*p*
Between-class variance in initial status	120.27	33.66	3.57	0.00 ***
Covariation in class-level initial status & linear growth	–1.02	1.89	–0.54	0.59 ns
Between-class variance in linear change	0.48	0.21	2.34	0.01 **
Between-student variance in initial status	174.20	15.89	10.96	<.0001 ***
Covariation in student-level initial status & linear growth	0.59	1.66	0.36	0.72 ns
Between-student variance in linear change	0.54	0.25	2.12	0.02 *
Residual	61.60	5.11	12.06	<.0001 ***
Total	357.09			

VARIABLES RELATED TO GROWTH IN ORAL ENGLISH LANGUAGE DEVELOPMENT (BEST)

The growth curve model for ESL literacy students' oral language skills measured by the BEST test showed that there was significant linear growth and a significant quadratic trend, meaning the linear trend tapers off over time. The mean BEST total scores started at 23.7 and increased at a rate of about 2.2 points per month for the first three months, or about 6.6 points. However, due to the growth deceleration, the model showed it would take the next six months to achieve the same amount of gain. The results of the growth curve modeling presented in Table 5.4 show that many student, class attendance, and instructional strategy variables were significantly related to linear growth in development of oral English communication skills.

Students' age had a small negative relationship to linear growth in oral English skills, as measured by the BEST. Younger adult ESL literacy students acquired English-speaking and listening skills at a slightly faster rate than their older counterparts. The model predicts that a twenty-year-old student would gain 0.4 more points more per month on the BEST compared to a forty-year-old student, all other variables being equal. Since younger students also tended to have slightly better oral English skills at the start of class, this age gap only widens over time.

Students with higher basic reading skills when class began, as measured by the Woodcock Johnson BRSC, were positively related to BEST scores initially (i.e., initial status) and were positively related with linear growth in oral English skills. This finding means that the better basic

Table 5.4 Growth in Oral English Skills (BEST): Predictors of Initial Status, Linear and Quadratic Growth Rates

Parameter	*Predictor*	*Co-efficient*	*se*	*df*	*t-ratio*	*p-value*
Initial Status	Intercept (Base Level)	30.37	4.92	26	6.17	<.001 ***
	Student Variables					
	Age	–0.09	0.06	832	–1.67	.09 $
	Formal Schooling in Home Country	–0.37	0.29	832	–1.28	.20 ns
	Employed	–0.42	1.59	832	–0.26	.79 ns
	Hispanic Student	–5.09	3.04	832	–1.67	.09 $
	Hmong Student	–3.83	3.84	832	–1.00	.32 ns
	Somali Student	–1.93	2.75	832	–0.70	.48 ns
	Basic Reading Skills (BRSC)	0.15	0.02	832	6.41	<.001 ***
	Attendance Variables					
	Attendance Rate	0.04	0.05	832	0.87	.39 ns
	Total Attendance Time (in hours)	–0.05	0.01	832	–4.26	<.001 ***
	Teacher Variable					
	Hispanic Teacher	–7.81	4.70	26	–1.66	.11 ns
	Class Variables					
	Length of Class (in hours per week)	–0.12	0.51	26	–0.24	.81 ns
	Mandatory Class	3.57	3.87	26	0.92	.36 ns
	Day Class	3.50	4.12	26	0.85	.40 ns
	Mixed Class	–4.13	3.31	26	–1.25	.22 ns
	Instructional Variables					
	Use of Native Language	–6.19	9.82	26	–0.63	.53 ns
	Varied Practice Strategy	–0.18	4.28	26	–0.04	.97 ns
	Connection to Outside Strategy	–3.98	3.11	26	–1.28	.21 ns
	Emphasis on Oral Communication	–63.20	16.40	26	–3.85	.001 ***

(continued)

Table 5.4 Growth in Oral English Skills (BEST): Predictors of Initial Status, Linear and Quadratic Growth Rates (continued)

Parameter	*Predictor*	*Co-efficient*	*se*	*df*	*t-ratio*	*p-value*
Linear Growth Rates	Intercept (Base Level)	1.92	0.27	832	7.21	<.001 ***
	Student Variables					
	Age	–0.02	0.01	832	–2.59	.01 **
	Formal Schooling in Home Country	0.04	0.04	832	1.00	.32 ns
	Employed	0.01	0.18	832	0.06	.96 ns
	Basic Reading Skills (BSRC)	0.01	0.00	832	2.11	.04 *
	Attendance Variables					
	Attendance Rate	0.01	0.01	832	2.24	.02 *
	Total Attendance Time (in Hours)	0.00	0.00	832	0.91	.36 ns
	Teacher Variable					
	Hispanic Teacher	–0.21	0.29	832	–0.72	.47 ns
	Class Variable					
	Length of Class (in hours per week)	0.21	0.06	832	3.34	.001 ***
	Instructional Variables					
	Use of Native Language	2.79	1.11	832	2.52	.01 **
	Varied Practice Strategy	0.56	0.26	832	2.13	.03 *
	Connection to Outside Strategy	–0.08	0.19	832	–0.41	.68 ns
	Emphasis on Oral Communication	4.50	1.02	832	4.41	<.001 ***
Quadratic Growth Rate	Intercept (Base Level)	–0.11	0.02	832	–4.82	<.001 ***
	Length of Class (in hours per week)	–.01	.01	832	–1.62	.11 ns
	Use of Native Language	–.05	.11	832	–.52	.61 ns

Note 1: se = standard error; df = degrees of freedom; ns = nonsignificant finding.
$ p < .10; *p < .05; **p < .01; ***p < .001

Note 2: The intercept is the estimate of the outcome measure when all the variables in the model take on a base or reference value. In other words, age = mean of 40.5, formal schooling = mean of 3.13 years, and so forth.

Variance Decomposition of Growth in Oral English Skills (BEST)

Variance Component	*Estimate*	*se*	*Z-value*	*p*
Between-class variance in initial status	94.07	25.77	3.65	0.00 ***
Covariation between class-level initial status & linear growth	−1.30	1.28	−1.02	0.31 ns
Between-class variance in linear change	0.25	0.13	1.87	0.03 *
Between-student variance in initial status	166.26	14.37	11.57	<.0001 ***
Covariation between student-level initial status & linear growth	−1.06	1.38	−0.76	0.45 ns
Between-student variance in linear change	0.70	0.20	3.42	0.00 ***
Residual	42.23	3.84	10.99	<.0001 ***
Total	303.51			

readers started higher and learned English oral skills faster than their less reading skilled peers.

As with the reading comprehension measure, rate of attendance was significantly related to positive growth in oral English. Other things being equal, including the length of class and the total amount of attendance hours, students who attended more regularly (i.e., with higher attendance rate) learned oral English at a faster rate than students who attended less regularly. The model also showed that the scheduled length of class in hours per week was positively associated with linear growth rate. In other words, the longer classes promoted faster growth in oral English acquisition.

The growth model revealed three instructional factors that were positively related to improvement in oral English. Students in classes where more time in instruction was spent on oral communication development activities (such as pronunciation practice, conversation practice, and dialogue drills) grew faster than students in classes where this type of instruction was provided less often.[7] The use of native language as instructional support also helped students learn oral English faster, as did increased use of the varied practice and interaction strategy.

We illustrate the growth curve model predictions for oral English-language development in Figure 5.2, using two of the instructional variables. Holding other variables constant, we compared the growth lines for low and high emphasis on oral communication instructional activities and low and high emphasis on the varied practice and feedback strategy. We also show the projected growth when both strategies are used at a high

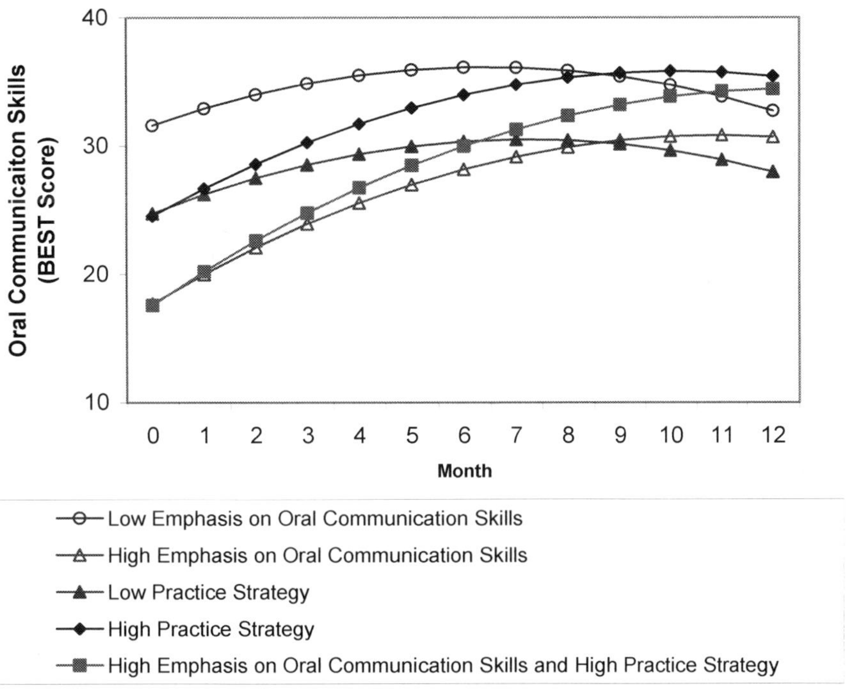

Figure 5.2 Effects of instructional emphasis on oral communication skills and practice strategy on growth in oral communication skills (BEST scores).

level, all else being equal. The curvilinear relationship is clearly evident, as are the relative effects of the two instructional emphases and the combined effect when both strategies are used.

SUMMARY AND DISCUSSION OF MAIN FINDINGS

The *What Works* study was successful in achieving its main goal of relating instructional strategies to student learning. Through the growth modeling approach, we found that three instructional strategies—connection to the outside world, use of the student's native language for clarification in instruction, and varied practice and interaction—were related to growth in student literacy and language learning. Table 5.5 summarizes the main findings related to instruction and program practices. These instructional strategies encompass a range of teaching activities, which we discuss following, along with an interpretation of why these strategies were effective. We conclude with a summary of findings related to student variables and suggestions for further research, including conducting longitudinal research.

Table 5.5 Key Findings Related to Instruction, Program Practices, and Students

Instructional Practices
• **"Bringing in the outside"**—students in classes where teachers made connections to the "outside" or real world had more growth in reading basic skills development.
• **Use of the students' native language for clarification**—students in classes where teachers used students' native language for clarification during instruction (e.g., to explain concepts and provide instructions on class work) had faster growth in reading comprehension and oral communication skills.
• **Varied practice and interaction strategy**—use of this strategy, where the teacher taught concepts in a variety of modalities and allowed student interaction, resulted in faster growth in oral communication skills.
• **Emphasis on oral communication**—students in classes where the teacher explicitly emphasized oral English communication skills in instruction had more growth in this area.
Program Practices
• **Scheduled class length** (in hours per week)—longer scheduled classes resulted in more growth in reading comprehension and oral communication skills, but *less* growth in basic reading skills. This suggests that teachers should not overemphasize basic reading skills for too long a time, but move on to higher level reading skills or other language skills.
Student Factors
• **Rate of attendance**—students who attended a higher proportion of scheduled time (in hours) had more growth in reading comprehension and oral communication skills.
• **Prior education and skills**—students with more years of education and higher incoming English-language and literacy skills had more growth, although the effect of years of schooling was limited to growth in basic reading skills development.
• **Age**—younger students developed basic reading and English oral communication skills faster than older students.

CONNECTION TO THE OUTSIDE: USING MATERIALS FROM EVERYDAY LIFE

Connecting literacy teaching to everyday life made a significant difference in reading basic skills development. To implement this strategy, teachers used materials from daily life that contained information that students wanted to know about or with which they had some experience. For example, a teacher might bring in grocery flyers from different stores and ask students to compare prices or use phone and electricity bills, letters from schools or immigration authorities, and other items that highlight literacy for adult contexts. Using authentic materials in this way, teachers can help build vocabulary skills, build background knowledge that helps students negotiate different types of document literacy, and increase reading comprehension skills. Work by Purcell-Gates and her colleagues has also demonstrated positive effects for using authentic materials in instruction (Purcell-Gates et al., 2002).

Activities of this sort might foster literacy development by linking new information to what learners already know and by engaging the learner in topics of interest. By starting with familiar materials that are of interest to learners and by creating situations for cognitive involvement, teachers can create interest, maintain high levels of motivation, engage students' minds, and through this process build literacy skills that have importance in the lives of adults.

USE OF STUDENTS' NATIVE LANGUAGE FOR CLARIFICATION

In classes where teachers used the native language as part of instruction to clarify and explain, students exhibited faster growth in both reading comprehension and oral communication skills.[8] Since the directions for a language and literacy task are sometimes more complex than the language required by the task itself (e.g., "write your name and the date on the upper right-hand side of the paper"), students who received clarification in the native language were able to focus on the task at hand, and the confusion and anxiety of not understanding the instructions were reduced.

Another reason why using both English and the native language in the classroom was effective may be that many of the learners, particularly along the U.S.–Mexico border, have become convinced that English can only be learned through a reliance on translation and are reluctant to use English outside of the classroom for fear of not understanding or not being understood and possibly being subject to ridicule. They may have lost confidence in their ability to get a point across in imperfect English or to understand a message if not all the words are understood. For these students, having a teacher who shares their language means being able to ask questions in a language they understand and having the security that access to the native language provides. Being in a classroom where the native language is used may provide less of a linguistic and more of a psychological advantage. Free from the anxiety of having to survive on English only in the classroom, these adults now have the opportunity to focus on learning and take in more information than otherwise possible. These explanations, however, must remain speculative, since we collected no data directly on these topics.

It seems clear that we cannot think critically in a language we cannot understand. Beginning ESL literacy students are not able to discuss options or articulate opinions to a deep level if they still struggle with even basic conversation in the new language. They may be able to understand a simple scenario presented to them, but they will be hard pressed to discuss the situation in detail or suggest more than the simplest course of action. Yet these types of situations present themselves daily to immigrants and refugees since the problems of real life do not wait for English to catch up: children have to be enrolled in school, supervisors need explanations, and newcomers get lost. By giving students a chance to use their own language in discussions, teachers can help students think about the situations that might confront them and can encourage them to work with others to brainstorm ideas, discover options, and think about consequences. By mixing the use of English with opportunities to use the native language where appropriate, the learning of English can be reinforced. This may be the process by which oral communication skills and reading skills improved, although again we can only speculate due to lack of data on this issue.

VARIED PRACTICE AND INTERACTION

The reason for the relationship of varied practice and interaction to oral language learning may be that learning how to communicate in English is a challenging process that requires different sets of knowledge: an understanding of sentence structure, grammar, and syntax; a good sense of how written language reflects oral language (phonology); the ability to interpret and use word endings that change the meaning of an expression; and a rich vocabulary. In other words, students need a good sense of "how English works" to understand what is being said and explain their ideas in ways that at least approximate Standard English. Communication requires a good sense of what is appropriate in any given situation, a sense of sociolinguistic competence.

While it is possible to learn English on one's own and slowly sort out the intricacies of the language, the process is aided by a teacher who draws students' attention to certain patterns and rules when appropriate and gives students a chance to talk in class without having to worry about accuracy at every step. This strategy is fundamental to direct teaching and explicit literacy instructional approaches, for example (Brown, Pressley, et al., 1996; Carell, 1985). However, while there is a place for direct teaching in the ESL literacy classroom, it is easy for students to become overwhelmed. Adults who did not study English formally in school often have difficulties understanding concepts such as "subject" or "direct object" and too academic terminology can frustrate both students and teachers. Setting time aside, however, to demonstrate to students how English works by highlighting patterns and to practice language in meaningful ways appears to pay off in terms of increasing oral proficiency.

STUDENTS AND TEACHERS

In examining the relationship of student background characteristics to English literacy and language development, we found that students' amount of formal education was related to growth in basic reading skills. While all of the students in the study had very little formal schooling, the more schooling they did have, the greater their development of basic reading skills—at least at first. This initial advantage of schooling faded over time. It may be that students with more prior schooling in their native language had some knowledge of basic reading that they were able to transfer to English, enabling them to learn faster. Students with less schooling struggled initially but eventually caught up to their more educated peers. This finding is consistent with the general research on the relationship of educational training in basic literacy to learning basic reading skills in a second language, the subsequent learning level, and second-language literacy development (Young-Scholten & Strom, 2006).

Students' English-language and literacy skills when they started class also were related to their subsequent learning. Students with higher basic reading skills (as measured by the WJRBRSC pretest) developed reading comprehension and oral communication skills faster than their peers. Similarly, students with higher initial English oral communication skills (as assessed by the BEST pretest) improved their basic reading skills faster.

Students' age was also an advantage to developing English oral communication and basic reading skills. Younger students developed these skills faster than older students. We cannot state from this research whether this finding is due to a cognitive advantage of age or higher motivation of younger students to learn English that older students. However, there was one assessment where older students had the advantage: the reading demonstration task. Older students tended to perform better over time reading the real-life, authentic materials (e.g., bills, labels, signs) used in this assessment than younger students. This intriguing finding may be due to the greater experience older students may have with these materials (see Condelli et al., 2003).[9]

We also looked at whether teacher background and training had an effect on adult ESL literacy student learning. We found that no teacher variables were related to any of the student outcome measures used in the study. However, the thirty-eight teachers in the study were relatively homogeneous. They were generally new, inexperienced teachers, and although most had bachelor or master's degrees, they had little training or professional development in teaching adult ESL or ESL literacy. These factors made it very difficult to find statistically significant effects for teacher characteristics.

SUGGESTIONS FOR FURTHER RESEARCH

Because the *What Works Study* did not employ experimental manipulation or a randomized design, we cannot state definitively which specific instructional practices will produce the outcomes we observed. For example, while our findings allow us to say that "bringing in the outside" teaching strategies are related to growth in adult ESL literacy students' basic reading skills, the study design does not allow us to say which *specific* instructional practices will *cause* these students' basic reading skills to improve. To make this type of inference, experimental research using random assignment is needed.

The study methodology and approach limited our ability to examine the effect of teacher characteristics on adult ESL student learning and to define the characteristics and behaviors of good teachers. Yet, the importance of a good teacher is widely acknowledged, and adult literacy students often identify their teacher as instrumental to their learning. In addition to focusing on instructional methods, a future study could identify teacher variables, such as ESL-specific training, educational background, and pedagogical approach of teachers that affect student learning.

While the study has demonstrated that instructional practices in adult ESL literacy class are related to language and learning growth, most students spend relatively little time in class. For example, we found our students attended an average of about 129 hours over sixteen weeks. Adults in these classes clearly rely on their environmental exposure to English and other methods of learning, in addition to classroom instruction, to acquire literacy and language skills. Such factors as the community in which learners reside, personal and family situation, employment, personal motivation, and literacy practices and needs also affect learning. The role of self-directed learning of adult ESL students has received little attention, but current research suggests it is an important factor in adult learning (Reder & Strawn, 2006).

Future research could also explore approaches to assessing adult ESL literacy students. We found that the assessments available for these students for instruction, research, and accountability purposes range from nonexistent to inadequate, particularly in the assessment of literacy skills in second-language learners. Empirical work to identify assessment approaches and to develop and evaluate new assessments would greatly benefit the field at all levels. Teachers need these tools to design appropriate instruction, researchers need them as outcome measures, and administrators need a gauge on student progress for accountability.

IMPLICATIONS FOR LONGITUDINAL STUDIES

This study is a first attempt to examine adult ESL literacy students' experiences with instruction and how literacy and language skills develop over time, using quantitative methods. Although the study period was only about nine months, our analytic model uncovered intriguing changes in learning development during this time. The HLM model for basic reading skills, reading comprehension, and oral English-language development each had a significant quadratic or nonlinear trend. In the modeling of basic reading skills, students with more schooling improved on these skills more quickly than less educated students, but this advantage diminished over time. The model also identified a changing learning dynamic for reading comprehension. Adult ESL literacy students who had higher basic reading skills showed significant growth in reading comprehension compared to students who had little or no basic reading skills, but this took time to appear. At first, students with low basic reading skills improved slightly but then later showed no growth in their reading comprehension skills. For oral English development, we observed rapid learning initially, which dropped off over time, according to the statistical model. Further research can study these learning changes over time and by student, only through longitudinal study.

Unaddressed in this study was the role of students' literacy and language learning outside the classroom. Students only attend for an average of about 129 hours over a four-month period, an insufficient amount of time for proficient language development, especially for students at the literacy level. However, students are exposed to English and negotiate American culture and daily life with varying degrees of success. Clearly, ESL students have other ways of learning English and supplementing their classroom experiences which longitudinal study could help reveal. Reder (this volume) has studied this issue with low-literate native English speakers.

As we found, however, longitudinal study of the adult ESL literacy population must contend with the reality that many students are new immigrants and refugees and move frequently to follow jobs and family. A substantial number of students from Mexico and Central America also return home frequently, maintaining ties on both sides of the border. Our strategy for maintaining contact with students was to use field staff from the community who developed personal ties with learners and maintained frequent, ongoing contact, which was somewhat successful when were able to implement it. For the second study cohort, we had a retention rate of almost 75 percent of learners. For the first cohort, however, which lasted over a summer, field staff lost track of many students and retention was just over 50 percent. This difference illustrates that continuous, ongoing student contact is essential to the success of any longitudinal effort.

NOTES

1. This paper focuses on the findings of instructional variables related to student learning gains measured by standardized assessments. For more detail on the descriptive analyses, results of qualitative assessments, and analyses of attendance patterns, and a fuller discussion of the statistical analyses and the research literature underlying the study, see the complete final report of the study (Condelli, Wrigley, Yoon, Cronen, & Seburn, 2003).
2. See Condelli and Wrigley (2004) for a broader discussion of these and other variables.

3. Results reported here use all available assessments, so comparisons across assessments are cross-sectional and do not account for differences in student characteristics due to study attrition. However, analyses using only students with all three assessments showed that they were on average 2–3 years older and 10–13 percent more were employed than students who dropped out. However, they did not differ significantly in prestudy ability, as measured by initial assessments. Employment was not found to be a significant predictor in any the growth models, and the age differences are too small to affect findings in the two analyses where age was identified as a predictor.
4. See the study final report (Condelli et al., 2003) for more information about this and other multivariate analyses of student attendance.
5. In addition, we tried several alternative models in preliminary analyses, including ones without a quadratic component. Invariably, models with nonlinear growth proved to be a better fit to the data, given the curvilinear nature of some of the relationships (see, for example, Figure 5.2). We acknowledge, however, that the use of three data points to estimate quadratic effects is a relatively uncharted area within HLM because, although the curvature parameter is fixed, we used covariates to predict its value. In some models, one or more predictors for initial status are also predictors of the random linear growth and/or quadratic growth parameters.
6. We also used fewer variables to predict to the linear growth rate than we did to compute initial status and even fewer measures to predict to the quadratic growth rate. This decision was made not only for statistical reasons (only a limited number of variables can be entered into the model), but also because there is no research or theory precise enough to predict quadratic patterns of growth among this student population.
7. Students in such classes not only grew faster on this measure but also started at a lower level. Students with lower oral English skills were more likely to be in classes with an oral communication emphasis, probably due to placement procedures of programs.
8. This finding occurred in classes where most learners and the teacher spoke Spanish. We did not have instances of other teacher-learner common languages in the study, nor did we include bilingual classes or native-language literacy classes, where the home language of the students is the language of instruction and the target is acquisition and improvement of literacy in a language other than English.
9. However, the HLM analysis using the reading demonstration assessment revealed no significant findings for instructional variables. Nor were any significant effects found for the writing assessment.

REFERENCES

Alderson, C. (1984). *Reading in a foreign language.* New York: Longman.

Bernhardt, E., & Kamil, M. (1995). Interpreting relationships between L1 and L2 reading: Consolidating the linguistic threshold and the linguistic interdependence hypothesis. *Applied Linguistics*, 16, 15–34.

Bransford, J. D, Brown, A., & Cocking, R. (2000). How people learn: Mind, brain, experience and school (expanded edition). Washington, DC: National Academy Press.

Brown, R., Pressley, M., Van Meter, P., et al. (1996). A quasi-experimental validation of transactional strategies instruction with previously low-achieving, second-grade readers. *Journal of Educational Psychology*, 88, 18–37.

Bryk, A. S., & Raudenbush, S. W. (1992). *Hierarchical linear models: Applications and data analysis methods.* Newbury Park, CA: Sage.

Carrell, P. L. (1985). Facilitating ESL reading by teaching text structure. *TESOL Quarterly*, 19, 727–52.

Carrell, P. (1991). Second language reading: Reading ability or language proficiency. *Applied Linguistics*, 12(2), 159–79.

Chisman, F., Wrigley, H., & Ewen, D. (1993). *Sparks of excellence: Program realities and promising practices in adult ESL*. Washington, DC: The Southport Institute for Policy Analysis.

Comings, J., Cuban, S., Bos, H., et al. (2003). *"As long as it takes": Responding to the challenges of adult student persistence in library literacy programs*. New York: MDRC.

Condelli, L., & Wrigley, H. S. (2004). *Identifying promising literacy interventions for adult ESL literacy students: A review of the literature*. Washington, DC: U.S. Department of Education (available from the American Institutes for Research).

Condelli, L., Wrigley, H., Yoon. K., et al. (2003). *What works study for adult ESL literacy students: Final report*. Washington, DC: U.S. Department of Education (available from the American Institutes for Research).

Fix, M., Passel, J., & Sucher, K. (2003). *Trends in naturalization* (Brief No. 3, Immigrant Families and Workers: Facts and Perspectives). Washington, DC: Urban Institute.

Greeno, J. G., Collins, A. M., & Resnick, L. B. (1996). Cognition and learning. In D. C. Berliner & R. C. Calfee (eds.), *Handbook of educational psychology* (pp.15–46). New York: Macmillan.

Purcell-Gates, V., Degener, S., Jacobson, E., et al. (2002). Impact of authentic literacy instruction on adult literacy practices. *Reading Research Quarterly*, 37(1), 70–92.

Quigley, Allan. (1996). *Rethinking literacy education: The critical need for practice-based change*. New York: Jossey-Bass.

Ramirez, J., Yuen, S., & Ramey, D. (1991). *Executive summary: Longitudinal study of structured English immersion strategy, early-exit, and late-exit transitional bilingual education programs for language-minority children*. San Mateo, CA: Aguirre International.

Reder, S., & Strawn, C. (2006). Self-study: Broadening the concepts of participation and program support. *Focus on Basics*, 8(C), 6–10.

Slavin, R. E., & Cheung, A. (2003). Effective reading programs for English language learners: A best-evidence synthesis. Baltimore, MD: Johns Hopkins University, Center for Research on the Education of Students Placed at Risk.

Thomas, W. P., & Collier, V. P. (2002). A national study of school effectiveness for language minority students' long-term academic achievement. Santa Cruz, CA: Center for Research on Education, Diversity and Excellence, University of California–Santa Cruz.

Van de Craats, I., Kurvers, J., & Young-Scholten, M. (2006). Research on low-educated second language and literacy acquisition. In I. Van de Craats, J. Kurvers, & M.Young-Scholten (eds.), *Low-Educated Second Language and Literacy Acquisition: Proceedings of the Inaugural Symposium—Tilburg 05*. Utrecht, The Netherlands: LOT.

Wrigley, H. (2001). Principles and indicators to assist in the development and evaluation of technology-based materials: A design framework for multimedia development in adult literacy. Retrieved from http://www.cyberstep.org/principles.html.

Wrigley, H., & Guth, G. (1992). *Bringing literacy to life*. San Mateo, CA: Aguirre International.

Young-Scholten, M., & Strom, N. (2006). First time L2 readers: Is there a critical period? In I. Van de Craats, J. Kurvers, & M. Young-Scholten (eds.), *Low-educated second language and literacy acquisition: Proceedings of the Inaugural Symposium—Tilburg 05*. Utrecht, The Netherlands: LOT.

6 Student Persistence in Adult Literacy and Numeracy Programs

John P. Comings

INTRODUCTION

Unlike children, who participate in schooling because of legal mandates and strong social and cultural forces, adult students choose to participate in educational programs. Adults must make an active decision to participate in each class or tutoring session and often must overcome significant barriers in order to participate in educational services. Student persistence, therefore, is an issue that all adult-education programs, including those focused on literacy and numeracy, must address. Persistence is comprised of two parts: intensity (the hours of instruction per month) and duration (the months of engagement in instruction). Persistence rates are stated as hours of instruction during a specific period of months, usually in increments of one program year.

The relationship between persistence and learning is supported by several studies. Two studies identified approximately one hundred hours of instruction as the minimum needed by adults to achieve an increase of one grade-level equivalent on a standardized test of reading comprehension (Sticht, 1982; Darkenwald, 1986). Comings, Sum, and Uvin (2000) found that, at 150 hours of instruction, adult students had a 75 percent probability of making a one (or greater) grade-level equivalent increase in reading comprehension or English-language fluency. Porter, Cuban, and Comings (2005) found that fifty-eight hours of instruction led to a .40 grade-level equivalent increase in reading comprehension. Rose and Wright (2005) examined the national reporting system (NRS) data of three states in the United States and found that at around 100–110 hours of participation, 50 percent of students were likely to show a one NRS level[1] increase or pass the General Education Development (GED)[2] test.

These studies point to one hundred hours of instruction as the point at which a majority of adult students are likely to show measurable progress, and, therefore, it serves as a benchmark that identifies an effective program. That is, if the majority of students are persisting for one hundred hours or more each year, the program is probably having a measurable impact on at least half of its students. Even one hundred hours of instruction is only about one-tenth of the time that formal school students spend in class during a year. A one grade-level increase, therefore, is a significant gain within this short period.

However, few adult students enter programs with goals that require only a few hours of instruction. In the United States, most adult students express the desire to improve their language, literacy, and math skills, acquire high school credentials, and move on to postsecondary education or skilled job training (Comings, Parella, & Soricone,1999; Reder, 2000). Program participation of one hundred hours, therefore, is probably inadequate for most adult students to reach their learning goals. Every adult education program, therefore, should help its students persist in learning long enough to reach their educational goals, which could require hundreds of hours over several years.

RESEARCH ON PERSISTENCE

By its very nature, persistence research requires a longitudinal perspective. The research reviewed in this chapter includes studies that followed students longitudinally, but it also includes studies that asked students to recount their past efforts to learn. This chapter will summarize the findings of four literature reviews, present findings from research published after these reviews, and describe the lessons learned in two connected studies. An analysis of these sources then presents advice to programs on how to improve persistence and to researchers on how to study persistence.

LITERATURE REVIEWS

Beder (1991) explores motivation as the force that helps adults overcome barriers to participation and then focuses more closely on those barriers. Beder suggests that adult education programs must change their recruitment and instructional practices to be congruent with the motivations and life contexts of adult students. If they did, more adults would enter programs, and they would persist longer. Beder also suggests that adults are weighing the perceived benefits and costs of participation and making decisions based on that analysis. In many cases, a decision to drop out may be justified if the costs outweigh the benefits.

The Beder review concludes by asserting that the system, at this time, probably only has enough resources to serve those who are eager to persist. However, it suggests that the effort could be more manageable for students if programs fit instruction to the needs and learning styles of adults and looked less like school and more like an activity in which adults would want to participate.

Drawing on the same sources as Beder did, Wikelund, Reder, and Hart-Landsberg (1992) comment on the reductionism of research and suggest that a useful theory of participation in adult education would incorporate the complexity of this phenomenon. The paper calls for broadening the definition of participation to acknowledge that adults engage in education in many ways that are not limited to participation in formal classes. The Wikelund et al. review criticizes the concept of "nonparticipant" because it implies that every adult who has low literacy and numeracy skills needs to enter a program, a situation that might not be true.

The review ends with the conclusion that research and theory, as well as practice, should break out of the framework of schooling. A new definition of participation would acknowledge that learning could take place outside of formal programs. With this new definition, programs could increase persistence by continuing to support learning at times when students cannot attend classes or participate in other formal arrangements.

In her review, Tracy-Mumford (1994) calls for programs to develop a commitment to and a plan for increasing persistence, which she suggests would send a message to students that the program is there to help them reach their goals. Since student goals can change, the program must be willing to make changes to accommodate new goals as they arise. For the commitment to be meaningful, the program should have a set of criteria for measuring persistence and a set of strategies that reduce dropout, increase student hours of attendance, improve achievement, increase personal goal attainment, and improve completion rates. Tracy-Mumford defines an effective persistence plan as one that both provides support to students and improves instruction.

The review ends with a summary of the findings of a large number of studies and descriptions of practice to provide a list of elements of a student persistence plan that weaves persistence strategies into all aspects of the program structure. These elements are:

- Recruitment methods should provide enough information for potential students to make an informed decision about enrolling.
- Intake and orientation procedures should help students understand the program, set realistic expectations, build a working relationship with program staff, and establish learning goals.
- Initial assessment tools should provide students and teachers with information on both cognitive and affective needs, should be integrated with instruction, and should form the foundation for measuring progress.
- Programs and teachers should establish strategies for formally recognizing student achievement.
- Counseling services should identify students at risk of dropping out early.
- Referral services should coordinate with social service agencies to ensure that all students are connected to the support services they need.
- A system for contact and follow-up that helps students who drop out return to the program and solicits information on ways to improve program services.

- Noninstructional activities should help form a bond between the program and its students and their families.
- Program evaluation should involve students in assessing, and offering advice on each aspect of the program.
- Childcare and transportation assistance should be provided.
- Instruction and instructional staff should be of sufficient quality to support effective learning.
- A student persistence team should coordinate dropout prevention activities, collect data on student persistence, and involve students and teachers in addressing this issue.

Tracy-Mumford's list is comprehensive, and it is useful to program staff because it translates theory into practical advice. Unfortunately, most adult basic education programs lack the funding required to implement all of these elements, but implementing some of them may contribute to increased persistence.

Quigley (1997) views persistence as significantly affected by the negative schooling experiences adult students had when they were younger and suggests the need to change programs so that they are different from schools. Quigley also describes three major constellations of factors that contribute to dropout, which he refers to as situational (influences of the adult's circumstances), institutional (influences of systems), and dispositional (influences of experience). He suggests that situational influences are largely beyond the control of adult education programs, though they receive most of the attention in the literature on dropouts. Institutional factors are areas that practitioners could affect and should work on continuously. However, he suggests that dispositional factors provide a focus for program reform that might affect persistence.

The review ends by suggesting that intake and orientation processes in the first three weeks are critical to improving persistence. Quigley suggests that intake should begin with goal setting and planning for success. Then, programs should place students in classes and with teachers that can meet their goals and learning needs. Since students are adults, they can take charge of this process, but they may need help in the form of careful questions and the provision of useful information for making these decisions.

RECENT RESEARCH

In recent research, Meder (2000) found that providing three opportunities for goal setting and discussion of the factors that might support or hinder adults in reaching their goals helped thirty-one students in her GED-level math course to persist in their studies. She had each student fill out a questionnaire on goals, barriers, and supports to persistence, and then discussed the answers with the students. This exercise took place at four weeks and eight weeks after the beginning of the class. She compared this treatment group to data available on the previous year's classes (in which this exercise did not take place). This study suggests that involving students in a process of thinking about their motivation to participate in adult education and ways to sustain that motivation helps them persist in learning.

Quigley (2000) followed twenty students who were judged to be at risk of dropping out in a quasi-experimental study of grouping students, classroom size, and support. This study sample was comprised of a control group of five subjects and three treatment groups, each with five subjects. The control group was in a classroom of fifteen students (large-group instruction), and none of this group persisted for three months or more. Among the treatment groups, three of the five adults placed in small-group instruction persisted for three months or more, two of the five adults placed in a learning environment with a teacher and a counselor persisted for three months or more, and one of the five adults in individual tutoring persisted for three months or more. Quigley also administered the Witkin Embedded Figures Test (Witkin, Oltman, Raskin, & Karp, 1971) and the Learning Style Assessment Scale (Flannery, 1993) to all students. He found that these subjects were highly field dependent (needing acceptance by peers) and global learners (who find it difficult to focus on topics that require linear sequencing). He suggests that small-group learning may be more appropriate for adult students who fit this profile.

Cuban (2003) reported on in-depth case studies of two women students in a library literacy program. The case studies make clear that adult life affects persistence in complex ways. The two women were caregivers for their various family members and put those responsibilities before their needs and desires. The students were interested in reading romance literature but this was not the focus of their classes. The study suggests that an approach to persistence support that is the same for all students might have a positive short-term impact but, for long-term persistence, programs may need to adapt their services to the lives of each student. This might require scheduling that can accommodate the changing demands on students' time and reading content that is of interest to students.

PERSISTENCE STUDY

In 1996, the National Center for the Study of Adult Learning and Literacy (NCSALL) began a two-phase study of the factors that support and inhibit persistence. The first phase interviewed and tracked the persistence of 150 adults in pre-GED[3] classes (Comings, Parella, & Soricone, 1999). The second phase studied the efforts of five library literacy programs as they attempted to increase student persistence over a three-year period and measured the relationship between persistence and achievement (Comings et al., 2001).

The first phase of the study employed a force-field analysis, which places an individual in a field of forces that are supporting or inhibiting action along a particular path (Gilbert, Fisk, & Lindzey, 1998; Lewin, 1999), as its theoretical model. This model expands the motivation/barriers and cost/benefit models to include a large number of forces on each side of the persistence equation. Understanding the forces, identifying which are strongest, and deciding which are most amenable to manipulation provide an indication of how to help someone move in a desired direction, such as reaching an educational goal.

Although all the negative and positive forces uncovered in the review of previous research could be important to persistence, some are probably more critical than are others. To explore this question, the research team interviewed 150 adult learners in nineteen pre-GED classes in fifteen adult basic education programs in five New England states. The research team then contacted subjects four months after the initial interview to find out if they had persisted in their studies.

Before the first interview with participants took place, each class participated in an orientation activity. The orientation had three objectives. The first objective was to start the study participants thinking about the forces that were supporting or hindering their persistence. The second objective was to explain the force-field analysis approach so that the study questions would make more sense to participants. The third objective was to build a friendly rapport between the students and the research team so that the interview could be relaxed, informal, and honest.

The orientation activity involved the study participants (either as a class or as individuals, depending on their program type) in a force-field analysis of a goal other than persistence in learning. The participants chose the topics, which included losing weight, owning a home, getting off welfare, and quitting smoking. The participants first brainstormed positive and negative forces, and then the class was broken into two groups, one representing the positive forces and the other the negative forces. One class member played the person trying to make the decision and the two groups tried to convince him or her by speaking from the point of view of the force they were representing. At the end of the exercise, the person charged with making the decision gave his or her answer. The students then talked about the experience.

Four months after the interview, the researchers contacted programs to see whether students had persisted. In addition, the research team attempted to contact each student to collect additional information that might be helpful to understanding their decision to persist or to drop out of the programs. All together 119 (79 percent) of the 150 students were contacted for the second interview.

Of the thirty-one students who could not be reached, attempts to locate seventeen participants resulted in no information. Of the remaining fourteen, one student was in a correctional facility and unavailable for interview; one was homeless and not reachable; one was "released by the program after multiple performance and attendance issues"; one was expelled from the program because of "getting in the way of other students' progress"; one moved and was hiding from her husband; one was having problems with her daughter and had to make family her priority; one had relapsed into using drugs and had been removed from the halfway house she was staying at; one had been sick and disappeared from a program; one had serious health problems and could not answer the questions; two had gone back to work and this presented a conflict with class time; and for three, several phone messages were left with people at their house or on a machine but these were never returned. This experience illustrates the difficulties of longitudinal studies of adult education students.

This study found that the many ways in which we can classify adult students (by gender, ethnicity, employment status, number of children, and educational background of parents or guardians) did not have a strong influence on persistence. The study did find that immigrants, those over the age of thirty, and parents of teenage or grown children were more likely to persist than others in the study. The greater likelihood of persistence by immigrant students in ESOL classes has been documented (Young et al., 1994a), and the findings of this study suggest that this effect continues as immigrants learn English and move on to pre-GED programs. Adults who are over thirty are more likely to have teenage or grown children than those under thirty. Their children may encourage their parents to join and persist in a program. These findings suggest that older students persist longer because they benefit from the maturity that comes with age, and they no longer have the responsibilities of caring for small children.

The study found that previous school experience (among U.S.-schooled students) was not associated with persistence. Of course, those potential students who are significantly affected by negative school experience may never enter a program or may have dropped out before the research team arrived. However, many of the study's subjects did describe negative school experiences, with most of the comments centered on high school. Respondents reported being ridiculed and even struck by teachers, bullied or intimidated by other students, and told that they were stupid and asked to leave school by administrators. Issues of class, race, and sexual orientation contributed to the negative school experience for some. Entering an adult basic education program may signal that a student has overcome the influences of negative school experiences and is ready to restart his or her education.

The study also found that prior nonschool learning experiences, particularly self-study focused on improving basic skills or studying for the GED, may be related to persistence. Attempts at self-study may be an indication of strong motivation, or some people may need several attempts at learning before they are ready to persist.

The students who persisted and those who dropped out talked about the forces both supporting and inhibiting persistence in the same way. Students mentioned four types of positive forces: relationships, goals, teacher and students, and self-determination. Relationships incorporate the support noted by subjects derived from their families, friends or colleagues, God or their church community, support groups, community workers, mentors or bosses, and their children. Goals included helping one's children, getting a better job, bettering one's self, moving ahead in life, attending college or some other academic institution, proving someone's assessment of the student's abilities wrong, or obtaining citizenship. The support provided by the people involved in their classes and their own determination to succeed were also important positive forces to which students pointed

Students mentioned three types of negative forces: life demands, relationships, and poor self-determination. Life demands included special childcare needs, work demands, transportation difficulties, the student's own or his or her family's health issues, age, lack of time, fatigue, bad weather, welfare and other official rules, unfavorable conditions at home, moving, and lack of income. Relationships included family members, friends, colleagues, community or welfare workers, and religious beliefs that were not in supporte of persistence, as well as fears about letting other people down by failing in a program. Lack of determination was indicated by comments such as "thinking negative thoughts," "my own laziness," and statements indicating a lack of confidence in students' ability to succeed.

How adults describe the positive and negative forces that affect them did not predict persistence, but this information is still valuable in that it gives practitioners input from adult students on what might be important. Adults in this study had much more to say about positive forces than about negative forces. Adding this information to the finding that negative school experience was not associated with persistence points to a conclusion that building positive supports may be more critical to increasing persistence than is the removal of barriers. If this is so, then understanding which positive forces are most important is essential to a model that supports persistence.

The study team summarized the implications of its findings by identifying four supports to persistence:

- *The first support to persistence is the establishment of a goal by the student.* The staff of an educational program should help potential adult students articulate their goals and understand the many instructional objectives that must be accomplished on the road to meeting those goals. Teachers should then use those student goals as the basis for instruction. The effort to identify goals must be continual as instruction proceeds because they may change over time.
- *The second support is to increase a sense of self-efficacy.* The self-determination mentioned by students must build on a foundation of self-efficacy, a feeling that they can reach their goals. Self-efficacy is the feeling of being able to accomplish specific tasks. According to Bandura (1986), self-efficacy is built when students meet people just like themselves who have succeeded at learning, accomplish learn-

ing objectives, learn to deal with negative emotional states associated with learning, and receive positive encouragement from friends, family, and program staff.

- *The third support is management of the positive and negative forces that help and hinder persistence.* Programs should help students develop an understanding of the negative and positive forces that affect their persistence. Building on that understanding, each student could then make plans to manage these forces so that persistence is more likely. The plans that come out of such an exercise should include strategies for persistence when the forces that affect peoples' lives cause them to drop out. These plans must be revised as adults persist in their studies and these forces change.
- *The fourth support is ensuring progress toward reaching a goal.* Since goals are important, adult students must make progress toward reaching their goals, and they must be able to measure that progress. Programs should provide services of sufficient quality that students make progress, and programs need assessment procedures that allow students to measure their own progress.

Phase 2. The second phase of the project observed ten library literacy programs in California, New York, and North Carolina that were attempting to increase student persistence and interviewed thirty of their students in depth about their history of participation and supports and barriers to their persistence. The study's research design relied on three components:

- A qualitative study followed both the individual programs and a number of their students over two years. Through many successive visits and in-depth interviews and conversations, the researchers developed a comprehensive picture of factors that influence persistence, including how both those factors and their influences develop over time. The qualitative study also documented how the individual programs attempted to improve student persistence. By visiting programs year after year, the researchers were able to distinguish deliberate program changes aimed at improving persistence from other program change that was happening due to changing funding environments, student needs, or other organizational and environmental issues affecting the programs.
- An analysis of quantitative program data documented how long students participate, what they do while they participate, and whether the duration and intensity of their participation changes as the programs implement persistence strategies. This analysis was designed as a cohort comparison. Such an analysis essentially compares participation outcomes for different cohorts of entering students in the various programs. The research team followed monthly cohorts of new students to establish whether later cohorts remain in the program longer

and participate for more hours, as one might expect if the programs' persistence strategies work.

- A longitudinal achievement study of a sample of 250 students collected two waves of a comprehensive battery of literacy tests, regardless of whether students continue to be active in the program or not. These achievement tests were carried out at the library or program offices or at another quiet location. The tests were administered by testers who were specially trained for this task and who remained in contact with the students between their two waves of testing, which took place fifteen to eighteen months apart. The battery of tests was specially designed to be used with readers at all levels of ability. It includes the Test of Word Reading Efficiency (TOWRE), the Picture Peabody Vocabulary Test (PPVT), the Adult Basic Learning Examination (ABLE), and, for ESOL students, the Basic English Skills Test (BEST). The test results for each student were compared and the relationship between gains on the test and the student's participation in literacy activities was analyzed using regression analysis. In this way, the potential effects of different degrees of student persistence on their achievement were measured with standardized tests.

Unfortunately, the quantitative data showed no impact of program changes on student persistence, and the longitudinal data showed no relationship between persistence and achievement. However, the qualitative data identified five persistence pathways that are determined by personal and environmental factors and ways that programs might support students on each pathway. Each pathway describes a pattern of persistence. However, any one student might begin a program on one pathway and change to another. In addition, these pathways might be arbitrary points on a continuum. These pathways, therefore, are guidelines that can help programs broaden the ways they help students persist rather than student types that could allow programs to identify the specific needs of an individual student. The five pathways are: Long-term, Mandatory, Short-term, Try-out, and Intermittent.

Long-term students participate regularly over a long period. Long-term students usually do not express specific goals but, rather, talk of education as an end in itself. Long-term students have managed the personal and environmental factors that support and inhibit their persistence. Presumably, they will persist in a program that is helping them meet their needs, is convenient for them, and that provides an enjoyable experience. In fact, this is the story told to the study team by long-term students. Most long-term students viewed their program as a comfortable and supportive community and talked about it as a family, a club, or a home base for learning. They referred to the program staff as friends or family members. Long-term students expressed a strong personal commitment to their programs and to their goal of becoming more educated.

Most of the long-term students identified through interviews were over the age of thirty. Adults over the age of thirty may no longer have child-care responsibilities and may have a stable income, housing situation, and set of relationships, while the younger students may not be as stable. The long-term persistence of older students may appear to be supported by their emotional maturity, but, in fact, it may be supported by stable personal and environmental factors related to children, partners, and employment.

For students who are able to travel the long-term pathway, an intake and orientation process that clearly sets out steps, with measurable objectives, along the path to reaching their often ambitious goals might help support their persistence. This study found that improving formal instruction (through training of teachers and volunteers, improving curriculum, and other program development inputs) and offering many different types of informal instruction appeared to increase hours of instruction for these students. The students on the other four pathways may need changes in program design, and even in the definition of persistence, in addition to these programmatic improvements.

Mandatory students must attend a program because they are required to do so by a public assistance or law enforcement agency. Their participation is usually regular and long term, and their goals are often those of the agency that is mandating their attendance. They look like long-term students while they are under the requirement to participate but usually leave abruptly once attendance is no longer mandatory and sometimes even while it still is a requirement.

Mandatory students overcome personal and environmental factors that constrain their persistence because they are required to do so. Since factors outside the program support their participation, programmatic improvements may not help these students to stay longer. However, if the program changes its services (making them more convenient, more useful, or more enjoyable), mandatory students might choose to participate for more hours or become long-term students. Counseling during intake and orientation and throughout instruction that focuses on helping mandatory students commit to learning as a way to improve their lives, understand how they learn best, find ways to enjoy learning, and build a support system to sustain their learning might help mandatory students persist after the mandate has ended. Additional hours of participation and persistence after legal mandates for participation have ended are probably good measures of impact for innovations meant to address the needs of mandatory students.

For students who are on the mandatory pathway, program intake and orientation must help students move past the required goals of attendance and begin to see learning as something they choose to do. This building of motivation probably requires identifying goals that are personal and an instructional process that helps students see that they can learn and that learning can be enjoyable. Literacy learning focused on family, work, personal interests, or even the problems that led to their legal or social service status might be a focus of instruction that supports persistence for adults on this pathway. An instructional process that involves discussions among a group of adults might provide a social network that supports persistence for mandatory students, and referral to support services (such as counseling, day care, and employment) might be necessary for these students to persist, even while they are under a mandate.

Short-term students enroll in a program and participate intensively for a short period in order to accomplish a specific goal. For some of these students, the short-term participation in a library literacy program meets their needs, but for some this participation leads to enrolment in a more suitable program. Though students on a short-term pathway may leave their programs after only a few weeks of instruction, some may persist in another program that more closely meets their needs. Since personal goals determine their length of participation, programmatic innovations may have little impact on the persistence of students on a short-term pathway. Transition into another program and accomplishment of a specific, limited goal are probably good measures of the impact of innovations meant to address the needs of short-term students.

For students who are on the short-term pathway, programs should be careful during intake and orientation to identify their specific goal. When transfer to another program is appropriate for reaching that goal, the program might be able to provide some learning opportunities that prepare these students to be successful in a more appropriate program. When new students have a specific short-term goal, programs should try to focus on it, possibly with an individual tutor, or make that goal the focus of their instruction in a more general learning environment.

Try-out students have barriers to persistence that are insurmountable and have goals that are not yet clear enough to sustain their motivation. These students end an episode of program participation quickly with neither goal achievement nor transfer to another program. Students on the try-out pathway are motivated to learn, and their decision to join program services is a positive step. However, they are not ready to make a commitment to program participation.

Program staff members believe that every new student can succeed and are usually opposed to counseling students to defer participation. However, admitting students who are likely to fail, particularly since most of these students have failed in education before, is probably not helpful to the student. Students on the try-out pathway who leave a program with a plan on how to address the personal and environmental barriers constraining their participation so that they can return at sometime in the future is probably a good measure of impact for innovations meant to address the needs of try-out students.

Helping try-out students during intake could improve program persistence rates by both lowering the number of students who drop out after very little participation and by providing more program resources to students who are on a different pathway. To do this, programs would have to design intake processes that identify try-out students, counsel them to delay entry, and help them design a program that would lead to successful participation some time in the future.

Intermittent students move in and out of program services. During the time that they are not attending program services, intermittent students may stay in contact with their programs, and their episodes of participation and nonparticipation may reoccur several times and take place in more than one program. Belzer (1998) found that students identified as dropouts in adult education programs often see themselves as still connected but temporarily unable to attend.

These students may have broad goals (such as improving language or basic skills ability) or specific (such as passing a citizenship test), but their goals require a long period of engagement to achieve. However, personal and environmental factors are limiting their ability to attend on a regular basis. Programmatic changes probably cannot have an impact on the persistence of these students unless program services change to fit a pattern of episodes of participation over a long period.

These program changes would redefine participation as connection to the program rather than hours of attendance in program services. This connection would have to be meaningful, not just a name in a database. An example of a meaningful connection could be monthly discussions between a program staff member and a student in which they review progress on a self-study plan. Programs would define any form of learning activity that serves the goals of the program and the student as participation. These activities would include classroom instruction or tutoring but might also include guided self-study at home or at the program venue. Length of continuous connection to the program and cumulative hours of engagement in learning might be good measures of impact for innovations meant to address the needs of intermittent students, regardless of whether they are formally participating in program activities.

The intermittent pathway may be the only one open to most students. Personal and environmental factors are always going to present barriers to long-term persistence, and most students have goals that require a good deal of study to achieve. Programs should accept this reality and look at ways to redesign their services to provide connected episodes of participation that use a multiplicity of learning resources. Program staff also need ways to help students maintain contact with the program and to continue to think of themselves as students.

CONCLUSIONS

Comings et al. (1999) proposed the term *persistence* because adults can persist in learning through self-study or distance education when they stop attending program services and sometimes can return to a program (not necessarily the one they dropped out of) after a lapse in attendance. The term *persistence* defines this phenomenon from the point of view of students who persist in learning inside and outside of a program until they have achieved their goals. Comings et al. (1999) redefined persistence as:

> *Adults staying in programs for as long as they can, engaging in self-directed study or distance education when they must stop attending program services, and returning to program services as soon as the demands of their lives allow.*

With this definition persistence becomes a continuous learning process that lasts until an adult student meets his or her educational goals, and persistence could start through self-study before a first episode of program participation.

The literature reviews see persistence as supported by motivation and constrained by barriers. The NCSALL persistence study broadened this equation to include a wide range of supports and barriers, but it also identified the limits on how much programs could do to address the barriers that students bring to programs. Most programs could make their services more convenient, enjoyable, and useful, but they may not have sufficient resources to address the factors that constrain persistence directly.

The insights from all of these sources were analyzed in order to make suggestions for how programs could support persistence for three chronological phases of program participation: Entrance into Services, Participation in Services, and Reengagement in Learning and for how research could study this phenomenon in ways that would provide better suggestions.

Entrance into Services includes recruitment, intake, and orientation, processes that prepare students to be successful in learning. Rather than trying to identify a student's pathway when he or she enters a program, program staff might just assume that all students are intermittent. That is, that students are prepared to participate in an episode of learning that, if it is short, might lead to additional episodes of learning that continue until they reach their goal.

The first step in this phase would be to help students express clear goals that represent their motivation for participation. The second step would be to develop a learning plan that includes both instruction and the support services a student needs to persist in learning to reach those goals. For try-out students, this phase would lead to their postponement of participation, but they would leave the program with a plan on how to prepare to enter services later. For other students, intake and orientation would lead to a plan for participation in the program. That plan would assume that students would engage in episodes of participation that lead to accomplishment of a specific goal, transfer to another program, or departure from the program followed by another episode of participation. Of course, students on a long-term pathway may only have a single long episode of participation.

The plan would be a written document that sets out the goals each student is trying to reach, the skills and knowledge the student needs to learn to meet those goals, and the services the student needs to complete that learning successfully. These services would include both instruction, self-study, and support services. The plan would also include a process that would allow the student to judge their progress.

Participation in Program Services includes both instruction and support services. General improvements in instruction and expansion of support services probably help support learning and persistence, but most students need more than just good services. They need instruction that fits their patterns of participation and support services that help them address their particular persistence needs.

A multiplicity of instructional modes (such as classes, tutoring, peer learning groups, technology, and print and media materials) provides students with ways to participate that do not always demand adherence to a regular schedule. However, these different modes would be more effective if they fit into a plan that both the student and the tutor or teacher follow.

The individual plan would allow a student who must stop instruction to continue learning (either at home or at another place such as a library) through self-study. When that student is ready to return to regular attendance, any tutor or teacher should be able to look at how far the student has been able to progress on his or her plan and start instruction there. Program services should include regular counseling that helps students meet their own needs for support services and identify the times when they will not be able to meet the instructional schedule and so begin the self-study part of their plan.

Reengagement in Learning is returning to program services after dropping out. Most students do not tell their program or tutor that they are stopping participation. Most just stop attending. Interviews with students uncovered that many of them believe that once they stop attending they cannot return (Comings et al., 2003). Programs should have a procedure for staying in contact with students who are not attending and for reengaging them in services, and this procedure should be explained during orientation. Former or even current students might be the best people to play this role, since they have addressed the same personal and environmental factors as the students who have dropped out. These new procedures require resources that are now being used to support instruction. However, if these procedures were successful, program services would improve as students who have had a single episode of instruction return to continue learning.

Further research could continue to explore the nature of persistence and the forces that support and inhibit it. This research would produce valuable knowledge, but now may be the time to test program models that incorporate the existing research. If these models prove to be sound, practitioners and policymakers could feel assured in making changes based on them, and further research could begin to build on this foundation.

NOTES

1. A one NRS level increase represents an improvement in a score on a standardized test from one defined level to another. Each NRS level is a range of approximately two grade-level equivalents, and so an increase could be as little as a single scale score point (a small segment of a grade level equivalent) to two grade-level equivalents.
2. Successful completion of the GED test leads to a high school equivalency certificate. A high school diploma represents twelve years of education and is usually completed when a student is eighteen years of age.
3. Pre-GED classes are comprised of adults whose literacy skills are between grade equivalents 6 and 8, which is probably insufficient to pass the GED test.

REFERENCES

Bandura, A. (1986). *Social foundations of thought and action: A social cognitive theory*. Englewood Cliffs, NJ: Prentice-Hall.

Beder, H. (1991). *Adult literacy: Issues for policy and practice*. Malabar, FL: Krieger.

Belzer, A. (1998). Stopping out, not dropping out. *Focus on Basics*, 2:A, 15–17.

Comings, J., Cuban, S., Bos, J., et al. (2001). *I did it for myself*. New York: MDRC.

Comings, J., Cuban, S., Bos, J., et al. (2003) *As long as it takes: Responding to the challenges of adult student persistence in library literacy programs*. New York: MDRC.

Comings, J., Parella, A., & Soricone, L. (1999). *Persistence among adult basic education students in pre-GED classes*. NCSALL Reports #12. Cambridge, MA: NCSALL.

Comings, J., Sum, A., & Uvin, J. (2000). *New skills for a new economy: Adult education's key role in sustaining economic growth and expanding opportunity*. Boston: Massachusetts Institute for a New Commonwealth.

Cuban, S. (2003). "So lucky to be like that, someone care": Two case studies of women learners and their persistence in Hasai'i literacy program. *Adult Basic Education*, 13(1), 19–43.

Darkenwald, G. (1986). *Adult literacy education: A review of the research and priorities for future inquiry*. New York: Literacy Assistance Center.

Flannery, D. (Ed.). (1993). *Applying cognitive learning theory to adult learning*. San Francisco: Jossey-Bass.

Gilbert, D., Fisk, S., & Lindzey, G. (1998). *Handbook of social psychology, Volume I* (4th ed.). Oxford, UK: Oxford University Press.

Lewin, K. (1999). *The complete social scientist: A Kurt Lewin reader* (Martin Gold, ed.). Washington, DC: American Psychological Association.

Meder, P. (2000). The effect of continuous goal setting on persistence in math classrooms. *Focus on Basics*, 4(A), 7–10.

Porter, K., Cuban, S., & Comings, J. (2005). *One day I will make it*. New York: MDRC.

Quigley, B. (1997). *Rethinking literacy education: The critical need for practice-based change*. San Francisco: Jossey-Bass.

Quigley, B. (2000). Retaining adult learners in the first three critical weeks: A quasi-experimental model for use in ABE programs. *Adult Basic Education*, 10(2), 55–69.

Reder, S. (2000). Adult literacy and postsecondary education students: Overlapping populations and learning trajectories. In J. Comings, B. Garner& C. Smith, C.,

The annual review of adult learning and literacy, Vol. 1. San Francisco: Jossey-Bass.

Rose, S., & Wright, M. (2005). *Using state administrative records for analyses of student attendance, student achievement, and economic outcomes: A three-state pilot study.* Washington, DC: U.S. Department of Education.

Sticht, T. (1982). *Evaluation of the reading potential concept for marginally illiterate adults.* Alexandria, VA: Human Resources Research Organization.

Tracy-Mumford, F. (1994). *Student retention: Creating student success* (NAEPDC Monograph No. 2). Washington, DC: NAEPDC.

Wikelund, K., Reder, S., & Hart-Landsberg, S. (1992). *Expanding theories of adult literacy participation: A literature review* (NCAL Technical Report TR 92–1). Philadelphia: National Center on Adult Literacy, University of Pennsylvania.

Witkin, H. A., Oltman, P. K., Raskin, E., & Karp, S. A. (1971). *A manual for the Embedded Figures Tests.* Palo Alto, CA: Consulting Psychologists Press.

Young, M., Fleischman, H., Fitzgerald, N., & Morgan, M. (1994a). *National evaluation of adult education programs: Patterns and predictors of client attendance.* Arlington, VA: Development Associates.

7 Teacher Attitudes Towards the "Skills for Life" National Strategy for Improving Adult Literacy and Numeracy Skills

Olga Cara and Jennifer H. Litster

In 1996 Tony Blair, MP, leader of the opposition Labour Party, declared that his three priorities for government would be "education, education, and education."[1] Early in its first administration (1997–2001), the new Labour government instructed a working group under the chairmanship of Sir Claus Moser to investigate the state of adult basic skills in the United Kingdom and to suggest ways of reaching adults with literacy and numeracy needs. Lord Moser's *A Fresh Start: Improving Literacy and Numeracy* (1999) concluded that up to seven million adults (one in five of the adult population) in England had difficulties with literacy and numeracy—a higher proportion than in any other Western country apart from Poland and Ireland.

In response to the Moser Report's recommendations, in March 2001 the Department for Education and Skills (DfES) launched a new national strategy for improving the literacy and numeracy skills of adult learners. Known as Skills for Life (SfL), the strategy's target is to improve the literacy, language, and numeracy skills of 1.5 million adult learners by 2007 and 2.25 million learners by 2010. SfL emphasizes the needs of priority groups at risk of exclusion, including: unemployed people and benefit claimants; prisoners and those supervised in the community; public sector employees; low-skilled people in employment; and younger adult learners aged sixteen to nineteen years.

In addition to boosting the demand for courses through a highly successful advertising campaign, and providing the courses to meet this new demand, a core component of the SfL strategy is to improve the quality of teaching and learning in literacy, numeracy, and English for Speakers of Other Languages (ESOL) provision through a new national learning, teaching, and assessment infrastructure. Through such infrastructural reforms, SfL has brought about a great number of changes in the working lives of all those who teach on the courses coming under the SfL banner. These include discrete and embedded courses in language, literacy, and numeracy at Level 2 or below on the National Qualifications Framework (NQF).[2]

The following chapter offers some findings and observations drawn from the first two waves of a large-scale longitudinal study examining the impact of the SfL strategy on teachers. From this rich data source, this chapter focuses on the attitudes of teachers of SfL subjects toward the strategy: what their attitudes are, what factors influence these attitudes, and what the implications of this may be for the strategy's successful implementation.

RESEARCH CONTEXT

In essence, SfL aims to transform the quality of language, literacy, and numeracy teaching and encourage a new professionalization in a sector where previously provision was marginalized and teachers and trainers operated without clear guidance on what was expected of them. With the strategy came consistent and robust national standards; screening and diagnostic assessments; new learning and teaching materials; and national tests which are now the benchmark for all literacy and numeracy achievement. The quality of teaching has been transformed by new national core literacy and numeracy curricula for adults, based on the national standards, which set out the specific literacy and numeracy skills that need to be taught and learned at each level within the National Qualifications Framework (NQF). The core curriculum for each subject area ensures consistency and continuity for the learner and helps teachers using focused teaching methods to meet the needs of individual learners. The introduction of this new approach was supported by induction and training courses for teachers, and during the life of the strategy thus far a range of Continuing Professional Development (CPD) programs and initiatives have been offered at regional, national, and local levels.

Prior to 2002 there was little available in the way of either initial teacher training or in-service provision for basic skills teachers. As part of SfL, the government developed mandatory teaching qualifications for new teachers, using a framework which recognised that adult literacy and numeracy were specialist subjects. For new teachers of adult literacy or adult numeracy, the requirement to gain one of these qualifications came into force on 1 September 2002 and for new teachers of adult ESOL on 1 September 2003. New teachers are required to have a generic teaching qualification such as a Postgraduate Certificate in Education (the PGCE is roughly the equivalent of a graduate diploma in education) *and* a Level 4 certificate in their subject specialization.[3] Existing teachers are also being encouraged to take these qualifications. The publication of *Success for All* in November 2002, a strategy document which articulated the government's vision for the future of post-16 education and training, reinforced the principle that all teachers should be qualified to teach. It is proposed that by 2010 only new entrants to teaching in the post-16 sector will be unqualified.

These changes to the teaching and learning infrastructure acknowledge 'that the goals of the SfL strategy cannot be accomplished by top-down government initiatives alone. Every relevant player on an individual or an organizational level can contribute by working to a common set of comprehensible objectives.

According to Fullan (1991), teachers are essential to the implementation of educational reform, because making changes to existing institutions, structures, and classroom practices is impossible without teacher participation. From this, Fullan (1999) argues that for an education reform to be successful, teachers must feel they have a personal investment in its outcomes. In effect, teachers are required to become agents of change. Teachers are best placed to know students and to understand students' needs (Barth, 1990). The knowledge, skills, and judgments that teachers bring to their role as "change agents" (Sanders, 1999) have critical roles to play in educational change.

Moreover, international research suggests that teachers' subjective perceptions of proposed change influence their behavior (McNess, Broadfoot, & Osborn, 2003; Hargreaves, 1998; Van den Berg, Vandenberghe, & Sleegers, 1999). Those seeking to instigate educational changes that teachers feel negatively about, or they exhibit behavioral resistance towards, face serious obstacles. Therefore, it is essential to consider teacher attitudes toward any educational reform they are involved in, and to identify the factors that are associated with the attitudes teachers express.

Fullan goes on to argue that if teachers "are to push for systematic change" (1999: 579), then time support and institutional support are critical. Currently, the SfL strategy requires teachers to improve their subject knowledge and pedagogy, to get new qualifications, and, in some cases, to engage in practitioner research. Teachers are also responsible for administration, for curriculum revision, and for outreach. But research has shown that teachers need *time* to try new approaches, learn new skills, develop new attitudes, and engage in professional development activities (Cocoran, 1995; Donohoe, 1993). For example, Turnbull (2002) found that teachers were more likely to support school reform when they had adequate training, resources, helpful support, participation in school decisions, and control over the implementation of reform in their own classrooms. Organizational environment overall is viewed as an important predictor of teacher attitudes and commitment (Hargraves, 1988; Reynes, 1992; Tam & Cheng, 1996).

This theoretical approach emphasizes the critical role of social-organizational factors, with "attitudes" being the individual's response to structures and processes (Reynes, 1990). However, "attitudes" can also be seen as an appropriate fit between individual personality needs and the opportunities and demands of the reforms. Because attitudes link people and their behavior to their context, it is important for research on attitudes to explore the impact of personal characteristics (Reynes, 1990). In this chapter both the organizational context and the personal characteristics of SfL teachers will be explored as they relate to the teacher's attitudes towards different aspects of the SfL strategy.

METHODS

The National Research and Development Centre (NRDC) was established in 2002 by the then Department of Education & Skills (DES) as part of the SfL strategy in order to conduct the further focused research the Moser Report called for. The NRDC comprises a consortium of university and other partners, who are specialists in the fields of literacy, numeracy and ESOL; the NRDC is based at the Institute of Education, University of London. The longitudinal study of the impact of the SfL national strategy for improving adult literacy and numeracy skills on teachers and trainers (referred to as "The Teachers Study") is one of the largest projects in the NRDC's portfolio and runs in tandem with a longitudinal study of the strategy's impact on learners. The Teachers Study is an exploratory research project designed to follow teachers working on a range of post-16 education and training programs and, as one of the first large-scale studies, allows policymakers to evaluate how the SfL strategy affects teachers.[4]

The Teachers Study is a panel study, with quantitative data gathered in three waves via a respondent questionnaire. These data are supplemented by in-depth qualitative interviews with a subsection of the respondent sample. Wave 1 quantitative fieldwork was carried out between May 2004 and August 2005, by means of a Computer Assisted Personal Interview (CAPI), for 309 respondents, and, for 718 respondents, a self-completion Web questionnaire.[5] The 1,027 valid questionnaires contain information on teachers' demographic profile; their qualifications; prior experience; employment details; professional development activities and training needs; organizational/work environment; reasons for entering the profession; job satisfaction; perceptions of their job roles; and attitudes towards the new SfL policies.

In Wave 2 (May–December 2006), quantitative data were gathered solely through a self-completion Web questionnaire. In the main, the second questionnaire is a truncated version of the first, designed to capture the changes taking place between survey waves; a very similar questionnaire will be used in Wave 3 (starts April 2007). Wave 2 fieldwork was ongoing at the time of writing, but to add a longitudinal dimension to this chapter, analysis has been carried out on a data set of six hundred Wave 2 questionnaires. Although this data set is partial, the missing data are random, and the observations are representative of the Wave 1 sample.

This said, readers should be aware of an important limitation of the Teachers Study. In the original study design, the sample was to comprise one thousand randomly selected teachers drawn from 245 adult-education programs in six different funding streams (Further Education [FE]; Prison Education; Adult and Community Learning; Work-Based Learning; JobCentre Plus; learndirect).[6] These programs were themselves randomly selected from a sample of eighteen (of forty-seven) Learning and

Skills Council (LSC) areas that represented all nine geographic regions of England.[7]

However, it proved extremely difficult to secure the participation of providers, managers, and teachers in the study, and especially for those in learning programs outside FE. The problem of recruiting participants was partly connected to the very complexity of the SfL sector: teachers on hourly paid contracts, for example, may have only periodic access to e-mail, and may have no fixed office or telephone at work to be contacted on, and agency staff are not always known to human resources departments. Teachers were therefore recruited on an opportunistic basic. To offset any reluctance on the part of busy teachers to complete what was a lengthy questionnaire, a Web version of the first survey was offered as an alternative to a face-to-face interview, and incentives in the form of book tokens were given to all those who completed the survey.

As a result, the sample of this study is now a volunteer (nonprobability) sample. It is nonrandom and thus nonrepresentative, although in many key areas the profile of our sample (see Tables 7.1 and 7.2) accords both with what available data from previous research tells us about similar kinds of teachers and with analysis of the Staff Individualised Records (SIR) data sets kept by the LSC.

In fact, it is worth noting that until recently *no* robust analysis of the SfL workforce profile had been carried out. The data used in all previous research were not based on representative, random samples of the whole population of SfL teachers, but on the substantial or less substantial numbers of teachers who agreed to participate. Moreover, the SIR analysis itself presents challenges. Firstly, the complexity of provision and the diverse nature of teacher contracts make precise identification of SfL teachers within the data set problematic, and it is also impossible to distinguish between literacy, numeracy, and ESOL teachers. Secondly, the SIR data include teachers working in FE colleges only, and no information is collected about community or work based learning organizations. Without robust evidence on the staff numbers and distribution that characterize the total population of SfL teachers, it is currently impossible to weight the findings from the Teachers' Study to make them more representative. By the time of our final analysis on the Teachers Study, however, this will be resolved, for in September 2006 Lifelong Learning UK (LLUK) was commissioned by the DfES to carry out a snapshot survey of the SfL workforce in England.[8] This survey is designed to find out the number and profile of specialist SfL teachers, tutors, or trainers, and our data will be weighted to reflect findings from the LLUK survey.

One other limitation of the Teachers Study is that although our questionnaire is detailed, it does not include any variables that measure the teachers' psychological characteristics. Early analysis suggests that such variables may explain the biggest variance in teacher attitudes.

Nevertheless, the data gathered in the Teachers Study are very rich and there are many possibilities for analysis. This chapter will focus on three main research questions:

- What are the attitudes of teachers towards certain aspects of the SfL national strategy?
- Which teacher or organizational characteristics are associated with which attitudes?
- How are these attitudes changing over time and what factors explain such changes?

By way of conclusion, this chapter discusses several important practical implications for policymakers suggested by these results.

INSTRUMENTS/MEASUREMENTS

In the Teachers Study quantitative questionnaire, attitude measurements are obtained through a 30-item scale developed by Giannakaki (Giannakaki, 2005). Each item was rated using a 5-point scale of agreement-disagreement, where responses were designated 5—"strongly agree"; 4—"agree"; 3—"uncertain"; 2—"disagree"; 1—"strongly disagree"; and questions were phrased to elicit both positive and negative views. An initial set of 77 measurements was piloted with a small sample of teachers (N = 178) in 2004 and was included in a reduced format in the main stage fieldwork. Giannakaki performed factor analysis to cut the number of attitudinal items for analysis and explore into which and how many groupings these attitudes could be divided. The seven main factors were extracted and labeled "Curriculum," "Inclusion," "Resources and Support," "National Tests and Learner Morale," "Teacher Qualification Requirements," "Inspections/Quality Assurance," and "Validity of Assessment."[9]

As Giannakaki acknowledges, a possible weakness in her analysis is that the number of cases (N = 178) is relatively small compared to the total number of items that were factor analysed (77). With this in mind, a separate factor analysis was carried out on the thirty-item attitude scale used for the 1,027 teachers sampled in Wave 1, which gave results broadly similar to those seen in Giannakaki's pilot study.

The thirty different attitudinal statements were next arranged in six bigger groupings identified with the factors.[10] For the purpose of this chapter's analysis, four of these six factors will be used as dependent variables: "Curricula and SfL Initiatives"; "Teacher Qualification Requirements"; "Validity of Assessment"; "National Tests and Learner Morale." The final scores of the scales are presented in Table 7.1.

Table 7.1 Descriptive Statistics of the Attitude Scales and Continuous Variables

Scale	*Mean Wave 1*	*Mean Wave 2*	*Range*	*Number of items*	*Cronbach's Alpha Wave 1*	*Cronbach's Alpha Wave 2*
Age	42.6 (16.64)					
Experience in post-16, years	9.57 (7.63)					
Years worked for the current employer	4.69 (4.67)					
Hourly wage, gross	15.23 (5.47)					
Paid weekly hours	15.25 (11.47)					
Contact hours	15.6 (9.58)					
Curricula and SfL initiatives*	29.90 (6.23)	29.34 (5.73)	9–45	9	0.88	0.85
Teacher Qualification Requirements	11.95 (3.22)	11.94 (3.14)	4–20	4	0.73	0.73
Validity of assessment*	15.47 (3.52)	14.92 (3.75)	5–25	5	0.75	0.77
National tests and learner morale	13.05 (2.83)	13.17 (2.48)	4–20	4	0.69	0.62
Job Satisfaction*	26.86 (6.37)	25.77 (6.52)	0–44	11	0.78	0.80
Managerial Support*	11.59 (3.19)	11.01 (3.51)	0–16	4	0.87	0.88
Collaboration with colleagues	7.82 (2.57)	7.66 (2.59)	0–12	3	0.60	0.65
Influence on decision making	15.99 (6.80)	15.99 (7.03)	0–32	8	0.86	0.87
Clarity of job role	16.84 (4.43)	16.87 (4.70)	0–28	7	0.77	0.81

*Notes: Figures in parenthesis are standard errors. Scales with * have significant change between two sweeps.*

On average, teacher attitudes towards the curricula and other SfL initiatives are quite positive. Attitudes towards the new teacher qualification requirements, to the validity of assessment, and to the national tests are more neutral with some negative qualities. In later analysis each of the six factors was used as a separate subscale. Scale scores were standardized and used as outcome variables in regression models to examine the effect of different individual and organizational characteristics on the attitudes of the SfL teachers.

Variables for individual teachers include: demographic characteristics (gender, ethnicity, and age); education and training (for example, new qualifications; highest qualification; professional development); and employment experience (for example, years of post-16 teaching experience; salary; contract type; main teaching subject; teaching program). Some relevant descriptive statistics on these characteristics are presented in Tables 7.1 and 7.2.

Table 7.2 Profile of the Teachers in the Sample, Wave 1

	Frequencies	*Percentage*
Female	814	79.3
Ethnicity: British white	895	87.1
British Asian	30	2.9
Black British	20	2.0
Other	82	8.0
Have or working towards a new Level 4 subject specialist qualification	480	46.7
Have a Cert Ed, PGCE, or BEd qualification	630	61.3
Have an ESOL, ESL, EFL qualification	249	24.2
Main teaching subject		
Literacy	406	40.0
ESOL	246	24.3
Numeracy	196	19.3
Other subject	166	16.4
Have permanent contract	791	77.0
Have a full-time position	598	58.2
Work in FE	728	70.9
Have curriculum manager experience	383	37.3

On an organizational level, five attitudinal variables were also added to the model. The first of these, *job satisfaction* (a measure based on eleven items), reflects satisfaction with different aspects of the teaching job, such as job security, benefits, professional development opportunities, learners, and so forth. The second variable includes four items on teacher perception of *help and support received from managers*. A third variable, again based on four survey items, shows *the extent to which teachers collaborate with colleagues*. The fourth variable, an eight-item scale, measures how teachers perceive their *degree of influence on decision making* in their employing organization. Finally, a fifth variable measures the *clarity of their professional role*, using answers to seven statements such as "I have clear, planned goals and objectives for my job" or "I know exactly what is expected of me." (Again, measurements for all items used a five-point agreement-disagreement scale.) These items were combined to produce an aggregate score for each relevant variable. (A summary of the descriptive statistics for these variables is presented in Table 7.1.) Scale scores were then standardized and used as independent variables in regression models to explore what association they had with teacher attitudes towards SfL.

FINDINGS

Correlation analysis revealed that salary, number of years teaching in the post-16 sector, number of days per year spent on CPD activities, length of time working for the employing organization, job satisfaction, influence on decision making, collaboration with colleagues, managerial support, and clarity of role all had a statistically significant association with teacher attitudes towards SfL core curricula. Moreover, other statistical tests showed that gender, main teaching subject, teaching sector (e.g., FE or work-based learning), holding a new Level 4 subject specialist qualification, or an ESOL, EFL (English as a Foreign Language), or ESL (English as a Second Language) teaching qualification, as well as a generic teaching qualification such as PGCE, showed a statistically significant relationship with teacher attitudes towards *Skills for Life* core curricula.

Our discussion of the multivariate analysis using multiple regression begins by examining the covariates of teacher attitudes towards the SfL curricula and other initiatives (see Table 7.3). These were entered into each analysis sequentially. When gender and qualifications were entered into the analysis, having a new Level 4 subject specialist qualification was related to attitude towards the SfL curricula and other initiatives. Gender and generic teaching qualifications were statistically insignificant, meaning that these two variables might be connected to new Level 4 qualifications or some other salient fourth variable, or that they explain the same variance in teacher attitudes as the new Level 4 qualification. In the next step, "has another ESOL qualification" was entered and was statistically significant. Then salary, years worked for the employing organization, and main teaching subject were entered. While all these variables, except salary, significantly predicted teacher attitude towards the SfL curriculum, their effect was not very strong in explaining the variance of the attitude. As a teacher's salary is often associated with the years worked for their employing organization, we do not find any significant effect of salary on teacher attitudes. When teaching sector and years of teaching experience in post-16 education were introduced as the next predictors in the regression analysis, only the coefficient of teaching sector was statistically significant. Years worked for the employing organization became insignificant when teaching sector was entered into the analysis. In the next step, the number of days per year spent in professional development per year was entered. Job satisfaction and clarity of job role had a statistically significant effect on teacher attitudes towards the SfL curriculum and other initiatives.

Overall, the analysis shows that teachers with the new Level 4 subject specialist qualifications and those teachers who have participated in a higher number of days of CPD are more positive about the SfL curriculum. Moreover, teachers who are more satisfied with their job, and teachers who understand their role more clearly, are also more positive about the SfL curricula and other initiatives. By contrast, those teachers in our sample who have other ESOL qualifications (such as the Certificate in English Language Teaching to Adults [CELTA]), ESOL teachers (compared to literacy teachers), and more experienced teachers are slightly inclined to have a more negative attitude towards the SfL curricula.

The next stage of our analysis examined teacher attitudes towards new teaching qualification requirements. Analysis of Variance (ANOVA) showed that the variables of having a Level 4 subject specialist qualification and main teaching subject had a statistically significant group effect on teacher attitudes towards new qualifications. The correlation analysis demonstrated that the degree to which teachers experience job satisfaction, experience managerial support, and have clarity about their job role is also associated with their attitude towards the new teaching qualification requirements.

All the control variables were associated with teacher attitudes towards the new teaching qualifications, with all measures being statistically significant, except for degree of managerial support (see Table 7.3). Those teachers who had, or were working towards, a new Level 4 subject specialist qualification had more positive attitudes towards these qualifications. If teachers were more satisfied with their job and had a clearer idea about their professional role, they were also more positive about the new teaching qualifications even when controlling for their teaching subject and job satisfaction. Teachers whose main subject was numeracy had slightly more negative attitudes compared to teachers whose main subject was literacy.

Table 7.3 Multiple Regression of Teacher Attitudes on Teacher Attributes

	CURRICULA AND SFL INITIATIVES	NEW TEACHING QUAL-IFICATIONS	VALIDITY OF ASSESSMENT	NATIONAL TESTS AND LEARNERS' MORALE
BME		−0.266		
		(0.113)*		
Male	−0.098	0.149		
	(0.080)	(0.088)		
Age				0.005
				(0.003)
Have or working towards new Level 4 subject specialist qualification	0.263	0.400		0.135
	(0.066)**	(0.068)**		(0.066)*
Have PGCE, CertEd, or BEd	0.107			
	(0.071)			
Have an ESOL, EFL, or ESL qualification	−0.289		−0.263	−0.027
	(0.106)*		(0.120)*	(0.108)
LogPay	−0.150		−0.314	0.173
	(0.133)		(0.144)*	(0.132)
Years worked for the employing organization	−0.013		−0.012	
	(0.007)		(0.008)	
(Reference Literacy) Numeracy	−0.155	−0.280	0.209	0.136
	(0.087)	(0.091)*	(0.093)*	(0.087)
ESOL	−0.278	0.110	0.076	0.277
	(0.114)*	(0.087)	(0.128)	(0.115)*
Other Subject	−0.029	−0.040	0.393	−0.235
	(0.096)	(0.100)	(0.103)**	(0.098)*
FE sector	−0.107		−0.062	−0.282
	(0.076)		(0.079)	(0.074)**
Years of experience in post-16 education	−0.011		−0.003	
	(0.005)*		(0.006)	
Managerial experience			0.182	
			(0.076)*	

(continued)

Table 7.3 Multiple Regression of Teacher Attitudes on Teacher Attributes (continued)

	CURRICULA AND SFL INITIATIVES	NEW TEACHING QUAL-IFICATIONS	VALIDITY OF ASSESSMENT	NATIONAL TESTS AND LEARNERS' MORALE
Number of days of professional development within the last 12 months	0.018 (0.008)*			
Attended 1 core curriculum seminar				0.272 (0.083)*
Attended more than 1 core curriculum seminar				0.204 (0.084)*
Job satisfaction	0.150 (0.042)**	0.095 (0.044)*	0.172 (0.046)**	0.167 (0.043)**
Clarity of job role	0.130 (0.041)*	0.139 (0.043)*	0.153 (0.045)*	0.049 (0.041)
Influence on decision making	0.068 (0.038)			0.018 (0.035)
Managerial support	0.033 (0.040)	–0.043 (0.040)	0.039 (0.041)	–0.031 (0.039)
Collaboration with colleagues	0.041 (0.037)			
Observations	773	836	744	894
R-Square	0.245	0.090	0.167	0.123
Adjusted R-square	0.226	0.082	0.151	0.109
F statistic	18.489	14.618	11.317	10.629
Pr (F-Stat)	0.000	0.000	0.000	0.000

*Notes: Standard errors in parentheses; * significant at 5%; ** significant at 1%; dependent variables are entered as Z scores.*

In the next stage, correlation analysis showed that the number of years respondents had taught in the post-16 sector and the number of years they had worked for a particular employer, as well as job satisfaction, degree of managerial support, and clarity of job role had statistically significant associations with teacher attitudes towards the validity of assessment. In addition, ANOVA showed that gender, ethnicity, main teaching subject, teaching sector, another teaching qualification, and experience of curriculum management had a statistically significant effect on teacher attitudes towards the validity of SfL assessment. It is interesting that when years of teaching experience was entered into the model, salary and years worked for the current employer became statistically insignificant predictors of attitudes towards the validity of assessment. The highest percentage of variance in attitudes was explained when attitudinal variables were entered.

The multiple regression analysis (see Table 7.3) showed that those respondents who taught numeracy, or whose main teaching subject was not an SfL subject, had more positive attitudes towards the validity of assessment than those who taught adult literacy, had managerial experience, a clearer understanding of their professional role, and higher job satisfaction. Having an ESOL or Teaching English as a Foreign Language (TEFL) qualification, coming from a black and minority ethnic (BME) background, and higher salary had negative associations with teacher attitudes towards the validity of assessment.

The last model examined teacher attitudes towards national tests, with a specific focus on how these tests affect learner morale (see Table 7.3). The statistical tests showed that the variables for main teaching subject, teaching sector, having a Level 4 subject specialist qualification, having an ESOL, EFL, or English as a Second Language (ESL) teaching qualification, and attendance at the SfL core curriculum training had a statistically significant group effect on teacher attitudes towards national tests and learner morale. Correlation analysis revealed that salary, age, pay, job satisfaction, managerial support, clarity of professional role, and influence on decision making had statistically significant associations with teacher attitudes towards national tests and learner morale.

Teacher characteristics did not explain a great deal of variance in attitudes in this model. Teachers with a Level 4 qualification or who were working towards one, those teaching ESOL (as compared to literacy), those who attended core curriculum training, and those with higher job satisfaction tended more than others to believe that national tests were connected with higher learner morale. A more negative attitude was found among those whose main teaching subject was not an SfL subject (as compared to literacy) and those working in FE.

We added a longitudinal dimension to our analysis by looking for any changes in teacher attitudes towards the SfL strategy between the two waves of fieldwork. An analysis of the partial data set from Wave 2 revealed significant changes in teacher attitudes towards the SfL core curricula and other initiatives and in teacher attitudes to the validity of assessment. In both cases, teacher attitudes became slightly more negative (see Table 7.1).

Since analysis of Wave 1 data revealed that a teacher's qualifications, and some attitudinal measures, had the greatest effect on teacher attitudes towards the SfL strategy, we wondered if these factors also influenced attitudes in Wave 2 and any attitudinal changes between the two waves.

Data analysis of Wave 2 showed that teachers who had, or were working towards, a Level 4 subject specialist qualification in Wave 1, and those who had gained this qualification or were working towards it by Wave 2, were more positive about the core curriculum and other SfL initiatives than those without the new qualification. There was no difference in attitudes towards the validity of assessment between these two groups of teachers. In addition, teachers with other ESOL teaching qualifications had a more negative attitude towards both the SfL curriculum and the validity of assessment. Here having an ESOL qualification could also serve as a proxy for being an ESOL teacher.

From the longitudinal analysis, firstly we observed that there was no significant attitudinal change among teachers without a Level 4 subject specialist qualifications and among those without generic teaching qualifications. Yet the analysis revealed that in Wave 2 the attitudes of teachers with Level 4 or generic teaching qualifications became slightly less positive compared to their attitudes in Wave 1. Teachers with ESOL qualifications, who in Wave 1 had slightly less positive attitudes towards the SfL curriculum and other initiatives, remained less positive in Wave 2 in comparison with the other teachers. Their attitudes towards the validity of the SfL assessment also became less positive. The downward change in attitudes towards the SfL curriculum of teachers with ESOL qualifications was smaller than among those without these qualifications: however, downward change in their attitudes towards the validity of the assessment was slightly greater.

Furthermore, combining data from both waves, teachers with generic teaching qualifications were also slightly more positive about the SfL core curriculum and other initiatives. Correlation analysis showed that job satisfaction, influence on decision making, clarity of professional role, managerial support, and satisfaction with organizational resources were associated with more positive teacher attitudes towards the core curriculum and other SfL initiatives in both waves. In addition, teachers who were more satisfied with their jobs and with organizational resources, who received more managerial support and were clearer about their role also had a more positive attitude towards validity of SfL assessment.

Furthermore, our data suggest that teachers were slightly less satisfied with their jobs and managerial support in Wave 1 than in Wave 2. There were also correlations among the attitudinal changes: teachers who became more negative about the core curriculum also became slightly more negative about the validity of assessment. There also was a statistically significant association between changes in teacher attitudes towards core curriculum and their job satisfaction. If teachers became more satisfied with their job, their opinion about the core curriculum became more positive. Even if change in evaluation of managerial support did not correlate with change

Table 7.4 Multiple Regression of Wave 2 Teacher Attitudes to Curricula Initiatives And Validity of Assessment on Teacher Attributes Taking Wave 1 Attitudes into Account.

	CURRICULA AND SFL INITIATIVES	VALIDITY OF ASSESSMENT
Attitude towards SfL curricula in Wave 1	0.107***	—
	(0.006)	—
Attitude towards SfL assessment in Wave 1	—	0.163***
	—	(0.012)
Have or working towards new Level 4 subject specialist qualification Wave 2	0.112*	–0.010
	(0.063)	(0.079)
Have PGCE, CertEd, or BEd Sweep 2	0.055	–0.006
	(0.078)	(0.101)
Have an ESOL, EFL, or ESL qualification Wave 2	–0.247***	–0.169*
	(0.076)	(0.101)
Job satisfaction Wave 2	0.017**	0.019**
	(0.007)	(0.009)
Clarity of job role Wave 2	0.009	–0.004
	(0.009)	(0.011)
Influence on decision making Wave 2	0.002	—
	(0.005)	—
Managerial support Wave 2	–0.006	0.002
	(0.011)	(0.014)
Collaboration with colleagues Wave 2	0.025*	—
	(0.014)	—
Organizational facilities and resources Wave 2	—	0.021
	—	(0.019)
Observations	450	420
R-Square	0.595	0.397
Adjusted R-square	0.586	0.385
F statistic	64.771	33.973
Pr (F-Stat)	0.000	0.000

*Notes: Standard errors in parentheses; *significant at 10%; **significant at 5%; *****significant at 1%; dependent variables are entered as Z scores.*

in teacher attitudes towards the SfL curriculum or assessment, it correlated significantly with change in job satisfaction. Teachers who became less satisfied with their job also evaluated the managerial support they received at their employing organization in a slightly less positive way.

After establishing the independent effect of our variables, we considered their relative effect and the extent to which one may account for the other. A multiple regression analysis was performed for two attitudes from Wave 2 where attitudinal changes were significant (Table 7.4). Teacher qualifications and attitudinal variables measured in Wave 2 were entered into the analysis as independent variables. Wave 1 attitudes were entered in the first step and qualifications and other attitudes in the second and third. Teacher attitudes in Wave 1 had the biggest effect on attitudes in Wave 2. However, even taking these attitudes into account, subject specialist qualifications, ESOL qualifications, collaboration with the colleagues, and job satisfaction were still associated with teacher attitudes toward the core curricula. After adjusting for attitude measurement in Wave 1, teacher attitudes towards the validity of assessment were related to job satisfaction and ESOL qualifications, but not to subject specialist qualifications.

DISCUSSION

This final section considers some of the methodological challenges we faced with this analysis before going on to look at patterns that run across our analysis of the four dependent variables measuring teacher attitudes towards different aspects of SfL strategy. We then offer some conclusions and policy discussion points about which personal characteristics and which organizational climates appear to have the greatest effect on attitudes.

The first important methodological challenge of this chapter is its longitudinal aspect. To add a longitudinal dimension, we used data from two waves of the panel survey. (The final response rate for Wave 2 was 74 percent.) A statistical analysis of the probability of a Wave 1 respondent being included in the Wave 2 sample suggests that the only statistically significant difference is for the sector of the main employing organization (that is, FE, work-based learning and so on). Taking account of this difference is complicated by the fact that teachers may change their employer, and their employer type, between waves. Overall, the analysis of attrition suggests that the missing data are random, and our observations are representative of the first wave sample for those teacher characteristics that we can account for.

Secondly, all the models we have used have quite low R-squared values, that is, they explained only a small proportion of the variance in the dependent attitudinal variables, especially when only sociodemographic or contextual variables were entered into the analysis. Adding in some attitudinal variables to serve as proxies for teachers' psychological profiles improved the explanatory power of the regression models, but the proportion of variance explained remained low, especially in the Wave 1 data analysis. In the Wave 2 analysis, the explained variance increased, but again this resulted from controlling for the Wave 1 teacher attitudes and thus for some factors that formed the later attitudes in the first place. This illustrates the complicated structure of attitude formation, where factors other than sociodemographic or contextual variables explain the greatest proportion of the variability.

Lastly, our main attitudinal scales were created using exploratory factor analysis as opposed to confirmatory factor analysis. Since our scales have not been used in any previous studies, and no other theoretical framework was available, we could not specify the number of factors and factor structure before our analysis for this chapter. When data for all three waves are gathered, we will look at a fit of the hypothesized factor structure using confirmatory factor analysis. At this interim stage of analysis (the final report on the Teachers Study will be submitted in December 2007) our main aim was to condense our thirty attitude items into a reduced factor structure of teacher attitudes and assess its internal validity.

Turning now to the early messages emerging from the Teachers Study, it is important to note that the SfL teaching population exhibits specific characteristics that may influence attitudes towards the strategy. From both SIR data analysis and previous research we know the workforce is predominantly female, and SfL teachers are, on average, older than compulsory education teachers.[11] A high proportion of SfL teachers work under part-time contracts (either fractional or hourly paid contracts) and historically there have been fewer full-time employment opportunities in this sector than in other FE programs or in the compulsory education sector. Moreover, a culture of change and new requirements contribute to a lack of stability in the workforce, a higher turnover of staff, and challenges for new entrants to the profession, all of which may bring about negative attitudes.

Although we investigated a range of sociodemographic characteristics for the teachers in our sample, few of these proved to have a significant effect on attitudes towards SfL. Only ethnicity had an effect on teacher attitudes towards the validity of SfL assessment procedures: in our sample, BME teachers tend to be slightly more negative about current assessment procedures. Gender and age *do* seem to be important in univariate analysis, yet when other variables are used as controls, they lose significance, indicating that gender and age are associated with certain other attributes, such as teaching qualifications and experience, or attitudes towards aspects of organizational climate that are related to the dependent attitude variable

In accordance with previous research (Avramidis, Bayliss, & Burden, 2000; Tsitouridou & Vryzas, 2003), which shows that more highly qualified teachers are more positive in general than their less qualified peers, the qualifications held by the teachers in our sample appear to be quite strongly associated with their attitudes towards SfL. Firstly, engagement in professional development activities, through attending a higher number of Continuing Professional Development (CPD) days or core curriculum training, had a positive effect on teacher attitudes. Secondly, the single strongest effect in all models concerning different aspects of the attitudes came from the variables concerned with the new Level 4 subject specialist qualification. Teachers who held, or were working towards, these qualifications had more positive attitudes towards SfL. However, teachers who held a different form of ESOL qualification had a slightly more negative attitude in both waves. Although on the face of it ESOL teachers seem to be the most qualified and experienced in our sample, this negative attitude may reflect uncertainty about the currency of these other ESOL qualifications in relation to the new teaching qualification requirements

This observation may also be connected to other issues specific to the ESOL field, including course funding uncertainties. A respondent's main teaching subject that is likely to be connected to teaching qualifications came out as significant in all models controlling for teaching qualifications. ESOL teachers had a tendency to be slightly more negative about the SfL core curriculum, but were more positive about the effect of national tests on learner morale. Also, numeracy teachers were less satisfied with the new teaching qualifications. One possible explanation for this is that when our Wave 1 fieldwork was conducted there was less availability of Level 4 subject specialist courses in numeracy than in the other subject specialisms, and a poorer quality of teacher training provision.

A lack of pedagogy in the Level 4 subject specialist courses is now being addressed and subject specialism courses will include subject specific pedagogy from September 2007. Our research supports the idea that a key factor in the successful implementation of SfL lies in encouraging teachers to gain these qualifications, which includes making sure that teachers are informed about the qualifications and about the availability of courses and funding opportunities, and that employing organizations support their staff. Moreover, practitioners should have a role to play in teacher training programs; the classroom experiences of "change agents" should feed into course design; for very often those studying for these new qualifications have many years of experience in the field. Practitioners also need to be kept up to date on changes to the teacher training courses and requirements and any impact these may have on the qualifications they currently hold. A recent focus group conducted as part of the qualitative strand of the Teachers Study highlighted the insecurities many teachers have about such changes: teachers need to know that the new qualifications they are encouraged to take will continue to have value, and not be rendered obsolete by further reforms to the qualifications framework.[12]

Consistent with other educational research, our analysis found that the characteristics of the strategy implementation context, that is, the features of the school, college, or other organization where a teacher works, influence attitudes towards educational reform and towards the changes it means for a teacher's professional life. It is notable that of the factual variables connected to employment, only teaching sector appears significant in explaining how teachers perceive national tests to influence learner morale: teachers working for employers other than FE colleges judge their learners to be more satisfied and positive about national tests.

A number of studies suggest that teachers are more positive about reforms and more motivated to put reforms into practice if their organization is characterized by a shared vision, by collaboration between staff members, support from management and administration, shared decision making, adequate resources, and training opportunities (Fullan, 1991; Reynolds, 1997; Tsui & Cheng, 1999; Geijsel, Sleegers, Van den Berg, & Kelchtermans, 2001; Turnbull, 2002). Our study also shows that the largest proportion of variance explained in all analyses comes from the attitudinal variables that address teachers' perceptions of the organizational climate and working environment. In the Teachers Study, job satisfaction and clarity of professional role had the strongest positive effect on teacher attitudes towards SfL. This could be explained in two ways. Firstly, of course, the explanation may lie in psychological characteristics not measured in this study, such as underlying optimism or pessimism. An alternative explanation is that having a clear and secure vision of a future teaching career path is associated with positive opinions about different SfL initiatives. If teachers feel involved, if they receive managerial support and their role is explained to them, if they have some autonomy over the hours that they work and over what and whom they teach, rather than being obliged to teach courses and classes according to organizational and funding pressures, job satisfaction is higher, and attitudes toward SfL are more positive.

SfL has brought many positive changes into the working lives of those who teach literacy, language, and numeracy to adults. Overall, teachers of SfL subjects appreciate that the new standards have helped to professionalize the SfL workforce. Although these new standards and systems have increased the teaching workload, particularly with regard to bureaucratic procedures and paperwork and, in many situations, instituted a more hierarchical structure between managers and teachers—factors often associated with decreased job satisfaction and more negative attitudes—national tests, new teaching and learning materials, and the core curriculum are viewed as helpful tools. When, in a series of qualitative interviews conducted with a subsection of our sample between September and November 2006, we explored the use of these tools in greater depth, teachers repeatedly reported that a main strength lay in their flexibility and the fact that tools could be adapted for different learners and different teaching situations. As one teacher remarked, "I don't use it prescriptively, but when we have learners they are always initially assessed, and we do the diagnostics, et cetera, and then we always use the curriculum for reference. Sometimes they have good ideas as well, and we think, 'Yes

I will do that.' It isn't a case of sticking to it to the letter. Basically you use it as background guidance really but you are guided by what the learners do in the group."

Teachers, then, are appreciative of the fact that they have the opportunity to be creative and use these standardized materials in an innovative way. This observation appears to be another factor key to the successful implementation of SfL. More often than not, teachers working in this field are motivated by a clear sense of purpose, even moral purpose, and a commitment to social justice. If teachers are to function as agents of change, and agents of the social, cultural, and, particularly, the economic transformations that the government trusts SfL will achieve, then they must not only feel *included* in the reform process, but share *ownership* of the reform initiatives. Retaining the power to feed a wealth of classroom experience into teaching and learning materials is one aspect of this; being offered the opportunity to take these "contact time" messages back to the strategy makers is another. As another respondent remarked when interviewed: "There has to be more bottom-up feedback. They have really got to shake things up and stop the top-down things coming down, and do bottom-up feedback. And really listen to what is happening and be prepared."

In offering some initial observations about the changes in attitudes between our first two survey waves, we acknowledge that SfL provision itself is continuously subject to change. More detailed analysis, and the results of our third wave research, will demonstrate whether the less positive attitudes reported here are sustained, unique to the SfL field, or part of a wider trend in post-16 education and training. As Roger Kline, head of equality and employment rights for the University and College Union (UCU), commented in a recent press interview: "I think [FE teachers have] lost ownership of their work. Nobody comes into further education to get rich. They came into FE with a view that they were going to have a fulfilling job in which they would really help other people. And I think they feel that has slipped away. They have less control over their work than even a decade ago. They don't feel that they are able to do the job they came into the profession to do. . . . The increased workload plus the diminished ownership and sense of vocation add up to a 'lethal combination' " (Kingston, 2006).

Our early research findings on SfL teachers certainly seem to suggest some points of convergence with Kline's view. As one respondent told us: "The difficulty I see in the system is you can get people who are very bright and very creative and that is kind of driven out of them because they have to conform, they have to meet targets, they have to fit figures, and that is the thing that matters to the system. [The system] stifles that creativity. . . ."

Of course time support and institutional support are critical to whether teachers are satisfied or not with the changes that have been made to the adult literacy and numeracy field. However, *what* these changes *are* may actually matter less in terms of teacher attitudes than whether teachers feel that they have ownership of these changes and believe that they have the power and the autonomy to continue "making a difference."

NOTES

1. Tony Blair, Speech to the Labour Party Conference, 1 October 1996.
2. A Level 2 qualification is equivalent to an A*–C grade in a GCSE qualification—the General Certificate of Secondary Education examinations taken in England and Wales at the end of compulsory secondary education (aged sixteen).
3. A Level 4 qualification is equivalent to a professional diploma, certificate or award.
4. To be eligible for the Teachers Study, teachers and trainers had to work in post-16 education and training in England and have responsibility for the teaching of speaking, listening, reading, writing, numeracy, or maths skills at Level 2 or below of the NQF. These include: (a) teachers and trainers of Adult Literacy, Language, ESOL, EFL, ESL, or Key Skills in Communication, (b) teachers and trainers of Adult Numeracy or Key Skills in Application of Number, (c) teachers and trainers of GCSE English or Maths, and (d) vocational teachers and trainers, who are involved in embedded provision of one or more of these subjects.
5. Although differences to the programming of the CAPI and Web instruments meant very slight differences between the two questionnaires, comparing the data collected by the two methods, we found some variation for the mean value of the key variables, but the differences are generally not statistically significant.
6. JobCentre Plus is the government-funded employment agency facility and the social security office in the UK; learndirect is a not-for-profit organisation created in 1998 to take forward the UK government's vision of a University for Industry (Ufi Ltd) in England, Wales, and Northern Ireland—it offers online *Skills for Life* courses.
7. The Learning and Skills Council is the nondepartmental public body responsible for planning and funding education and training for everyone in England other than those in universities.
8. The research is being carried out on behalf of LLUK by NRDC, the London Strategic Unit for the Learning and Skills Workforce (LSU) and independent consultants SQW Limited.
9. Reliability coefficients for each scale were calculated and ranged from 0.70 to 0.84.
10. Alpha coefficients of reliability across subscales ranged from 0.63 to 0.88.
11. "Forecasting future skills needs in the development of literacy, language and numeracy skills provision in the post-16 sector" (Host Policy and Research, 2001); "New initial teacher education programs for teachers of literacy, numeracy and ESOL 2002/3: an explanatory study" (NRDC, 2004b); "Qualifications of Staff in LSC-funded provision" (Host Policy and Research, 2004); "Recruitment and retention in the post-16 learning and skills sector" (York Consulting, 2005) ; "Skills for Life core curriculum training programmes 2001/03: characteristics of teacher participants" (NRDC, 2004a); "Survey of staffing in literacy, numeracy and language provision" (NIACE and TES, survey return date 10 March 2006).

12. The qualitative strand of the Teachers Study consists of a focus group and sixty-three in-depth interviews with teachers from the quantitative sample.

REFERENCES

Avramidis, E., Bayliss, P., & Burden, R. (2000). A survey into mainstream teachers' attitudes towards the inclusion of children with special educational needs in the ordinary school in one local education authority. *Educational Psychology*, 20(2), 191–211.

Barth, R. (1990). *Improving schools from within.* San Francisco: Jossey-Bass Foundation.

Cocoran, T. (1995). *Transforming professional development for teachers: A guide for state policymakers.* Washington, DC: National Governors' Association.

Donahoe, T. (1993). Finding the way: Structure, time and culture in school improvement. *Phi Delta Kappan*, 75(4): 298–305.

Fullan, M. G. (1991). *The new meaning of educational change.* New York: Teachers College Press.

Fullan, M. G. (1999). *Change forces: The sequel.* London & Philadelphia: Falmer Press.

Geijsel, F., Sleegers, P., Van den Berg, R., et al. (2001). Conditions fostering the implementation of large-scale innovation programs in schools: Teachers' perspectives. *Educational Administration Quarterly*, 37(1), 130–66.

Giannakaki, M. S. (2005). Using mixed-methods to examine teachers' attitudes to educational change: The case of the Skills for Life strategy for improving adult literacy and numeracy skills in England. *Educational Research and Education*, 11(4), 323–48.

Hargreaves, A. (1988). Teaching quality: A sociological analysis. *Journal of Curriculum Studies*, 20, 211–31.

Hargreaves, A. (1998). The emotions of teaching and educational change. In A. Hargreaves, E. Lieberman, M. Fullan, & D. Hopkins (eds.), *International handbook of educational* change (pp. 558–70). Dordrecht, The Netherlands: Kluwer.

Kingston, P. (2006, November 21). Shock tactics. *The Guardian*. Retrieved November 26, 2006, from http://education.guardian.co.uk/print/0,,329637434–108283,00.html.

McNess, E., Broadfoot, P., & Osborn, M. (2003). Is the effective compromising the affective? *British Educational Research Journal*, 29(2), 243–57.

Reynes, P. (1990). Organisational commitment of teachers. In P. Reynes (ed.), *Teachers and their workplace: Commitment, performance, and productivity* (pp. 143–62). Newbury Park, CA: Sage.

Reynes, P. (1992). *Preliminary analysis of teacher organizational commitment: Implication for restructuring the workplace.* Washington, DC: Office of Educational Research and Improvement, U.S. Department of Education. .

Reynolds, D. (1997). Linking school effectiveness knowledge and school improvement practice. In A. Harris, N. Bennett, & M. Preedy (eds.), *Organizational effectiveness and improvement in education* (pp. 251–60). Buckingham & Philadelphia: Open University Press.

Sanders, R. (1999). Exploring obstacles to educational reform: Observations from "Finding a Way." *Professional Geographer*, 51(4), 578–85.

Tam, W. M., & Cheng, Y. C. (1996). A typology of primary-school environments: Synergetic, headless, mediocre and disengages. *Educational Management and Administration*, 24, 237–52.

Tsitouridou, M., & Vryzas, K. (2003). Early childhood teachers' attitudes towards computer and information technology: The case of Greece. *Information Technology in Childhood Education Annual*, 2003(1): 187–207.

Tsui, K. T., & Cheng, Y. C. (1999). School organizational health and teacher commitment: A contingency study with multi-level analysis. *Educational Research and Evaluation*, 5(3), 249–68.

Turnbull, B. (2002). Teacher participation and buy-in: Implications for school reform initiatives. *Learning Environment Research*, 5, 235–52.

Van den Berg, R., Vandenberghe, R., & Sleegers, P. (1999). Management of innovations from a cultural-individual perspective. *School Effectiveness and School Improvement*, 10, 321–51.

White, J., & Barber, M. (1997). *Perspectives on school effectiveness and school improvement.* London: Institute of Education, University of London.

8 The Impact of the "Skills for Life" Learning Infrastructure on Learners

A Summary of Methods and Findings

John Vorhaus, Ursula Howard, Greg Brooks, Ann-Marie Bathmaker, and Yvon Appleby

Note on authorship: This chapter was written and compiled by John Vorhaus and Ursula Howard from reports on the three strands of the Learners study. Ann-Marie Bathmaker wrote the report on the existing quantitative data strand, [1] *Greg Brooks is the principal author of the report on the new quantitative strand,*[2] *and Yvon Appleby is the principal author of the report on the qualitative strand.*[3]

INTRODUCTION

Context: Policy and Purpose

The idea for a longitudinal study of learners in England emerged in 2003 among the policy community responsible for Skills for Life (SfL), the government-led strategy launched in 2001 to address the literacy and numeracy levels of the adult population. Alarmingly low levels of literacy and numeracy in the United Kingdom had been revealed in the OECD survey of adult literacy, which was published in 1997 (Carey, Low, & Hansbro, 1997). The new Labour administration set up a working party to investigate the extent of the problem further: academics worked alongside policymakers and senior practitioners, including economists and educationalists, with understanding and experience of longitudinal studies and their ability to inform and generate policy development. The report of the working group chaired by Sir Claus Moser, *A Fresh Start for Literacy and Numeracy* (DfEE, 1999), recommended that the government establish a research centre as part of a national strategy to address adult literacy and numeracy.

When the National Research and Development Centre for adult literacy and numeracy (NRDC) was set up in 2002, longitudinal studies were at the heart of its efforts to gather and analyze evidence, theorize the field of adult literacy, develop practice and "inform, refresh and help take forward" the national strategy. In this sense, some longitudinal studies conducted by NRDC were also designed as evaluations of how policy interventions are introduced, and then develop and establish themselves in practice over time, exploring the extent and cumulative impact and effectiveness of policy to inform future developments.

THE PURPOSE OF THE STUDY AND ITS COMPANION STUDY ON TEACHERS

Two key NRDC longitudinal studies in particular had these multiple purposes. The first was a study of "The impact of SfL on Teachers and Trainers" (the Teachers Study), which began in 2002 (Cara & Litster, this volume). This was followed in 2003 by a study explicitly commissioned by the (then) Department for Education and Skills (DfES) in England as a sister study. This study of the "Impact of the SfL Learning Infrastructure on Learners" (The Learners Study) is the subject of this chapter. The SfL strategy developed a wide-ranging set of reforms to teacher, tutor, and trainer qualifications and professional development, setting ambitious targets to "professionalize" a service in which tutors were often not qualified teachers, and the use of volunteers underpinned what had been called a "Cinderella" service. The argument was that those adult learners with the greatest learning needs deserved the most knowledgeable and best-trained teachers.

With regard to learners and learning, the teacher education reforms were matched by the creation of a new set of standards, comprising five levels of achievement or learning outcomes: Entry levels 1, 2, and 3 and Levels 1 and 2, with Level 2 broadly equivalent to the UK (age sixteen) school-leaving certificate (the General Certificate of Secondary Education [GCSE] or the U.S. General Educational Development [GED] tests). National core curricula were introduced for literacy, numeracy, and English to Speakers of Other Languages (ESOL). To replace and simplify a plethora of qualifications and forms of recognition of achievement and accreditation, nationally recognized multiple-choice tests were introduced. Other developments included initial assessment guidance and tools, individual learning plans, and a range of materials to support teaching and learning. Together, these became known as the SfL learning infrastructure.

THE FOCUS OF THE LEARNERS STUDY

The Learners study focused primarily on the impact of this learning infrastructure on the experiences and achievements of adult literacy, numeracy, and ESOL learners. It also examined how the infrastructure was introduced and developed over time. The focus of SfL is both economic and social and gives particular attention to the government's "priority groups"—unemployed people, benefit claimants, prisoners, workers in low-skilled jobs, and other groups at risk of social exclusion.

The study comprised three strands of research activity, two quantitative and one qualitative, across the fields of literacy, language (ESOL), and numeracy (LLN). In the first quantitative strand, existing data from the Individualised Learner Records (ILRs) gathered by the national Learning and Skills Council (LSC) in England were analyzed. In the second quantitative strand, new data were gathered from two samples of learners in 2004/05 and 2005/06, using literacy and numeracy tests, attitude questionnaires, and Profiles of learners' background characteristics. In the qualitative strand, case studies were conducted in six areas of England, using in-depth interviews, observations, and focus groups.

The Learners and Teachers studies were designed to allow for triangulation as follows. The learners who participated in the new quantitative data strand of the Learners study were all students of teachers and trainers participating in the Teachers Study. The intention was to combine quantitative data from the two studies to explore the link between teacher characteristics, attitudes, and practices on the one hand and such learner outcomes as motivation, persistence, and achievement on the other. Qualitative comparisons would also be made between teachers' and learners' perceptions of, and attitudes towards, the various elements of the new learning infrastructure.

THE LONGITUDINAL NATURE OF THE LEARNERS STUDY

A longitudinal study follows the same individuals over time The study of teachers and managers of LLN programs, which followed a significant number of respondents through three sweeps covering nearly four years, was therefore clearly a longitudinal study. However, it was discovered early on that the Learners study in the quantitative strand could not be wholly longitudinal in the conventional sense of the term (see next section). The other two strands had to take account of the fact that adult LLN learners do not routinely study over long periods of sustained attendance. Attendance is episodic, flexible, and often short. The learner population is more difficult to track and hold in a study over time than a professional group such as teachers or college managers. These two strands, therefore, were longitudinal to the extent that we recontacted learners after a few months rather than any longer interval. In other words, the learners' study was longitudinal in two strands, in the sense of repeated investigation of the same individuals over time. However, all strands were able to offer significant insights as an evaluation of the impact of the implementation of the SfL policy over a four-year period.

In what follows we describe the methods used in each of the three strands, point up their strengths and limitations, and offer some concluding remarks. We also provide brief summaries of findings.

USING EXISTING QUANTITATIVE DATA

Introduction

The Learning and Skills Council's (LSC's) Individual Learner Records contain data on adult basic skills learners in England at three points: enrollment, completion, and achievement. Originally it was thought that it would be possible to estimate the progress which LLN learners make by using both pre- and postcourse assessments of their attainment. But the ILRs contain information on learners' achievements only at the *end* of a course, and no estimate of their attainment at the beginning. Without a pretest, no measure of progress was possible. Analysis of the ILRs continued, but limited to numbers of enrollments, courses completed, and qualifications obtained. However, it did prove possible to analyze trends over time in those areas between 2000/01 and 2004/05.

THE LSC DATA

The LSC gathers data on all learners participating in LSC-funded provision, for funding, monitoring, and audit purposes. The chief recipients of LSC funding are further education colleges. SfL courses form a small part of the overall data sets.

The ILR data used in this strand of the study were provided by the Learning and Skills Council to the University of Sheffield in October 2006 and deemed correct by the LSC on that date. The data covered the complete population of LSC-funded learners in England over a period of five academic years, from 2000/01 to 2004/05. At the time of working on this study, no further years of data were available for analysis. However, the available data covered the first years of the SfL strategy, and included 2004, the year of the first government target of 750,000 learners improving their skills.

ENROLLMENT, COMPLETION, AND ACHIEVEMENT OF LEARNING AIMS

Another key feature of the ILR data sets is that the entries comprise not the population of learners but of their learning aims. A learning aim is the goal that a learner is aiming to achieve at the outset of a program of learning. Each learner may have more than one learning aim. Data are collected at three key points: enrollment (learning aims taken up), completion (learning aims completed), and achievement (learning aims achieved).

The learning aims data were analyzed to show, first, trends and patterns in participation, completion, and achievement between 2000/01 and

2004/05. Secondly, the numbers for enrollment, completion, and achievement were used to create percentages showing: completion as a proportion of enrollment (completed/enrolled), achievement as a proportion of enrollment (achieved/enrolled), and achievement as a proportion of completions (achieved/completed).

FINDINGS: PARTICIPATION AND ACHIEVEMENT IN SfL

There was a considerable increase in overall figures for enrollment, completion, and achievement in SfL provision over the five years. Figure 8.1 shows the total number of learning aims taken up, completed, and achieved for each academic year from 2000/01 to 2004/05 (left-hand scale).

Enrollments more than doubled from 1,043,087 to 2,180,253, whilst achievements rose by an even greater percentage and almost tripled from 441,364 to 1,284,531. However, although the number of enrollments, completions, and achievements increased each year, there was a considerable difference every year between the number of enrollments in SfL provision and the number of completions, and again between the number of completions and the number of achievements recorded.

In fact, the line graphs show (right-hand scale) that the number of completions as a proportion of enrollments stayed almost the same throughout the period, rising from 71 percent in 2000/01 to 73 percent in 2004/05. At the same time, the proportion of enrollments that resulted in achievement

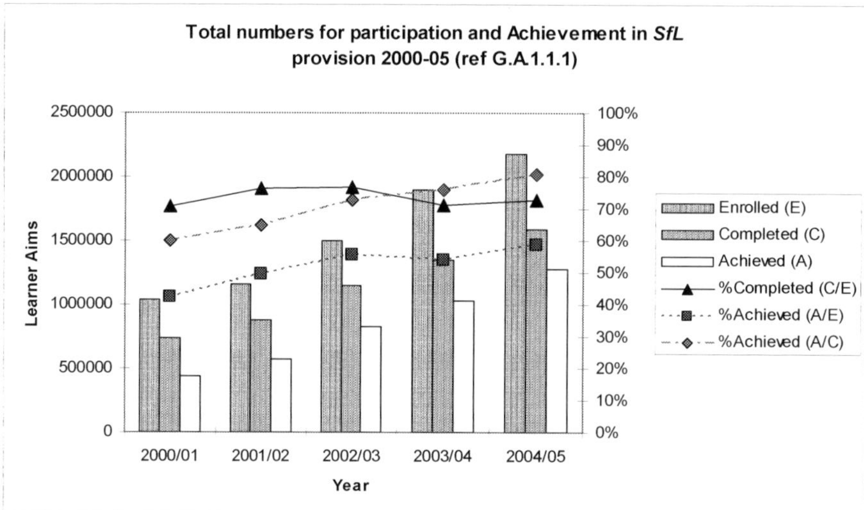

Figure 8.1 Total participation and achievement in SfL provision between 2000 and 2005.

was much lower, and whilst this figure rose, particularly in the first three years, from 42 percent of enrollments leading to achievement in 2000/01 to 59 percent leading to achievement in 2004/05, a 17 percentage point increase in total, this trend was flattening out in the last two years, rather than continuing to rise.

There were much higher rates of achievement where programs of learning were completed. The number of achievements as a proportion of completions rose steadily over the five-year period, by a total of just under twenty-one percentage points, and in 2004/05 the proportion of completions that resulted in achievement was 81 percent. The area where there was little improvement was in completion rates. At just over 70 percent throughout the period, the retention rate was high for adult learners participating in adult basic and key skills provision. At the same time, since completion rates would appear to have a direct impact on overall achievement rates, this is an important area for further investigation.

What level of qualification do learners achieve?

SfL provision is offered at three levels, starting with Entry Level, progressing to Level 1 and finally to Level 2 within the National Qualifications Framework for England. Level 2 is the equivalent of GCSEs at grades A*–C, which is considered to represent successful completion of qualifications by sixteen-year-olds at the end of compulsory schooling. Level 2 is also the first level of qualification defined in government policy on workforce skills as representing an adequate base level of skill in the context of a competitive, globalized economy (HM Treasury, 2006; DfES, 2005).

Figure 8.2 shows both the overall figures for achievement at these three different levels (shown as bars in the graph) and the proportion of all achievements by level (shown by the line graphs). The bar graphs show that overall achievements were highest for Level 1 qualifications throughout the five years. They were lowest for Entry Level at the start of the five-year period in 2000/01, but Entry Level figures increased substantially in 2002/03, overtaking the number of achievements at Level 2. From this year on, Level 2 achievements represented the lowest number for all three levels.

The line graphs show the proportion of all achievements by level. Here it becomes clear that the proportion of the total made up by Level 2 achievements fell between 2000/01 and 2002/03 and remained steady from then on at 17 percent of all achievements. The proportion of the total made up by Entry Level rose between 2000/01 and 2002/03 and fell for Level 1. By 2002/03 they shared an almost identical proportion of achievements at 41.6 percent for Entry Level and 41.8 percent for Level 1. The proportions then reduced for Entry Level, to 29.1 percent of the total by 2004/05 whilst for Level 1 they rose to 54.2 percent of the total in the same year.

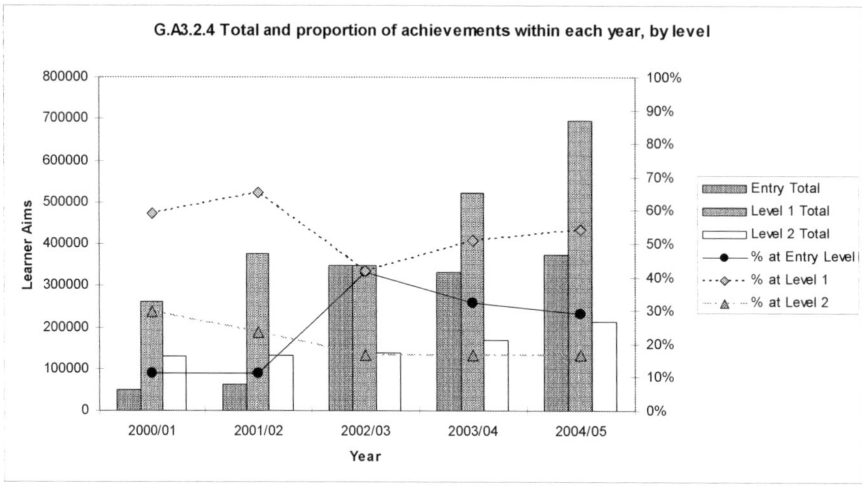

Figure 8.2 Comparison of achievements in SfL by level.

WHAT TYPE OF QUALIFICATION DO LEARNERS ACHIEVE?

As with other areas of education and training, there is a range of different qualifications which accredit achievement in the skills of literacy, numeracy, and ESOL. One of the things that SfL has done is to regulate which qualifications are funded as part of LSC provision. This has changed over the period under examination as qualifications have been revised over time and brought into line with SfL policy. ESOL qualifications were the last to be brought into line, and there is now a list of recognized qualifications provided by the LSC. The qualifications fall into three broad types: basic skills, key skills, and GCSEs. Since 2001, basic skills and key skills share the same national test. Key skills qualifications require a portfolio in addition to the achievement of the test. The GCSE is a quite separate qualification, representing the standard qualification in English and maths taken by school students at sixteen, and also available to adults, usually through a course at a further education college.

Figure 8.3 shows enrollments by type of qualification and graph four achievements by type of qualification.

Basic skills made up the highest number of all enrollments, with 504,050 enrollments in 2000/01 rising to 1,413,086 in 2004. Because these numbers increased so considerably, this also meant that the proportion represented by basic skills rose from just over 48 percent in 2000/01 to just under 65 percent in 2004/05.

The figures for achievement followed this pattern even more strongly (Figure 8.4).

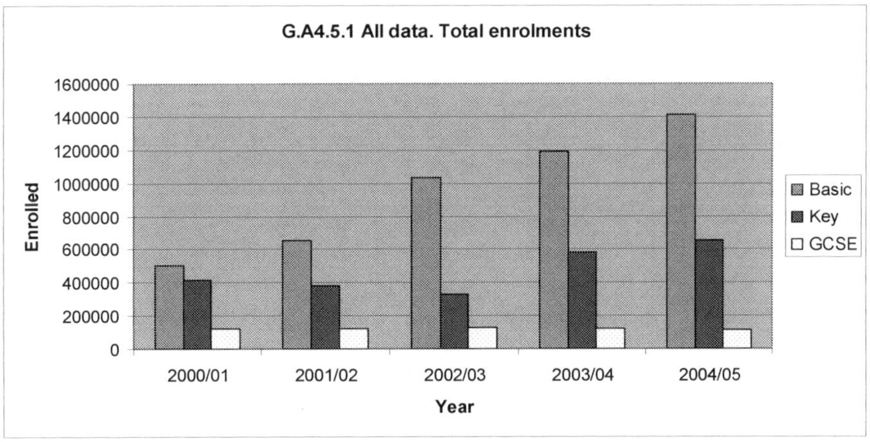

Figure 8.3 Enrollments in SfL by type of qualification.

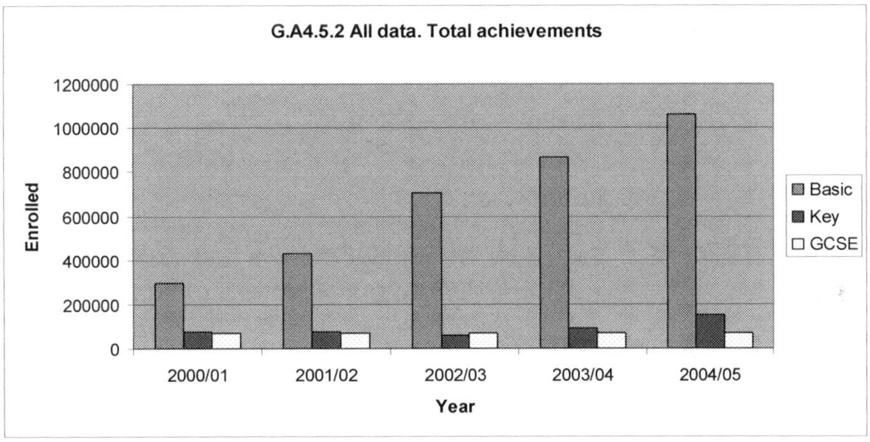

Figure 8.4 Achievements in SfL by type of qualification.

Figures for achievement in basic skills rose from 296,743 in 2000/01 to 1,062,409 in 2004/05. This meant that basic skills qualifications made up an even greater proportion of all achievements, rising from 67 percent of overall achievements in 2000/01 to 84 percent in 2002/03, and then remaining almost steady for the following two years. Over the same period, there was a small but steady decrease in the proportion of enrollments and achievements represented by GCSEs.

LIMITATIONS OF THE DATA

Although the LSC data sets are the most comprehensive available, there are limits on using them to measure progress towards the SfL targets and to evaluate progress and progression. This is not a criticism of the LSC Individualised Learner Record, for it was not set up to serve these purposes. However, these limits need to be taken account of when using LSC data to make claims about the SfL strategy.

Numbers in the ILR comprise "learning aims" rather than individuals, whilst the SfL targets are concerned with individual learners rather than learning aims. For example, the 2001 SfL target of 750,000 denotes 750,000 different learners; if a learner achieves in more than one basic skill, in literacy and numeracy, or in numeracy and ESOL, then that learner will only count once towards the target. Moreover, if a learner achieves at Level 1 and then moves on and successfully completes a qualification at Level 2, that learner will again only be counted once.

In order to calculate the number of individual learners whose achievements count towards the SfL target, a formula was devised which converts the number of learning aims into a figure representing the number of individual learners. In 2003/04, for example, the figure of 1,066,085 enrollments was translated into 240,758 learners, based on an average of 4.4 learning aims allocated to each learner. Over the period as a whole, an average of approximately four learning aims was allocated to each learner. These figures, however, are based on an earlier incomplete data set provided by the LSC in 2006. The full data set used for our final report did not permit identification of individual learners and the analysis was therefore confined to learning aims.

The data for the ILR are collected by providers and returned to the LSC; they are therefore dependent on whether providers interpret data fields correctly and collect data accurately. Both the LSC and practitioners in colleges expressed concerns about the robustness of the data.

There is a further limitation to record. It is not possible to speak with any certainty about the distance traveled by a learner, because there is no systematic assessment of the level of a learner's achievement on entry to a learning program. In other words, there was no direct evidence that achievement of a certificated outcome actually represented progress by an individual learner from a previous level of skill, as distinct from representing a level of skill the learner had started out with and barely moved from.

Nor is there accurate information about subsequent progress following the achievement of a stated learning aim. There are no unique numbers that follow learners wherever they go. Only learners who remain with the same provider can be easily tracked from one course or year to the next, and even then this often applies only to further education provision which is funded by the LSC.

A further area that goes unrecorded in the LSC database is progress in the broader sense of benefits gained by learners as a result of developing their basic skills. One gain repeatedly identified is the positive effect on self-image, confidence, and self-esteem (see Brooks, Davies, Duckett, Hutchison, Kendall, & Wilkin, 2001). This may be linked to a further benefit, the decision to continue with education and pursue further courses. In addition, family literacy programs have been found to enhance children's learning as well as that of adults, by encouraging and enabling adults to help children with language, literacy, and numeracy development and by encouraging them to become involved in their child's school (Brooks, Gorman, Harman, Hutchison, Kinder, Moor, & Wilkin, 1997; Brooks, Gorman, Harman, Hutchison, & Wilkin,1996; Hannon & Bird, 2004). None of these benefits can be analyzed using the ILRs.

NEW QUANTITATIVE DATA

Data Gathering

In this second, longitudinal, strand of the Learners Study, data were gathered from 1,649 adult literacy, ESOL, and numeracy learners in academic years 2004/05 and 2005/06. All completed a background characteristics profile, an attitudes questionnaire, and an assessment during a first interview near the beginning of the academic year. Those who agreed to be, and were, recontacted completed an attitudes questionnaire and an assessment during a second interview on average six months later.

Data from the Learner Profiles were analyzed to provide details of the samples of learners. Data from the questionnaires and assessments of those learners who were interviewed twice were used to estimate whether their skills and/or attitudes had improved.

INSTRUMENTS

The learners' background details were gathered via Learner Profiles. A common profile was used with literacy and ESOL learners, and a similar but slightly longer one with numeracy learners. Both profiles were devised by researchers working on other NRDC numeracy and literacy projects and were based on those used in those projects. At each stage (pre- and post-) numeracy learners completed an attitudes to numeracy questionnaire and a numeracy assessment; literacy and ESOL learners completed an attitudes to literacy questionnaire and one or two literacy assessments. The attitudes questionnaires were devised by researchers working on other NRDC numeracy and literacy projects and were based on questionnaires used in those projects.

The numeracy researchers also devised the numeracy assessment, which consisted of a subset of the numeracy items used in the SfL needs survey in 2002/03 (Williams, Clemens, Oleinikova, & Tarvin., 2003). The literacy assessment used, titled the "Go test," was that developed for NRDC by the National Foundation for Educational Research (NFER) in 2003 (see Evans, Waite, & Admasachew, this volume). It is a secure instrument, in the sense that its use is restricted to NRDC projects and it is not publicly available. The test consists of:

- A Locator booklet containing seven reading comprehension items based on a short passage contained in the booklet.
- Two parallel writing assessments (Forms 1 and 2), each based on a simulated glossy magazine.
- Two parallel reading assessment instruments (Forms 1 and 2) based on the simulated glossy magazines and each having both an easy version designed for learners at Entry level (version a) and a less easy version designed for learners at Levels 1 and 2 (version b).
- Each form of the writing assessment contains three tasks: writing two or three sentences giving an opinion about the magazine, writing a letter requesting further information about a career mentioned in it, and stating an opinion about an issue featured in it.

Each form of the reading assessment contains thirty questions based on passages in the relevant magazine. Some are common to the easy and less easy versions, while others are unique to one or the other. A table supplied by NFER and based on data from the piloting of the tests is used to convert raw scores on all four versions to a common 0–100 scale having a mean of 50 and a standard deviation of 10.

ALLOCATION OF LEARNERS

All literacy and ESOL learners in the study completed the Locator booklet and then the attitudes to literacy questionnaire. Approximately two-thirds were allocated to complete a reading assessment, and the rest a writing assessment. While each learner completed the attitudes questionnaire, the administrator marked the Locator booklet and then allocated those who were to complete a reading assessment to either the easy or the less easy version, based on their score on the Locator booklet.

However, ESOL learners whose scores on the Locator booklet would have indicated allocation to version (a) of the reading assessment were not in fact asked to attempt it. It was felt that for many the level of demand would have been inappropriate, and it would have been impossible to establish whether low scores on it were due to lack of understanding of the reading passages or to lack of fluency in spoken English. Also, some learners who were said to be below Entry Level 1 were not asked to attempt the reading assessment. No such exemption was applied for the writing assessments or to the (b) versions of the reading assessments: all those allocated to complete these were asked to do so, whether English was their first or an additional language.

In the first interview, literacy and ESOL learners were allocated in equal numbers to complete Form 1 or Form 2 of the reading or writing assessment (though, as stated earlier, some ESOL learners did not complete a reading assessment). Then at the second interview each learner who was seen again completed the other form. This allocation cut across those described above.

DATA ANALYSIS

The assessment results for numeracy, reading, and writing were processed by different routes:

- The numeracy instrument contained all objective items requiring only a marking key to score, with no judgments about correctness of answers being required of the markers. These tests were therefore scored by the organization which conducted the interviews.
- The reading test forms were collected by interviewers and sent to NFER for scoring, using the detailed marking guide supplied with the assessment instrument by NFER, supplemented by experience in the Effective practice in reading project (Brooks, Burton, Cole, & Szczerbiński 2007a, 2007b). The raw scores were then converted to scaled scores for analysis.
- The writing scripts were collected by interviewers and sent to the Learning and Skills Network (LSN) for scoring. Each script was marked independently by two members of a small panel using a detailed guide based on that supplied with the assessment instrument by NFER, supplemented by LSN colleagues' experience of using it in other NRDC projects. When all the scores were available, small corrections were made to the scores of a couple of markers, for either severity or leniency, and the two scores for each script for each occasion were averaged.

All three forms of data were analyzed at the University of Sheffield.

FINDINGS

Returners and Nonreturners

In order to have confidence that the results for learners who took the assessments at both stages ("returners") are reliable, and that any gains they made were not due solely to their being different from the nonreturners to start with, it is necessary to check that the two groups did not differ significantly on their preassessment scores. This check was conducted separately for numeracy, reading, and writing scores: in all cases there was no statistically significant difference. It was concluded that the returners were sufficiently representative of the full original samples.[4]

EFFECT SIZES

An effect size is a statistical measure of the impact of a program based on the gain, that is, the difference between a group of learners' average scores at pre- and posttest (here, the first and second interviews), and four purposes is standardized by dividing by the standard deviation (10) producing values in the range of 0 to1. Ideally, calculation of this statistic also requires data from a control or comparison group of learners who did not receive the teaching. In this study this was impossible: there were no such learners. Alternatively, where a standardized test is used, data from the standardization exercise can be used, on the assumption that the people who provided those data can be treated as an "unseen" control group. In this study this was possible for the scaled scores on the reading assessment, which was standardized by NFER during development, but not possible for the numeracy and writing assessments or for the raw scores for reading.

The usual rule of thumb for interpreting effect sizes is that those below 0.25 are very small and possibly not educationally significant; those between 0.25 and 0.5 are small; those between 0.5 and 0.8 are medium; and those above 0.8 are large.

NUMERACY

Pre- and postscores on the numeracy assessment were available for 239 learners (56 percent of those pretested) across the two cohorts. The average scores are given in Table 8.1. A statistical test showed that *the gain was highly statistically significant* ($p < 0.001$). However, statistical tests of differences in average gain between subgroups (e.g., men/women; those with English as first/additional language) showed that none of those differences were statistically significant.

The average gain of 3.3 score points was equivalent to about 5 percent of the maximum score of 60. The pretest average score would put a learner with this score towards the upper end of Entry Level 3, and the posttest average score would put a learner with that score just over the threshold into Level 1. Though modest, the average gain therefore seems worthwhile.

Table 8.1 Numeracy Learners' Average Scores (N = 239)

Pre-		*Post-*		*Gain*	
Average	(SD)	Average	(SD)	Average	(SD)
22.9	(11.0)	26.2	(12.4)	3.3	(8.1)

READING

The analyses and interpretations reported in this section are based as appropriate on either the learners' raw scores or the scaled scores and National Qualifications Framework (NQF) levels derived from them by using the NFER conversion tables. Analysis for the full sample is based on the scaled scores because using raw scores for this would have failed to account for the fact that a particular score on the easier, (a), versions of the assessment indicates less competent reading than the same score on the less easy, (b), versions. However, interpretation against NQF levels can only be based on the raw scores, and those are therefore presented for the (a) and (b) forms of the assessment separately.

Pre- and postscaled scores on the reading assessment were available for 186 literacy learners (65 percent of the 284 pretested on reading) and 135 ESOL learners (55 percent of the 245 pretested on reading) across the two cohorts. The average scores are shown in Tables 8.2 and 8.3, respectively.

Table 8.2 shows that the literacy learners' pre- and postaverage scaled scores were both slightly above the national average of 50.0. A statistical test showed that *the gain was highly significant* ($p < 0.001$). However, statistical tests of differences in average gain between subgroups showed that none of those differences were statistically significant for literacy learners. Also, the effect size was not very large because the average gain was just four score points. This does not contradict the statistical significance of the gain—a small difference can be highly statistically significant if the sample size is large, as here.

More important is the question of what the gain implies for these learners' ability in reading. This can be judged better from the average raw

Table 8.2 Literacy Learners' Average Scaled Scores for Reading (N = 186)

Pre-		*Post-*		*Gain*		*Effect Size*
Average	(SD)	Average	(SD)	Average	(SD)	
51.4	(23.2)	55.4	(20.6)	4.0	(13.5)	0.4 (small)

Table 8.3 Literacy Learners' Average Raw Scores for Reading (N = 186)

Assessment Form	N	*Pre-*		*Post-*		*Gain*	
		Average	*(SD)*	*Average*	*(SD)*	*Average*	*(SD)*
b (less easy)	128	22.9	(8.4)	24.2	(7.3)	1.3	(5.8)
a (easy)	57	19.7	(8.1)	22.3	(6.7)	2.6	(4.8)

scores, which are shown in Table 8.3 separately for learners who took the two forms of the assessment.

Both gains in raw score were statistically significant (Form b, p = 0.015; Form a, p < 0.001). The average gains of 1.3 and 2.6 raw score points were equivalent to about 4 percent and 9 percent of the maximum score of 30, respectively. The pretest average raw score for the literacy learners who took Form b would put a learner with this score quite near to the upper end of Level 1, and the Form b posttest average score would put a learner with that score almost at the top of Level 1. The pretest average raw score for the literacy learners who took Form a would put a learner with this score towards the upper end of Entry Level 3, and the Form a posttest average score would put a learner with that score just over the threshold into Level 1. Though modest, the average gains therefore seem worthwhile.

ESOL

It is possible to show both scaled and raw reading scores for the ESOL learners in one table (Table 8.4) because they all took the same form of the assessment, the (b) Form, as explained on page 210.

The pre- and postaverage scaled scores were both well below the national average of 50.0; indeed, the preaverage score was almost two standard deviations below the norm, and therefore within the bottom 5 percent of the distribution. Statistical tests showed that *both measures of the gain were highly significant* (p < 0.001). Statistical tests of differences in gain between subgroups showed that only one such difference was statistically significant: ESOL learners aged sixteen to nineteen made significantly more progress than other age groups. The average gain of 2.7 raw score points was equivalent to about 9 percent of the maximum score of 30. The pretest average raw score for these ESOL learners would put a learner with this score near to the upper end of Entry Level 3, and the posttest average score would put a learner with that score just into Level 1. Again, this suggests that the modest gain was worthwhile.

Table 8.4 ESOL Learners' Average Scaled Aand Raw Scores for Reading (N = 123)

	Pre-		*Post-*		*Gain*		*Effect Size*
	Average	*(SD)*	*Average*	*(SD)*	*Average*	*(SD)*	
Scaled	32.7	(15.3)	38.9	(18.3)	6.1	(14.3)	0.61 (medium)
Raw	11.4	(6.7)	14.1	(8.2)	2.7	(6.0)	n/a

WRITING

Across the two cohorts, ninety-six literacy and 119 ESOL learners completed a writing assessment both pre- and post-. The average scores for the literacy learners are shown in Table 8.5. (For the ESOL learners, see following.)

The maximum score is twenty-nine, so the average pre- and post-scores were just above half marks. The gain was only about 2.5 percent of the maximum score, and was not statistically significant ($p = 0.15$). Only one subgroup difference was statistically significant: white learners made significantly less progress than learners of other ethnicities (it must be remembered that all the learners in this analysis had English as their first language).

The writing results for ESOL learners are shown in Table 8.6.

Here both average scores were slightly below half marks. The gain was about 4 percent of the maximum score, and *was statistically significant* ($p = 0.01$). None of the subgroup differences were statistically significant.

The ESOL and literacy learners' writing scores were compared statistically. Both pre- and post-, the literacy learners' average scores were significantly higher than the ESOL learners' scores ($p < 0.001$), but the difference in the gains was not statistically significant ($p = 0.52$).

SUMMARY

There were modest but worthwhile gains in numeracy for numeracy learners, in reading for both literacy learners and those taking courses in ESOL (English for Speakers of Other Languages), and in writing for ESOL learn-

Table 8.5 Literacy Learners' Average Writing Scores (N = 96)

Pre-		*Post-*		*Gain*	
Average	(SD)	Average	(SD)	Average	(SD)
17.1	(8.3)	17.8	(7.3)	0.7	(4.9)

Table 8.6 ESOL Learners' Average Writing Scores (N = 115)

Pre-		*Post-*		*Gain*	
Average	(SD)	Average	(SD)	Average	(SD)
13.2	(7.4)	14.4	(6.8)	1.2	(4.8)

ers (but not literacy learners). Literacy learners' overall attitudes to literacy and self-confidence improved, as did ESOL learners' self-confidence.

There were only two statistically significant differences in amount of progress made by different subgroups—ESOL learners aged sixteen to nineteen made significantly more progress in reading than other age groups, and white learners made significantly less progress in writing than learners of other ethnicities. On the whole, this means that provision was working equally well for many different groups of learners, and few groups were being left behind.

QUALITATIVE DATA

Aims

This strand explored the impact of the SfL infrastructure on the experiences of learners, focusing on the government's priority groups—unemployed people, benefit claimants, prisoners, workers in low skilled jobs, and other groups at risk of social exclusion. (The other two strands were unable to focus on special groups because they are not distinguished in the relevant data sets.) This strand also sought to assess how policy had been implemented and adapted through the infrastructure to achieve the aims of the SfL strategy.

The qualitative strand team looked at the experiences of learners, asking about their reasons for attendance, what they were hoping to learn from their programs, the importance of qualifications to their motivation, and what their longer-term learning and life aspirations were.

METHOD

To understand the impact on all SfL stakeholders, including managers, coordinators, tutors, and learners, a 'staircase' model of policy implementation was used (Saunders, 2006; see Appendix 8.1). This model also enabled understanding of the roles and responsibilities of each stakeholder: each group on the staircase was involved in implementing the policy, through the infrastructure, which shaped the provision the learners engaged with. Those highest up the staircase engaged more directly with policy, those in the middle were interpreting and implementing the strategy, and those further down worked within the infrastructure as it affected their experiences of teaching and learning. The staircase illustrates an interrelated and dynamic relationship, and one that, whilst largely invisible to learners, nevertheless shapes their experience of the infrastructure.

A total of 562 people took part in this strand: 416 adult literacy, numeracy, and ESOL learners, and 146 managers, coordinators, and tutors. The six research sites (West Yorkshire, Birmingham and Solihull, Cheshire and Warrington, Wiltshire and Swindon, West London, and Cumbria and Northumberland) were chosen to represent geographically diverse areas. The sites provided a variety of examples of policy implementation, often relating to local issues and constraints such as organizing and supporting provision in rural or urban settings. Together they contributed to a general picture of how a national policy became adapted and embedded over time.

In the first phase an initial sweep of the six sites was conducted to obtain baseline information from local coordinators and managers. A minimum of twenty managers and coordinators were interviewed in each site by a project researcher. These interviewees were selected to represent, where possible, a range of provision that included work-based, prison, college, voluntary, and community contexts, and which also covered literacy, numeracy, and ESOL. The interviews took, on average, an hour and were transcribed and used to generate initial findings, enabling identification of questions and priorities for the next stage of the research. Approximately 20 percent of the managers and coordinators in each site were reinterviewed in the second phase, by phone, e-mail, or face-to-face.

The first interviews asked informants details of their teaching or managing background, their current roles and experience of the infrastructure, and their views on the strengths and constraints brought about by the implementation of the infrastructure. The second interviews tracked changes in individual career trajectories, enquiring into changes in experience of provision and views about changes in the impact of the infrastructure on learners.

Interviews were conducted with 416 learners (a minimum of fifty per site) across all six sites. Learners were also selected to represent, where possible, a cross section of provision across the same range of contexts, to cover literacy, numeracy, and ESOL and to represent learners of different ages working at different levels from Entry Level 1 to Level 2. Most interviews lasted on average of about twenty minutes and were recorded and transcribed. Where learners were unhappy with this, comprehensive notes were made of their responses. Interviews were carried out both individually and in small groups, depending upon people's confidence and language skills. Many ESOL learners, and learners from a mental health group, opted for a group interview as offering more language support and being less intimidating than a one-to-one interview. Other learners, such as those in prison, had little choice other than to be interviewed on their own because of the constraints of their environment. Learners were asked why they had joined the class, what they were getting from it, whether it was helping them in their life, whether it was affecting their confidence, and what they wanted to do after the course finished.

Second interviews were carried out with approximately 20 percent of learners per site: an overall total of 135 across all six sites. These interviews were conducted mainly by phone, but where communication and confidence were an issue they were conducted face-to-face (as, for example, with groups of ESOL learners and those attending a mental health group). The second interviews asked learners about their current activities, what they had taken from the course they attended, whether it had helped them, whether they felt they had made progress, and whether they had achieved what they set out to. Information on learners who were not reinterviewed was collected from tutors and other students.

Early emerging findings suggested that 90 percent of learners were satisfied with their experience of learning. This was at variance with some of the difficulties in implementing the infrastructure being described in the first interviews by managers, coordinators, and tutors. In discussion forums it emerged that tutors were significant in making the infrastructure work, in spite of the difficulties, largely owing to a commitment to their learners.

FINDINGS

Positive benefits were found to include an increased national profile given to literacy, language, and numeracy, new resources, increased access to learning opportunities, an increased sense of professionalism, and improved quality of provision. Many learners and some employers valued national tests and certificates, the national curriculum was valued by many teachers (particularly those new to the profession), there were clear routes for progression for teaching and learning, and there were new areas of entitlement for learning.

The direct consequences of policy changes were thought to include a greater emphasis on embedded provision, a strengthening of vocational learning by increasing the age of basic skills entitlement, a greater emphasis on work-based learning, and flexible methods of delivery. The indirect consequences were thought to include some movement away from provision initially characterized by its diversity, including generous funding for nontraditional learners, towards provision that is in some cases less diverse; some decrease in flexibility in the curriculum and in methods of delivery, particularly for "at-risk" and "hard-to-reach" groups; and difficulties in funding time required by slower learners to achieve their learning aims.

FINAL THOUGHTS

Together with its companion, the Teachers Study, the Learners Study has contributed to an evaluation that charts in detail the impact of a government strategy on all the most affected stakeholders, and it has permitted the tracking of variations in those changes over time. It is a data set that enables us to specify in detail the profile of SfL cohorts and their changing composition. We can analyze enrollment, participation, and achievement rates for the SfL population as a whole; for literacy, numeracy, and ESOL learners and by age, ethnicity, and gender; and we can interrogate the data to draw out relationships between these and related variables. From the two longitudinal strands, we have evidence on learners' progress in literacy, ESOL, and numeracy and on changes in their confidence and attitudes to learning. And our qualitative study gives evidence of how a wide range of individuals and groups, variously placed within the SfL context, have been affected by this unprecedented national strategy. We learned a good deal about the implementation of policy and the dependence of policy on its reception and adaptation by principal stakeholders.

At the same time, difficulties with the design of the study are plain to see: previously existing data were found wanting in more than one respect—the lack of a unique learner identifier in the LSC data set being of special significance—and the learners were sometimes distinguished by their moving in and out of provision and by being in other ways hard to keep track of. The terrain of SfL does not always provide the most congenial conditions for longitudinal study. Indeed, only two of the three strands adopted this method, and these only partially. Integrating and analyzing data across the three strands thus presents a challenge for several reasons: not only are we working with qualitative and quantitative data; the data are arrived at by different methods, and both existing data sets and those we created ourselves are far from complete. Any research program must overcome difficulties of this sort and adapt itself to the varying demands and limitations that go with widely varying SfL territories.

The question arises: how far is the context of SfL as whole, including what we know about patterns of enrollment and participation, and the variously endowed contexts of learning, congenial to a longitudinal research design of the kind intended? Perhaps some contexts—further education colleges—are better suited than others—prisons, adult ,and community learning organizations? Or should we rather say that some contexts—including the larger, securely funded and more elaborate organizations—are better equipped to serve as sites for a longitudinal quantitative study; whilst others, being smaller, less well resourced, with learners whose learning patterns are less predictable, are better suited to case study—biographical interview, ethnographic investigation, and so on? If we accept the first suggestion we concede that there is a limit on how far such a design can be applied across the context of SfL as a whole. The second suggestion acknowledges that different learning contexts may require different approaches to data gathering, raising the question of how the data thus gathered are integrated and analyzed. And that question is especially pressing if, as in the present case, respondents selected for qualitative analysis are not a subsample of those making up the quantitative sample. In fact, in one of our four studies that used the LSC data set, the unit of analysis was not even individual learners but their earning aims.

These are the challenges presented by longitudinal research on SfL learners. But none of this should take away from the fact that the project we have reported on remains a uniquely extensive and illuminating study of a major national initiative to improve literacy, language, and numeracy, and one that should serve as an exceptional evidence base for future generations of researchers.

APPENDIX 8.1

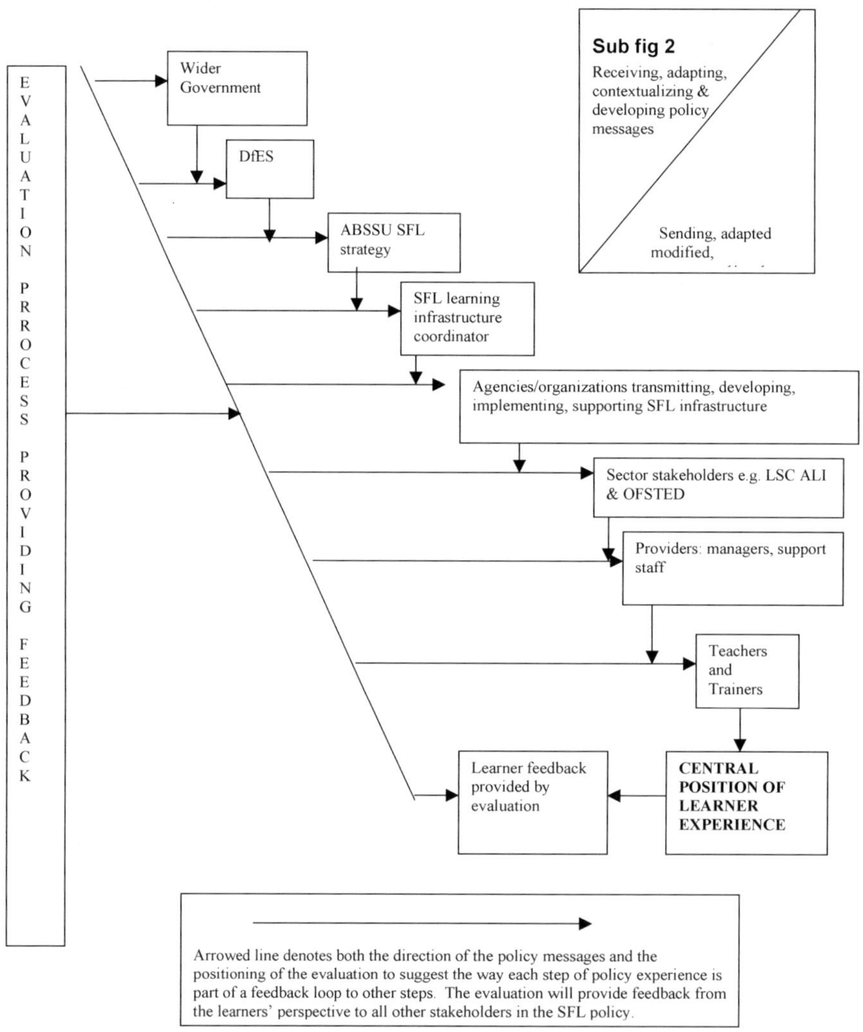

Figure 8.A.1 Implementation staircase: evolution of the SfL infrastructure.

NOTES

1. With statistical help from Sammy Rashid and Mark Pilling (University of Sheffield).
2. Of which he and Mark Pilling are the authors, with statistical help from Sammy Rashid.
3. Of which she, Ann-Marie Bathmaker, Pam Cole (University of Sheffield), and Linda Pearce (University of Lancaster) are the authors.
4. Note on statistical significance: The statistical significance of the difference between two scores is expressed as a p (probability) value. For example, $p < 0.05$ means that the result could have occurred by chance only once in 20 times at most (0.05 = 5% = 1 in 20). This is usually taken as the minimum level of statistical significance that is considered reliable. The other two levels often indicated are $p < 0.01$ (less than 1 chance in 100 that the result was due to chance), often described as "very significant," and $p < 0.001$ (less than 1 chance in 1,000 that the result was due to chance), often described as "highly significant."

REFERENCES

Brooks, G., Burton, M., Cole, P., & Szczerbiński, M. (2007a). *Effective teaching and learning: Reading.* London: NRDC.

Brooks, G., Burton, M., Cole, P., & Szczerbiński, M. (2007b). *Effective teaching and learning: Reading. Summary report.* London: NRDC.

Brooks, G., Davies, R., Duckett, L., et al. (2001). *Progress in adult literacy: Do learners learn?* London: Basic Skills Agency.

Brooks, G., Gorman, T. P., Harman, J., et al. (1997). *Family literacy lasts: The NFER follow-up study of the Basic Skills Agency's Demonstration Programs.* London: Basic Skills Agency.

Brooks, G., Gorman, T. P., Harman, J., Hutchison, D., & Wilkin, A. (1996). *Family literacy works: The NFER evaluation of the Basic Skills Agency's Family Literacy Demonstration Programmes.* London: Basic Skills Agency.

Carey, S., Low, S., & Hansbro, J. (1997). Adult literacy in Britain, J0024129, C15, 9/97, 5673. London: The Stationery Office.

Department for Education and Employment. (1999). *Improving literacy and numeracy: A fresh start.* Report of the Working Group chaired by Sir Claus Moser. London: HMSO.

Department for Education and Skills. (2005). *14–19 education and skills.* White Paper. London: HMSO.

Hannon, P., & Bird, V. (2004). Family literacy in England: Theory, practice, research and policy. In B. H. Wasik (ed.), *Handbook of family literacy.* Mahwah, NJ: Lawrence Erlbaum Associates.

HM Treasury. (2006). *Prosperity for all in the global economy—world-class skills: Final report.* London: HMSO.

Saunders, M. (2006). The 'presence' of evaluation theory and practice in educational and social development: towards and inclusive approach. *London Review of Education,* 4(2), 197–215.

Williams, J., with Clemens, S., Oleinikova, K., & Tarvin, K. (2003). *The Skills for Life survey: A national needs and impact survey of literacy, numeracy and ICT skills.* Norwich, UK: The Stationery Office. Retrieved 02/14/08 from http://www.dfes.gov.uk/research/data/uploadfiles/RR490.pdf.

Part III

The Impact of Policy and Programs

9 Outcomes for Basic Skills Learners

A Four-Year Longitudinal Study

Hilary Metcalf and Pamela Meadows

BACKGROUND

The Skills for Life (SfL) program in England is designed to improve the literacy, numeracy, and language skills of adults and of young people (aged sixteen to seventeen) who have left full-time education and who do not have a Level 2[1] qualification in English or mathematics. The evaluation, commissioned by the Department for Education and Skills (now the Department for Innovation, Universities and Skills), was undertaken jointly by the National Institute of Economic and Social Research and BMRB (British Market Research Bureau) Social Research. It has followed through a group of literacy and numeracy learners who were taking courses in further-education colleges leading to qualifications in 2002 or 2003 and compared their progress on a range of outcomes with a comparison group of people who also had low or no literacy or numeracy qualifications but who did not take an SfL[2] course. Both groups were interviewed for the first time when the learners were on their courses and were reinterviewed three times at roughly annual intervals. In the final year of the project there was a qualitative study of a small subsample of the learners in order to gain a deeper understanding of what the course meant to them.

The evaluation did not attempt to measure any skills that learners gained as a result of their attendance on a literacy or numeracy course. Rather, the aim was to assess the impact of participation in SfL courses on wider economic and social outcomes. A central objective was to estimate the rate of return to literacy and numeracy courses. This approach means that any identified outcomes cannot be attributed to the development of literacy or numeracy skills. Rather, they reflect the impact of going through the process of attending a literacy or numeracy course in a college.

The sample

A sample of SfL courses was drawn from the Learning and Skills Council list of courses funded under the program, supplemented by a sample of learners whose courses were funded by the European Social Fund (ESF).

Within each course learners were sampled at random (sometimes with the assistance of course tutors), although young people under the age of nineteen (who represent the majority of college learners) were undersampled in order to produce adequate numbers of adult learners. In the event, only sixty-two ESF learners were included in the study. The learners generally had their first face-to-face interview at the college where they were studying, although they could choose to be interviewed at home if they preferred this. All follow-up interviews (and all comparison sample interviews) took place in the respondent's home. The original sample contained 2,012 learners.

The comparison group were selected from three sources:

- Respondents to the National Adult Learning Survey (NALS) (a representative sample of all adults) who were identified as having few or no qualifications and likely to have literacy or numeracy problems.
- Respondents to the SfL Survey (SfL) (again a representative sample of all adults) who had taken a test which revealed that they had literacy or numeracy problems.
- A booster sample selected from within the wider population based on their lack of qualifications and a short screening test.

The original comparison sample comprised 2,255 nonlearners. This sample was not representative of the population with literacy or numeracy difficulties. Rather, the aim was to have a sample that shared as far as possible the characteristics of the learner sample. In particular, the incidence of literacy and numeracy difficulties is known to be higher among older age groups (Organisation for Economic Co-operation and Development [OECD], 2000), but older people are underrepresented among adult learners and therefore in the comparison group. The numbers interviewed from each of the sources were: NALS 429, SfL 759, freefind 1,067. Further details of the sampling can be found in Metcalf and Meadows (2005).

ATTRITION

Longitudinal surveys are prone to attrition. People may refuse at the end of the first interview to be contacted again in future. Those who are willing to be contacted may subsequently refuse, or they may have moved and be untraceable. In this study most of the attrition took place between Wave 1 and Wave 2. Only just over half those who were interviewed at Wave 1 were interviewed a second time. In each subsequent wave around two-third of those who had been interviewed at the previous wave were reinterviewed. In general the main personal characteristics of those interviewed in Waves 2 to 4 (including age, gender, family status, number of children, literacy and numeracy test scores and self-assessed competence at Wave 1, school-leaving age, and experience of school) did not differ significantly from those interviewed in Wave 1. There

Table 9.1 Number of Interviews Conducted at Each Wave

	Wave 1	*Wave 2*	*Wave 3*	*Wave 4*
Learners	2,012	1,094	682	461
Proportion of Wave 1 sample		54%	34%	23%
Proportion of sample in previous wave		54%	62%	68%
Nonlearners	2,255	1,122	674	468
Proportion of Wave 1 sample		50%	30%	21%
Proportion of sample in previous wave		50%	61%	69%

were two characteristics where there were some differences: people of white ethnic origin were more likely to respond to the follow-up surveys, so that they were 80 percent of the Wave 4 sample but only 75 percent of the Wave 1 sample. Those who were economically active at Wave 1 were slightly less likely to respond to subsequent waves. At Wave 1, 49 percent were economically inactive, while among Wave 4 respondents 53 percent had been economically inactive at Wave 1. Taking all characteristics together, however, the differences were relatively minor, and are unlikely to be a source of bias.

CHARACTERISTICS OF LEARNERS

These SfL learners were not typical of the general population. They were predominantly female (58 percent), single (65 percent) and white (75 percent—proportionally less than in the general population). They were spread across the age range, although more concentrated in the nineteen to forty-four age group, with an average age of thirty-seven. Those from ethnic minorities were spread fairly equally across the main ethnic minority groups. A minority had children aged under sixteen (29 percent), although many of those with children were lone parents (11 percent of all SfL learners).

Illness or disability was very common amongst SfL learners. Forty-one percent reported a long-standing illness or disability at Wave 1. Previous research has found that people with literacy skills below Level 1 are twice as likely to be in poor or very poor health as people with skills at Level 2 or above (Bynner & Parsons, 2006). The survey did not identify whether respondents had learning difficulties, including dyslexia. However, as many of the respondents in the qualitative research referred to being dyslexic or appeared to have learning difficulties, it seems likely that learning difficulties were common amongst SfL learners. Previous research has found a high incidence of dyslexia among people with literacy skills below Level 1 (Bynner & Parsons, 2006).

MEASURING OUTCOMES

Overall, the approach taken in the evaluation of SfL was essentially a form of "black-box" evaluation: the impact on literacy and numeracy was not measured, but final outcomes such as employment, health, and social participation were. As with all black-box evaluations, this means that the process by which adult literacy and numeracy training might affect wider outcomes is unknown. It might be through improvements in skills, but it might be through improvements in motivation or self-esteem.

The study was designed not only to enable the measurement of what had happened to the learners but also to provide an assessment of what would have happened to them if they had not taken their literacy or numeracy courses (the counterfactual). Establishing the counterfactual is important, since some indicators (earnings, for example) are liable to change over time as a result of increased experience and changes in the overall economic environment. Others (such as moving into paid employment) are likely to be affected by life-course changes. Without a counterfactual, there is a risk of wrongly attributing movements in these indicators to having taken a literacy or numeracy course. It is therefore important to compare the outcomes for those who have done courses with the outcomes for others in similar circumstances.

In terms of methodology, there are two broad ways of achieving this: an experimental approach, where those who express an interest in taking a literacy or numeracy course are randomly assigned to doing or not doing the course, and a nonexperimental approach, where people who have the same general problem but who have not done a course are used to represent the outcomes that would have occurred for the learners if they had not done their course.

In the present study an experimental design was not feasible, not least because all adults without a Level 2 qualification are entitled to take courses leading to a qualification at that level, and there would have been no practical means of preventing those assigned to any control group from taking courses at other colleges. This meant that a comparison group approach was the only feasible option.

The standard sample selection problem associated with using a comparison group is that any observed differences in outcomes will partly reflect the true impact of doing a literacy or numeracy course and partly reflect other differences between those who do courses and those who do not. People who choose to go on literacy or numeracy courses are not a random selection from the population. They have differing characteristics and (often) experience different circumstances. Thus, one could observe different proportions in paid employment for those who had done a course relative to those who had not even if the courses themselves had no impact, perhaps because individuals who take courses had better health or motivation, for example (Heckman, Smith, & Clements, 2007).

In this study the learners group was representative of learners on courses leading to qualifications in either literacy or numeracy. However, the comparison group of nonlearners cannot be regarded as representative of any underlying population. Rather, they are just comparators for the learners group.

Although the comparison group was selected to match as far as possible the characteristics of the learners, when the learners group was compared with the nonlearners after their first interview, there were some differences between the two groups, which might lead to bias in the measurement of outcomes. The analysis therefore used two statistical methods to reduce the impact of bias: propensity score matching and difference-in-differences.

PROPENSITY SCORE MATCHING

Instead of trying to directly match the learners and the comparison group on a range of characteristics such as age, education, number of children, local labor market conditions, and so on, propensity score matching develops a single composite indicator, and the matching is done on the basis of that indicator (Rosenbaum & Rubin, 1983). In this study, as in many studies of the outcomes of training and other labor market interventions, the composite indicator is the probability of taking part in a SfL course.

The propensity score is calculated using logistic regression for each individual in both samples taken together based on their observed characteristics before starting the course. The details of the equation used can be found in Metcalf, Meadows, Rolfe, Dhudwar, Coleman, Wapshott, & Carpenter (2007). The Wave 1 variables used were:

- age
- number and ages of children
- gender
- marital status
- ethnicity
- previous educational attainment
- literacy and numeracy levels achieved at Wave 1
- whether English is first language
- whether English is spoken at home
- self-perceived problems with literacy and numeracy
- employment status
- indicators of attitudes towards education and training
- index of employment commitment
- local unemployment rate as at November 2002

All those who had been interviewed as "nonlearners," in that they had not been sampled at colleges, but who in fact had been doing an SfL course, were excluded from the propensity score matching process. There were some 280 of these, which reduced the size of the comparison sample. The reason for the exclusion is that the evaluation was trying to measure the impact of doing an SfL course, and thus the comparators should all be people who have not done a course.

The average estimated propensity to take part in learning for the learners was 0.651 and for the nonlearners it was 0.635. Of course, the actual propensity to take part in a course for the learners is one and for the nonlearners it is zero. The purpose of the propensity score matching is not to predict these outcomes. Rather, it is to find comparators whose characteristics mean that their propensity to take part in a course is similar to that of the learners. The matching process had a significant impact on the measured bias in the case of almost all the variables involved. For the majority of variables the estimated reduction in bias was over 80 percent. Once the propensity score was calculated, the learners were matched with nonlearners based on their predicted propensity to take a course. The aggregate comparisons were then based on the sum of the outcomes for these matched pairs.

DIFFERENCE-IN-DIFFERENCES

A potential source of bias from using a comparison group approach lies in the unobserved characteristics of individuals. To offset this, a difference-in-differences approach was used. For both learners and nonlearners the value of an outcome indicator at Waves 2, 3 or 4 was compared with the value of the same indicator at Wave 1. Thus, the comparison between the two groups is of the average *change* in an outcome indicator for those individuals compared with the initial value of that indicator for the same individuals. The effect of this is to take account of bias as a result of any initial differences in the starting points of the two groups (Heckman, Smith, & Clements, 1997).

WHAT THE IMPACT EVALUATION MEASURED

The changes between Wave 1 and subsequent waves were measured for a range of outcomes. These related to paid work and to personal, social, and family issues. Differences between the two groups were treated as statistically significant if $p < .05$. All outcomes which appeared to be only marginally insignificant at this level were reestimated using $p < .1$, but only one outcome was significant at this level and insignificant at $p < .05$. This is reported explicitly following. Those that were significant at $p < .01$ are also noted.

For most indicators there were no statistically significant differences between the learners and the comparison group. In part this reflects the

Table 9.2 Summary of Statistically Significant Outcomes

Outcome	*No. of waves*
Commitment to education and training	3
Self-perceived literacy improvement	3
Self-perceived numeracy improvement	3
Self-esteem	2
Take-home pay	1
Satisfaction with promotion prospects	1

fact that some potential outcomes were only measured for a minority of respondents, which meant that the number of people for whom the outcome measure was available was too small to detect significant effects. For example, the survey asked a series of questions about reading stories to children and helping them in various ways. However, these questions could only be answered by survey respondents who had children living in their household. At Wave 1 only 29 percent of respondents were living in households where there were dependent children.

Table 9.2 summarizes the outcomes where there were statistically significant differences between the learners and the comparison group in at least one of the three follow-up waves. More detailed information is provided in Table 9.3.

SELF-ESTEEM

At Wave 2 and Wave 4 there were statistically significant differences in self-esteem between the learners and the nonlearners. However, the difference at Wave 3 was not significant. Self-esteem was measured in this study using the shortened and simplified version of the Rosenberg self-esteem scale (Rosenberg, 1965) developed by Smith, McVie, Woodward, Shute, Flint, and McAra. (2001). The Rosenberg SES is the most widely used measure of self-esteem in the social sciences and has been extensively validated (Blascovich & Tomaka, 1993). It is based on ten questions which are completed on paper by the respondent. Because the sample in this study had literacy problems, all questions had to be asked by an interviewer, and the use of only six questions both reflected limitations on the time available for the interview and also the risk of upsetting or embarrassing respondents. However, offsetting this is the disadvantage that as the modified scale has not yet been widely used, it is not as well validated as more commonly used scales. Respondents were asked to agree or disagree with a series of six questions, each of which had five possible answers. In each case high self-esteem was scored with a value of 5 and low self-esteem with a value of The six questions were:

1. I like myself.
2. I often wish I was someone else.
3. I am able to do things well.
4. I don't think much of myself.
5. There are some good things about me.
6. There are lots of things about myself I would like to change.

Self-esteem was therefore measured on a scale of 6 to 30, and changes between Wave 1 and subsequent waves on a scale of–24 to +24.

The average wave 1 value of the scale was 23.6 for the learners and 22.8 for the nonlearners. At Wave 2 learners had increased their self-esteem by 0.5 points compared with Wave 1, while nonlearners had reduced theirs by 0.1 points. This difference was statistically significant. At Wave 3 the change in the learners' self-esteem was similar to that at Wave 2 (an increase of 0.5 points) while that for the nonlearners was similar to the Wave 1 level, so that overall their self-esteem had improved slightly between the two waves. The difference between the two groups was not statistically significant. At Wave 4 learners' self-esteem was 0.7 points higher than it had been at Wave 1 (and therefore higher than at Waves 2 and 3), while for the nonlearners self-esteem was 0.5 points lower than it had been at wave 1 (and therefore below the level found at Waves 2 and 3). Although not statistically significant at $p < .05$, it was significant at $p < .1$.

In the qualitative interviews respondents frequently referred to improvements in their confidence. This is particularly noteworthy given that interviewees were not asked if they felt more confident but were responding to questions about whether the course had changed them and whether anything had been different because of the course

Respondents identified the following factors as affecting their confidence and self esteem:

- having qualifications, often for the first time;
- skills, in literacy, numeracy and Information Technology (IT);
- improved skills in communication, particularly among those for whom English is not their first language;
- being less dependent on others for everyday tasks;
- feeling able to use Information Technology, and particularly the Internet, for activities such as information searches, family tree research, and eBay;
- being able to write formal letters and deal with officialdom.

Activity in these areas had the effect of making participants feel less like outsiders. Many talked of the effects of their newfound confidence in

Table 9.3 Key Outcomes: Difference in Differences After Propensity Score Matching

	wave 2				*wave 3*				*wave 4*			
	Difference	*Sig*	N = learners	N = *non-learners*	*Difference*	*Sig*	N = learners	N = *non-learners*	*Difference*	*Sig*	N = learners	N = *non-learners*
Labor market and work												
Change in employment status (net increase/decrease in proportion of sample)	–0.7%		1020	1022	1.4%		623	616	1.7%		425	435
Change in take-home pay (nonemployed = 0)	–£268		1020	1022	£1,272	**	590	587	–£27		424	435
Change in satisfaction with promotion prospects (scale–4 to +4)	–0.17		998	969	0.3	**	201	285	0.14		117	197
Self-esteem												
Change in self-esteem (scale–24 to +24)	0.63	**	1020	1022	0.48		623	616	1.18	*	419	432
Education and training												
Change in commitment to education and training (scale–16 to +16)	0.94	**	1020	1022	0.63	**	623	616	0.43	**	390	426

(continued)

Table 9.3 Key Outcomes: Difference in Differences After Propensity Score Matching (continued)

	wave 2				*wave 3*				*wave 4*			
	Differ-ence	*Sig*	N = learners	N = *non-learners*	*Differ-ence*	*Sig*	N = learners	N = *non-learners*	*Differ-ence*	*Sig*	N = learners	N = *non-learners*
Self-perceived literacy and numeracy **+**												
net proportion reporting self-perceived improvement in literacy in past year	34.5%	**	1020	1022	19.2%	**	623	616	18.6%	***	425	435
Net proportion reporting self-perceived improvement in numeracy in past year	39.2%	**	1020	1022	27.1%	**	623	616	14.6%	**	425	435

*significant at 10% level; **significant at 5% level; ***significant at 1% level.

+Outcomes based on reported levels not difference in difference.

going out to work, progressing further in education, using public transport, dealing with shopping transactions, and mixing socially. Success in these areas appeared to have a cumulative effect, leading to more confidence and greater independence and participation.

LABOR MARKET OUTCOMES

Adults with poor literacy and numeracy levels have significantly worse labor market outcomes than those who have better literacy and numeracy levels. Only around half of British adults with poor literacy skills are in paid employment, while around three-quarters of those with good literacy skills are (Leitch, 2005). More generally, unemployment, temporary work, and chequered job histories are more common among those with poor basic skills and earnings are lower (Bynner & Parsons,1997; Dearden, McIntosh, Myck, & Vignoles, 2000; Ekinsmyth & Bynner, 1994; McIntosh & Vignoles, 2001). The impact analysis assessed changes in several types of labor market outcomes:

- Net change in the proportion of the sample in paid employment
- Change in net earnings
- Change in satisfaction with pay
- Change in employment commitment
- Change in satisfaction with promotion prospects

As shown in Table 9.3, the net changes in the proportion of the two samples in paid employment were not statistically significant at any of the three follow-up waves. However, the difference between the two groups grew between Wave 2 and Wave 3 and between Wave 3 and Wave 4. At Wave 2 there had been a growth of 3.7 percentage points in the proportion of learners in paid work, but a growth of 4.3 percentage points for the nonlearners. By Wave 3 learners' employment rates had increased by 5.6 percentage points while the nonlearners had remained at 3.8 percentage points above the Wave 1 level. At Wave 4 the proportion of learners in paid employment had increased by 6.7 percentage points, while the nonlearners had increased by 5 percentage points.

The fact that there were not statistically significant changes in employment rates does not necessarily imply that there are no employment benefits from taking a literacy or numeracy course. However, if there are benefits, they appear to take more than three years to emerge. Moreover, the qualitative interviews indicated that some were still pursuing education to achieve their long-term employment goals.

The qualitative interviews found examples of a direct link between taking a basic skills course and subsequently finding work. A number of respondents had read job advertisements in their local paper for the first time or had been able to complete an application form. For these individuals it is clear that the course made a difference both to the process of finding paid work and to the kind of work they felt able to apply for. However, members of the comparison group also moved into paid work without receiving this support.

The analysis also considered changes over time in the net annual earnings of the learners group and the comparison group. Comparisons of earnings can often be problematic when only the positive earnings of those in paid work, and not the zero earnings of those who are not working, can be observed (Heckman, 1979). In this study, those with zero earnings at either Wave 1 or a follow-up wave have been included in the comparisons. Moreover, because the impact is based on the average of the increases and reductions in earnings experienced by the same individuals over time, and not on the relationship between individual earnings and other characteristics, the problem of unobserved characteristics, which besets cross-sectional studies of earnings, is also avoided.

Among those in paid employment there were both high and low earners, and there were movements into and out of paid work. This means that the changes for some individuals are quite large, which in turn means that the 95 percent confidence intervals around the average changes are also large. Between Wave 1 and Wave 2 the learners had an average increase in net annual earnings of £124 while the nonlearners had a fall of £144. The net difference was therefore £268, but the 95 percent confidence interval was between–£489 and +£1,327, so the difference was not statistically significant. Between Wave 1 and Wave 3 the learners had increased their net annual earnings by £558 while the nonlearners had had a reduction of £713. The difference between the two groups was £1,272 and the 95 percent confidence interval was between £576 and £2,114. Thus, this estimate was just statistically significant. Between Wave 1 and Wave 4 the learners' increase in earnings was similar to that found in Wave 3 (£560) while the nonlearners' gain was similar to that of the learners (£542). The difference (£27 a year) was not statistically significant.

Across the three follow-up waves the variations in earnings within the two groups were many times greater than the differences between them. The one statistically significant difference observed at Wave 3 has a very large confidence interval, and it is only just possible to rule out the difference being zero, as it is at Waves 2 and 4. Other UK studies also find that the rate of return to adult basic skills courses appears to be zero (Blanden, Draca, & McIntosh, 2005; Dearden et al., 2000).

Only one of the changes in other work-related outcomes was significantly different between the learners and the nonlearners. At Wave 3 learners were more likely than nonlearners to believe that their promotion prospects had improved. However, this did not persist, and by Wave 4 the difference was not statistically significant.

Whilst SfL was not found to improve perceived promotion prospects, the qualitative research suggested that for some former learners it may have improved job competence and increased the scope of work undertaken. A number of qualitative interviewees said that they were now able to do their job better as a result of their new skills. These included a health care assistant who no longer avoided answering the telephone for fear of being asked to write messages and a hotel kitchen porter who was promoted to head steward. This supports the evidence of employers that if employees' basic skills improve, some aspects of their work, such as customer service and their ability to take advantage of training opportunities, also improve (Ananiadou, Jenkins, & Wolf, 2003).

PERCEPTIONS OF IMPROVEMENTS IN LITERACY AND NUMERACY

As Table 9.3 shows, there were large and statistically significant differences in the extent to which former learners and comparison group members perceived that their literacy and numeracy skills had improved. When they were first interviewed, respondents were not asked what they thought had happened to their literacy and numeracy over the previous year. These questions were only asked in the three follow-up waves. This meant that there was not a baseline from which to calculate differences in self-perceived literacy and numeracy improvements. Thus, only the perceptions themselves can be reported. As a consequence there may be some bias in the measurement of these impacts. Nevertheless, the sizes of the effects are very large and appear to be persistent.

Respondents were asked whether they felt that their literacy and numeracy had shown (a) definite improvement, (b) some improvement, or (c) no improvement in the last year. Thus, there were three possible answers for literacy and three for numeracy scored at 2, 1, and 0, respectively. Between Wave 2 and Wave 3, learners' assessments of their progress in numeracy over the previous year was an average of 0.83 points, while the nonlearners' assessments of their progress over the previous year was only 0.37. This difference is statistically significant. Between Wave 3 and Wave 4, learners again recorded an average gain of 0.69 points over the previous year, while nonlearners recorded an average of 0.42 points. Again, this difference was statistically significant.

For literacy the perceived improvements were larger for both groups: 1.1 for the learners and 0.7 for the nonlearners between Waves 2 and 3, and 1.06 and 0.7 between Waves 3 and 4. Again, both these results were statistically significant .

An alternative approach to measuring perceived literacy and numeracy outcomes is to examine the proportion reporting any improvement. At Wave 2, 59 percent of learners and 20 percent of nonlearners reported an improvement in their numeracy over the previous year. At Wave 3, 56 percent of learners and 29 percent of nonlearners reported further improvement. At Wave 4, 47 percent of learners and 32 percent of nonlearners reported further improvement. In all three years the difference was statistically significant.

More than three-quarters of learners (76 percent) reported at Wave 2 that their literacy had improved over the previous year, as did 42 percent of nonlearners. At Wave 3, 68 percent of learners but only 19 percent of nonlearners reported that their literacy had improved. At Wave 4, the proportion of learners reporting a continued improvement in literacy was similar to that in Wave 3 (69 percent), but the proportion of nonlearners was 50 percent. These figures too are statistically significant at all three follow-up waves ($p < .05$ Waves 2 and 3; $p < .01$ Wave 4).

Many respondents in the qualitative research described changes in the use of literacy or numeracy. Improvements in literacy appeared to have the greatest impact on respondents' day-to-day activities. Many referred to reading novels and newspapers as a result of newly acquired reading skills, expanded vocabulary, and understanding of punctuation.

With regard to writing skills, respondents reported particular improvements in spelling and punctuation, which had been problematic for many before taking courses. Writing skills were put to use in everyday life, for activities such as writing shopping lists or letters. Many commented on their new ability to write formal letters, for example, to their child's teacher or to the local authority. This was a great source of satisfaction and pride. With regard to numeracy, improvements were reported largely with mental maths. Improved skills at addition, subtraction, and calculating percentages were reported to be particularly useful when shopping. However, whilst respondents reported that their literacy and numeracy skills had improved by taking the basic skills courses, in many cases this had taken some time. This was particularly apparent for those with learning disabilities and with dyslexia.

EDUCATION AND TRAINING

There were large and statistically significant differences in the change in commitment to education and training between the former SfL learners and the comparison group at each of the three follow-up waves, although the effect was strongest in Wave 2 and declined thereafter.

Respondents were asked for the extent to which they agreed with the following four statements:

1. You are more likely to get a better job if you do some learning.
2. Learning new things makes you more confident.
3. To get a job, who you know is more important than what you know.
4. Getting qualifications takes too much effort.

Each question had five possible answers. These were added together to produce a scale of commitment to education and training whose value could vary between 4 and 20 (with 20 being the maximum level of commitment). The change between waves could therefore vary from–16 to +16.

At Wave 1 learners had an average value of the scale of 14.8 and the nonlearners had a value of 14.3. For the learners there was little difference in their commitment to education and training at Wave 2 compared with Wave 1, but the nonlearners had a fall of 0.9 points. At Wave 3 the earners had an average increase in the value of their index of 0.9, while the comparison group members had an increase of 0.3. At Wave 4 learners' commitment to education and training was lower than it had been at Wave 1 (by 0.2 points), but the nonlearners' commitment had fallen by a larger amount (by 0.7 points). These differences were statistically significant at all three follow-up waves ($p < .05$ for Waves 2 and 3 and $p < .01$ for Wave 4).

The qualitative research suggested that commitment to learning from participants in basic skills courses was influenced by their enjoyment of the learning process and the courses themselves, the skills gained and benefits gained from putting them into practice, and the possible employment benefits of gaining further qualifications.

Another significant difference between the two groups was the extent to which they had gone on to take a new education or training course. The analysis excluded all comparison group members who had been taking a course at Wave 1. Thus, any course members of the comparison group were taking at subsequent waves was a new course. For the learners, taking a course at Wave 2, 3, or 4 might represent either a continuation of an existing course or the start of a new course.

Overall, across the three follow-up waves more than three-quarters of learners (78 percent) but only 19 percent of nonlearners had taken a new course. This difference is statistically significant ($p < .01$). Among the learners, about half those who had taken a new course at some point were taking a new course at Wave 4 (40 percent). Among the nonlearners the proportion taking a new course at Wave 4 was the same as the proportion who had taken a new course at any wave. Again, the difference in the proportion taking a new course at Wave 4 was statistically significant ($p < .01$).

This finding is particularly important in that it provides evidence to support the hypothesis that the main value of improving literacy and numeracy skills for adults is that it opens the way to further learning opportunities, some of which might help to develop skills which are directly relevant to work. Literacy and numeracy, while they are useful in themselves, may have their real value in enabling people to progress to further education and training (including training provided by employers) and to develop skills that more directly influence their employment and earnings prospects. The fact that a large majority of the learners were using their literacy and numeracy courses as a stepping-stone to other learning may be part of the explanation for the fact that there were no differences in employment outcomes between the learners and the comparison group at any of the three follow-up waves.

The qualitative research found respondents continuing with education courses for a number of reasons. In some cases, they had planned progression to further courses from the outset but had needed to improve their basic skills first. Others had less specific goals but had progressed to further courses to improve their job prospects. It was also apparent that some continued in learning because attending college had become an important part of their lives. These included participants with learning disabilities and individuals with caring responsibilities. It was also apparent that for some individuals with learning disabilities or with dyslexia, progress had been very slow: in a number of cases, participation in basic skills courses predated the start of the research in 2002 and progression to Level 2 was still not in sight.

SOME ISSUES ARISING FROM THE RESEARCH

SfL learners' lives had been circumscribed and their quality of life diminished by lack of basic skills. There were many ways in which their SfL course had enabled them to lead a fuller life: to travel unaccompanied, to budget and manage their own finances, to write letters, to go shopping. SfL also appeared to have reduced isolation. These benefits were not measured in the impact analysis and few would have been quantifiable in cost-benefit terms. However, this does not mean that they are neither of benefit nor of economic benefit.

Perhaps the most important conclusion of the research is that for those with poor literacy or numeracy skills, who often have other disadvantages such as poor health or learning difficulties, it can take a long time (longer than the three years in this study) before they are able to move into paid employment. Literacy and numeracy alone rarely deliver the skills needed for the workplace, but without literacy and numeracy skills people are unable to do the kind of mainstream education and training courses which do enhance employability.

NOTES

1. In the English qualification system Level 2 equates to a grade A–D in the GCSE examination taken by most sixteen-year-olds.
2. Although English for Speakers of Other Language courses are part of the SfL program, they were excluded from the evaluation because of the practical difficulties of interviewing the learners.

REFERENCES

Ananiadou, K., Jenkins, A., & Wolf, A. (2003). *The benefits to employers of raising workforce basic skills levels: A review of the literature.* London: NRDC

Blanden, J., Draca, M., & McIntosh, S. (2005). *The Economic and social returns to FE colleges: A report for the Department of Education and Skills,* http://www.dfes.gov.uk/furthereducation/index.cfm?fuseaction=content.view&CategoryID=20&ContentID=19.

Blascovich, J., & Tomaka, J. (1993). Measures of self-esteem. In J. P. Robinson, P. R. Shaver, & L. R. Wrightsman (eds.), *Measures of personality and social psychological attitudes* (pp. 115–60; 3rd ed.). Ann Arbor, MI: Institute for Social Research.

Bynner, J., & Parsons, S. (2006). *New light on literacy and numeracy.* London: NRDC.

Bynner, J., & Parsons, S. (1997). *It doesn't get any better: The impact of poor basic skills on the lives of 37-year-olds.* London: Basic Skills Agency.

Dearden, L., McIntosh, L., Myck, M., et al. (2000). *The returns to academic, vocational and basic skills in Britain.* DfEE research report no. 192. Nottingham, UK: DfEE.

Ekinsmyth, C., & Bynner, J. (1994). *The basic skills of young adults.* London: Basic Skills Agency.

Heckman, J. (1979). Sample selection bias as a specification error. *Econometrica,* 47(1), 153–61.

Heckman, J., Smith, J., & Clements, N. (1997). Making the most out of social program evaluations and social experiments: Accounting for heterogeneity in program impacts. *Review of Economic Studies,* 64(4), 487–535.

Leitch Review. (2005). *Skills in the UK: The long-term challenge. Interim report.* London: The Stationery Office

McIntosh, S., & Vignoles, A. (2001). Measuring and assessing the impact of basic skills on labour market outcomes. *Oxford Economic Papers,* 53(3), 453–81.

Metcalf, H., Meadows, P., Rolfe, H., et al. (2007). *Evaluation of the impact of* SfL *learning: Longitudinal survey of learners, final report.* London: DIUS.

Metcalf, H., & Meadows, P. (2005). *Evaluation of the Impact of* SfL *Learning: Report on Wave 1.* London: DfES, http://www.dfes.gov.uk/research/data/uploadfiles/RW50.pdf.

OECD. (2000). *Literacy in the information age: Final report of the international adult literacy survey.* Paris: OECD.

Rosenbaum, P. R., & Rubin, D. B. (1983). The central role of the propensity score in observational studies for causal effects. *Biometrika,* 70(1), 41–55.

Rosenberg, M. (1965). *Society and the adolescent self-image.* Princeton, NJ: Princeton University Press.

Smith D. J., McVie, S., Woodward, R., et al. (2001). *The Edinburgh study of youth transitions and crime: Key findings at age 12 and 13,* http://www.law.ed.ac.uk/cls/esytc/findingsreport.htm.

10 Enhancing "Skills for Life"? Workplace Learning and Adult Basic Skills

Karen Evans, Edmund Waite, and Lul Admasachew

The educational and labor market policy context in England has generated an unprecedented level of state support for workplace basic skills programs as part of the national Skills for Life (SfL) strategy. Substantial government funding, through the channels of the Learning and Skills Councils (LSCs), Regional Development Agencies (RDAs), and Trade Union Learning Fund (ULF), has given rise to a wide range of literacy, numeracy, and ESOL provision across all sectors of the economy and public sector. These include discrete literacy, numeracy, and ESOL courses in the workplace, SfL embedded in IT, SfL embedded in vocational and job-specific training as well as "LearnDirect" SfL courses undertaken in online learning centers in the workplace. The absence of evidence about the effects on learners and organizations of participation in such provision underlined the need for large-scale longitudinal research into "Adult Basic Skills and Workplace Learning." This research, part of the United Kingdom's Economic and Social Research Council's Teaching and Learning Research Program (TLRP) and cosponsored by the National Research and Development Centre for Adult Literacy and Numeracy (NRDC), has provided a framework within which the realities of adult basic skills learning accessed through the workplace can be explored longitudinally.

The project "Adult Basic Skills and Workplace Learning":

- aims to assess the effects on individuals and on organizations of engagement in workplace basic skills programs.
- is designed longitudinally to gain longer-term perspectives and deeper insights into both the trajectories of learners and the features of the organizations and workplaces than are possible through short term evaluations.

Workplace basic skills courses successfully implemented under the SfL strategy have typically provided a standard, initial thirty hours of instruction in or near the actual work site; have focused predominantly on literacy; are often built around the use of computers ('Laptops and Literacy' is a typical title); and use teaching material that is generalist rather than directly related to occupations.[1] The research is asking about what happens to the employees that may be related to their learning experiences, and what happens in the company/organization that may be related to the existence of the learning program. The research extends to 2008 and is therefore at the stage of interim findings. Our objective is to identify when and how workplace programs are effective in improving adults' basic skills, as well as their effects on other life-course variables (employment stability, promotion, enrollment in further educational programs); and to examine the impact on enterprises of sponsoring such programs, in terms of potential improvements in productivity and changed attitudes or commitment to the organization. Analyses combine quantitative analysis using a range of outcomes (measured progress in basic skills, employment, attitude inventory changes, etc.) with qualitative analyses of transcribed interviews and in-depth studies carried out in selected organizations. The underpinning conceptual and theoretical framework draws on economics (especially human capital theory), sociology (including social capital and sociocultural theories), management theory, and psychology.

THE USE OF MIXED METHODS IN THE "ADULT BASIC SKILLS AND WORKPLACE LEARNING" PROJECT

The "Adult Basic Skills and Workplace Learning" project employs structured interviews with 564 learners (as well as managers and tutors) from fifty-five organizations in a variety of sectors. Interviews were conducted at three times, T1 (2004), T2 (2005/6), and T3 (2007) over a five-year period. Each learner was also assessed early in the course using an assessment tool that has been especially designed to take account of small changes in literacy development. We also conducted in-depth interviews with a subsample of sixty-four learners from ten of the sites. In upholding the potential for qualitative and quantitative methods to have "complementary strengths and non-overlapping weaknesses" (Brewer & Hunter, 1989: 17) and opposing the concept of an *intrinsic* link between a particular method and epistemological worldview, our approach to mixed methods complies broadly with what Tashakkori and Teddlie (1998) have termed methodological "pragmatism." The rationale for employing in-depth interviews, in addition to structured questionnaires, includes the need to collect more detailed, narrative-based biographical data, to allow for the analysis of more subjective aspects of the learning experience (e.g., effect of previous learning experience, motivation/confidence, stigma surrounding poor basic skills, social capital, etc.), and to pursue the broader organizational context in terms of organizational case studies. The in-depth interviews will also add to the longitudinal dimension of the research insofar as they ask the respondents to think about past, present, and future in their learning experiences and use of skills at work, and will be followed up with a further in-depth interview in the final stages of the research.

The in-depth interviews allow for more detailed exploration of how "basic skills," particularly those centered on literacy, are actually employed in differing contexts. This allows us to take issue with assumptions underpinning SfL discourses about the existence of large basic skills deficiencies that inevitably impact negatively on performance at work. Instead, we pursue a more nuanced approach that illustrates the diverse range of techniques that are employed in literacy practices, whilst highlighting those cases where skills deficiencies exist. The respondent-led "emic" feature of in-depth interviews also allows for the exploration of attitudinal changes that cannot be so easily explored within the confines of a structured questionnaire. Finally, in-depth interviews with learners, managers, and tutors at selected sites allow for the exploration of key organizational processes that impact on the usages of literacy in different contexts. These include the impact of an "audit culture" manifested in increasing report writing, the effects of delegating responsibility to lower level employees, and a general trend towards textualization of the workplace (Scheeres, 2004).

The qualitative aspects of the project benefit from the quantitative insofar as we are able to consider the findings from the in-depth sample against the broader sample of structured interviews both in terms of learner profile as well as key themes and trends identified from in-depth interviews. Table 10.1 shows how we are able to benchmark the employees whose cases are used in this chapter to exemplify important findings against the wider data set.[2] The quantitative aspects of the project benefit from the more detailed exploration of attitudinal issues, as well as more detailed exploration of organizational issues that are pursued through in-depth interviews.

The employment of the Go literacy measure[3] provides us with an independent measure of literacy which can track small changes of literacy level over time. In-depth interviews then allow us to explore the individuals' own perceptions of their literacy skills, including the question of whether they are coping/struggling in the workplace. The importance of these various research instruments in allowing for a more complete understanding of literacy practices in the workplace is underlined in the cases that follow, where independent assessments of measured literacy level and insights into how the employee manages to carry out the job combine to circumvent incorrect assumptions that can stem from overreliance on only one form of evidence.

PARTICIPATION IN WORKPLACE BASIC SKILLS PROGRAMS

Of the 564 participants surveyed, almost two-thirds were male with an average age of just over 40. Almost all were in permanent full-time employment at the time of the first interview. The average length of employment with the

Table 10.1 Employee Data Benchmarked Against the Full Sample

	Kathleen Croft	*Victoria Appiah*	*Tracy Beaumont*	*Bill Williams*	*Trevor Woodford*	*Faiza Anwar*	*Melanie Taylor*	*Abdul Nazif*	*Bob Murphey*	*Fareen Ahmed*	*Laila Rahman*	*Overall Sample*
Gender	F	F	F	M	M	F	F	M	M	M	F	65% Male3 5% Female
Age at first interview	50	54	38	45	22	27	36	33	42	35	52	Mean age: 42 18–24 3.5% 25–34 23% 35–44 32% 45–54 24% 55–64 15% 65 + 1.5%
Age at leaving school	16	16	16	15	16	16	16	16	16	15	16	Mean: 17.2 Mode: 16 Median: 16
Qualifications on leaving school	CSEs including English and Maths	None	None	5 CSEs (not Grade 1)	None	D/K	GCSEs	none	none	Indian A level	none	None: 55% Some qualification: 44%
Literacy Level	Level 2 or above	Level 1	Level 1	L2 or above	Level 1	L2 or above	L2 or above	Level 1	Entry3	L1	Entry3	Below E2: 5.5% E2: 3.4% E3: 19.6 L1: 45.6 L2 or above: 25.7
Likes/Dislikes Job (on a scale of 1-7 with 1=dislike, 7=like a lot)	7	7	7	5	6	5	6	7	7	5	7	5.8

(continued)

Table 10.1 Employee Data Benchmarked Against the Full Sample

	Kathleen Croft	*Victoria Appiah*	*Tracy Beaumont*	*Bill Williams*	*Trevor Woodford*	*Faiza Anwar*	*Melanie Taylor*	*Abdul Nazif*	*Bob Murphey*	*Fareen Ahmed*	*Laila Rahman*	*Overall Sample*
Likes/Dislikes the Course (on a scale of 1-7 with 1=dislike, 7=like a lot)	5	7	7	7	6	Missing	7	6	7	5	7	6.7
Likelihood of doing a further course at work	Not very likely	very likely	very likely	Very Likely	n/k	Very likely	D/K	n/k	Quite likely	Very likely	Quite likely	Not at all likely 4.6% Not likely 4.9% Quite likely 22.4% Very likely 55.6 D/K-not sure: 12.5%
Likelihood of doing a further course outside of work	Very likely	not at all likely	Not at all likely	Not at all likely	Quite Likely	Quite Likely	Very likely (currently doing course)	Quite likely	Quite likely	Not at all likely	Not at all likely	Not at all likely 23.3% Not likely 16.5% Quite likely 29.2% Very likely 17.4% D/K-not sure 6.8%
ELLI growth orientation/ challenge	3.5	3.8	3.1	3.1	2.6	3.1	3.2	3.4	3.5	2.6	3.6	Mean: 3.0
ELLI dependence/fragility	1.9	2.2	1.7	2.0	2.4	2.2	2.2	2.67	2.8	2.2	2.5	Mean: 2.3
ELLI imagination/creativity	3.3	3.4	2.7	1.5	2.0	2.9	3.5	3.3	3.0	1.9	3.1	Mean: 2.6

*Mean age is 42. (18–24, 3.5%; 25–34, 23%; 35–44, 32%; 45–54, 24%; 55–64, 15%; 65 + 1.5%).

current employer was almost eight years. Fifty-four percent had left full-time education with no qualifications; 14 percent were qualified to Level 3 or above when they left FT education; 23 percent acquired further qualifications after leaving formal education; and 3 percent acquired further qualifications at Level 3 or above after leaving FT education. Basic skills program participants in the full sample were 35 percent ESOL; in the follow-up sample, 25 percent ESOL.[4] Our overall rate for successful recontact is just under 60 percent.[5]

Two intertwined themes have emerged from qualitative analysis of the sixty-four in-depth cases, undertaken in parallel with quantitative analysis of the survey results. The first theme is the diversity of motivations and outcomes reported by employees, many of these personal, some directly related to the job. The second theme is the shifting attitudes towards learning apparent in many of the participants. These themes, taken together, point to the potential of workplace programs to respond to shifting attitudes and their capacity to compensate for previously negative educational experiences.

The themes emerging from qualitative analysis were consistent with broad patterns found in the survey results. Both structured and in-depth interviews have indicated a complex set of factors behind learners' engagement in courses, beyond merely the wish to develop job-specific skills. Only 34.3 percent of learners from the broader sample initially selected the need to "increase skills for current job" as a reason for joining the course.

Almost one half gave reasons for participation that were primarily related to current or future job; others had either more general or home/family–related reasons. As Figure 10.1 shows, 40 percent and 54 percent of

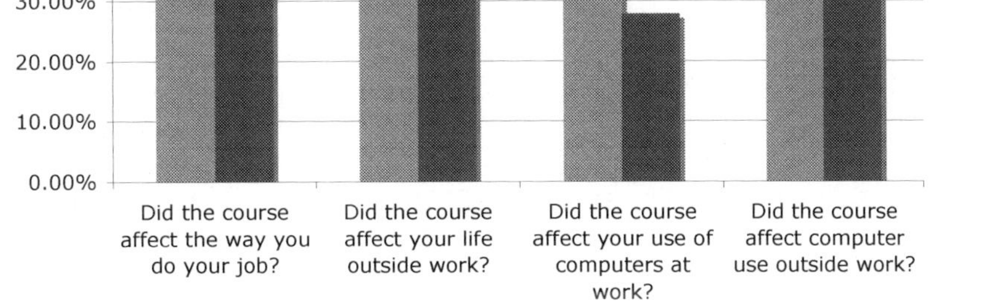

Figure 10.1 How the training has affected learners.

the respondents had indicated that the course had affected the way they do their work and the life they had outside work, respectively.

Shifting perceptions of benefits were highlighted by the follow-up survey (Time 2) which showed that only a quarter of respondents identified the same benefits as they had when first asked, prior to taking the course (Time 1).

Most Popular Choices

- TIME 1
- Acquire new skills
- Improve chances of promotion
- TIME 2
- Acquire new skills

- Get a better job
- Help do current job better
- Help do current job better
- Meeting new people
- Making work more interesting

This is consistent with our qualitative finding that employees who participate in these programs often manifest shifting attitudes to learning (see following) and perceive more opportunities in their work than previously. Among the 40 percent who reported an effect on the way they do their current job, almost two-thirds ascribed this primarily to the skills they had gained and approximately a quarter to increased confidence. Approximately a third of the respondents identified an impact on their computer use at work and outside work. Higher proportions of ESOL participants than non-ESOL reported positive effects on all items except "use of computers at home," where fewer reported impact. Given our interest in the linkages between workplace-linked courses, workplace activities, and learner biographies, we have been especially interested in how shifting attitudes to learning relate to the employees' wider sense of themselves and their work and life priorities, as well as the features of the workplace programs themselves.

INITIAL MOTIVATIONS

In-depth interviews allowed us to pursue in a more detailed and nuanced manner some of the diverse reasons for participation in workplace courses. These factors ranged from "curiosity" to wanting to make up for missed earlier educational opportunities; from wanting specific help with job-relevant skills to wider career aims; from a desire to help children with schoolwork to wanting self-improvement.

Where personal or "nonwork" reasons were highlighted, this was often to do with wanting to make up for previous "failures" or disruptions in education combined with a sense of challenge—wanting to know what one might be able to achieve. Faiza Anwar's early education was very much impeded by her father's approach to education. Victoria too had missed out as her parents moved around in her early life in Ghana, before coming to England and raising her own family. Both Faiza Anwar and Victoria Appiah signed up primarily for the challenges of learning, and could see resulting benefits in both their work and personal lives.

For some, accounts move swiftly from their old identities of themselves as people who "failed," or missed out on education in their early lives, to their current identities as parents and workers. Kathleen Croft's parents did not encourage her to work hard at school, and she was consequently keen to adopt a more involved approach with her own children (aged eight and eleven in 2004). For Kathleen, the main motivation has been to "better herself," and the company scheme was seen as offering the means to do that. This was not rooted in any sense of having inadequate job skills. Having managed to gain a few CSEs[6] in "English, Maths, Sociology and Commerce" at school, Kathleen felt that she had sufficient literacy skills to cope with her current job. She also had a very positive view of the company: *"I do enjoy my job, I like the fact that it's so varied and it's so intense, and that you've got to meet deadlines. . . . They are a people company, I think they are supportive of people. . . . I think there's vast amount of opportunities here and because it's such a big establishment, you can move within the same umbrella and you know, better yourself."*

Others, by contrast, are motivated by workplace difficulties that stem from self-acknowledged problems with literacy. Born in 1971, Abdul Nazif left school in Nigeria without any qualifications and has been working as a resident caretaker since 1999. In his current job he needs to keep an account of incidents in the buildings. *"I know my problem is spelling and writing and expressing myself in writing which I find very difficult still to be honest with you . . . I wanted to better my position at work . . . I don't always want to be doing manual work."*

Tracy Beaumont also had a specific "weakness" in spelling in mind in signing up for the course at the learning centre, which allowed her to study at home on a computer that she bought for her daughter five years ago. Tracy likes the flexibility associated with learning on computers, which allows her to learn at her own pace. She would not have had the confidence to sign up for a college course. She attributes her low confidence to bad experiences at school which have conditioned the nature of learning as an adult. In Tracy's case, the process of helping her child with her homework had given her a greater sense of urgency in developing her own literacy skills. Unlike Abdul, she does not feel that her level of literacy and numeracy have held her back in the workplace, although she admits to feelings of panic in work situations such as meetings, for example, when she has to read things out.

Kathleen and Tracy's accounts revealed a gendered interweaving of work and personal factors. While Kathleen used the perceived "affordances" of the company for learning, which has both personal and work benefits, Tracy was seeking to overcome "weaknesses" that she felt as a parent and as an employee. Abdul and Bill perceived very clear boundaries between work and home, but while Abdul's motivations were almost entirely work motivated, Bill's main reason for signing up for the courses at the food processing company (Coopers) was that he was *"just curious . . . I mean I left school with no qualifications to speak of, just CSE*[7] *things, which are probably in museums now . . ."* Though registered as a learner at the company learning center, Bill did most of his studying away from the workplace, with the help of the tutor who gave him a disk to take home. This type of learning suited Bill partly because it fitted with his identification of the company exclusively with work: *"I'm old fashioned, once I finish work I want to go home. Learning is nothing to do with work, its still on the premises. I take this home (i.e. the disk) and I'll do this at home. . . . "*

The main motivation for Bill undertaking the courses was personal interest as well as the goal of helping his child with homework. Others, such as Melanie Taylor, had already taken the first steps in adult learning through courses at a center for Life Long Learning. She first heard about the workplace literacy course from her manager. She had no initial fears about undertaking the course, due partly to her previous experience of learning. She enjoyed the course and appreciated the chance to refresh and consolidate her existing skills: *"I found that really interesting because we covered a lot of things that I did like years ago at school and it was just like completely forgotten and it made me realize just what I didn't know."*

Unlike Bill, she saw immediate uses in the bakery in which she worked, which she valued and saw as improving her performance rather than overcoming specific difficulties at work. These cases serve as a caution against assuming that those with literacy needs will inevitably struggle in the workplace.

Some do struggle in particular workplace activities, though, as we have seen in Tracy's experiences with activities that involved reading out in public. For Trevor, his main use of literacy in the workplace consisted of writing reports about accidents or incidents of graffiti on the estate. By his own admission, he struggled with this aspect of his work. Trevor undertook the "communications course" for three hours a week over a five-week period. He did not admit to any feelings of anxiety about embarking on such a course, which he regarded as "something extra to help us." All members of the mobile care team attended, which had the advantage of creating a "we're all in the same boat" feeling. He regarded the familiarity with other students as a key advantage of a workplace course. Environments as well as individual dispositions play a part in feelings about the course, as shown by Evans et al. (2006). The supportive features of the environment that helped in Trevor's case were not experienced by Tracy, who continued to feel exposed and humiliated.

BENEFITS WITHIN AND BEYOND THE WORKPLACE

Workplace benefits that stem from participation in various forms of "basic skills" learning range from applications in the immediate job to wider career horizons.

For example, Trevor Woodford was very specific about the immediate and medium-term benefits. He worked in a mobile team of caretakers which rotated between different flats. Trevor felt that the course had improved his report-writing capabilities: "*Before (I was) just guessing where it went really. It's sort of the same for me but it's a lot better now I know where to put things. . . . I'm a lot better at writing letters.*" The course has also improved his abilities in oral communication, for example, in undertaking formal telephone conversations. He also gave examples of using these abilities outside his paid work; for example, in writing an e-mail to the council complaining about the tiles in his kitchen: "*Without putting it nastily I wrote a decent sort of email to them and I got a reply back with the same sort of manner I wrote the email in. As before I suppose I wouldn't have even bothered you know, I would have just phoned them. . . . I basically stuck up for myself . . . [because] . . . I knew what to say to them in a way . . . how to word something.*"

Trevor Woodford is planning to apply for a housing officer post, an office-based job working with tenants. He links this directly to improved skills: " *. . . the course helped me with the spelling of letters and writing letters and telephone manner, so I just thought well that would help if I did decide to go for this job in a couple of years time . . .*" Bob, in the same organization, also exemplifies how a SfL course can have a wide-ranging transformative effect on an individual's life. He had self-declared literacy needs which were severely impeding his capacity to do his job and indeed posed a threat to his overall job security. Developing literacy skills both improved these aspects of his work and boosted his confidence in ways that permeated through to his life outside work. It was noticeable that Bob attributed his progress to the longer and more intensive course he was currently taking, rather than to the initial thirty-hour course. Initial indications are that substantially more than the standard thirty hours is necessary for effecting significant change, as shown elsewhere (e.g., Alamprese, this volume; Comings, this volume; Comings & Soricone , 2005; Rose, this volume).

Victoria, an employee in a transport company, has persisted with learning. She passed the English course and is now taking a math course with a view to a further ICT course at a later stage. She gives examples of how improved spelling ability has helped her at work, and how she is better able to fill in various worksheets and records. But she emphasizes repeatedly her growth in self-confidence as the most significant and multifaceted personal outcome. She also talks about how she sometimes acts as a "negotiator" helping to resolve conflicts or help people who are feeling stressed—this is something that she does at work and in her life outside (as a "befriender" for her local church). The personal gains of increased confidence and self-belief can affect all activities whether paid, unpaid, day-to-day personal, or leisure activities. Victoria struggles to explain what she means by confidence in a more general sense, saying that the word is inadequate to express fully what she means: *"You know I wish I could describe it stronger than what I'm saying, it's like you've been caged and set free."* So confidence for Victoria is also about feelings of being freed from constraints, from things that held her back.

Increased confidence is not, though, a guaranteed outcome of improved literacy or vice versa. Individual dispositions, personal histories, and workplace environments all have their part to play in supporting or undermining the confidence that can come from exercising new abilities (see Billett, 2004; Evans et al., 2006).

Tracy reports that the courses have had some effect in improving confidence. However, her confidence is still low despite having completed the literacy Level 2 test: *"But I'm still not there you know what I mean just because I've got it (ie the Literacy L2). I still haven't got the confidence."* She still feels that she struggles with reading, especially in public situations: *"Its like if you go in a meeting and you read things I panic, I panic, you know what I mean I'm really like conscious about it . . . because like, a lot of people take the mickey because you can't read, and now I'm really self conscious of it."* Tracy has undertaken an impressive range of courses in literacy, numeracy, and ICT. Though she deals with graphs at work, she

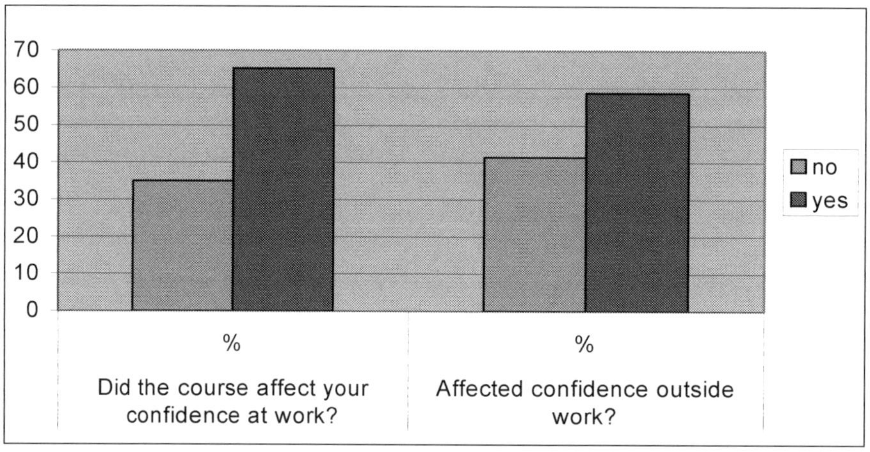

Figure 10.2 Shows that the majority of respondents have indicated that the course has affected their confidence at work (66%) and outside work (59%) at Time 2.

perceives that there is no use of maths in her job: *"I just get the computer to add them up for us."* She learned how to use this technology at work without going on a formal course. The courses have had the effect of marginally boosting confidence but have not been sufficient to alleviate a deep-seated crisis in confidence, originating from school and compounded by incidents of humiliation in working life. Despite this, she doesn't feel that her level of literacy and numeracy has actually held her back in the workplace: *"I think if I'd have gone for a job like . . . like a paper or something like that then yeah I would say yes definitely but like all this kind of work no. . . ."* Despite self-acknowledged problems with reading and writing, it is noticeable that, as in the case of many other individuals in our sample, Tracy feels that she has still managed to cope effectively with the demands of her job.

For Bill, whose motivation for undertaking the numeracy courses was mainly for personal interest, work benefits were not apparent. *"I wouldn't say there's anything I actually learnt off the numeracy that I use day to day that I didn't already know."* The literacy courses have helped him with writing letters and have improved his capacity to help his child with homework but have not impacted substantially on the workplace: *"As I say paperwork, you have sheets to fill in every day, but there's very little writing involved, it doesn't matter if its grammatically correct or not."* He also carries a portable spell-checker. Bill is planning to stay with his current company in more or less the same job: *"I've only got 16 years to go until I'm 60 . . . at my age and my qualifications, I'm getting double the minimum wage . . . I'm relatively happy in what I do. . . ."* For Melanie Taylor, report writing was a more significant part of the job. She emphasized how the workplace course had improved her capacities for report writing: *"You're more aware of how you're going to word things."* The process of undertaking this course, as well as the other courses undertaken over the last few years, have had a major effect in boosting Melanie Taylor's confidence, which has allowed her to gain promotion: *"I just didn't have any belief in myself I don't know, and I think all these courses has sort of helped me to have more confidence and feel better about myself. . . ."*

Different effects of workplace learning on job horizons are also exemplified by our cases. While Bill has no aims for advancement, Trevor Woodford, Bob Murphey, and Abdul Nazif all had concrete plans, and Kathleen Croft has more vague intentions to go for a better job stemming from newfound abilities. Faiza Anwar had already been promoted (after taking the course) to the position of "hygiene coordinator" in the bakery and currently supervises approximately twenty people. As part of her job, she monitors products, writes "incident report forms," undertakes audits, and uses email. Her use of ICT in the workplace includes power-point presentations to staff on quality issues. She also uses maths in relation to meeting key performance indicators (KPI's). Faiza Anwar felt the course impacted positively on her work in the bakery: *"because there is a lot of written work in our job, so . . . it did help."* She proceeded to take various online computer courses but feels that the literacy did not necessarily facilitate progression to these courses: *"I probably would have done them anyway. I mean I went for the computer course, simply because when I applied for this job, there was quite a lot of computer work involved you know like graphs and things like that and spreadsheets."* Brightland Bakeries has a policy of making 70 percent of promotions internally; so this is an example of how company policy and environment count as well as individual initiative.

Ahmed, in the same company, claims that the course helped him progress to a Level 3 NVQ in food and drink, sponsored by his employer, by improving his written and spoken English and confidence in communicating: *"I'm also doing another course NVQ in bakery, at Thames Side college and the company is sponsoring me so that has helped me in my writing skill as well because I have to do assignments so it helped me."* He is positive about the opportunities that are available for him in the company: *"It's a very big group, and there are a lot of different fields here, I don't need to go and look at other jobs, all the jobs are here, supervisor, management, sales . . . quality—everything is here so that's why I'm doing that course, (NVQ 3) and then I'm certainly looking for a supervisors job . . ."* Laila Rahman also emphasizes the positive impact the course has had on her job and future promotion prospects in Brightland Bakeries. As in the case of many other ESOL learners involved in our project, she sees a very direct and tangible link between the course and the skills needed at work. The process of a "flattening out" of management structures means that Laila is increasingly required to take on more responsibility that also entails increased paperwork. Her case underlines some of the advantages and disadvantages of workplace learning: such training offers accessibility but also can potentially be negatively affected by pressure from managers/supervisors on employees to miss learning sessions in order to fulfill their duties in the workplace (as appears to have occurred to several employees on Laila's course).

Paid employment is only one form of work. Indeed, Livingstone's (1999) research has drawn attention to the fact that much of the work in economies is unpaid work. The uses of literacy in work activity outside paid employment[8] have already been suggested in the cases so far. Household, caring, and community work are as readily used as examples by some research participants—for example, Victoria, who talks of "befriending" work for her church; Trevor, who talks of work in his house; Bill, who is primarily motivated by supporting his child; and Kathleen Croft, who sees the most important practical outcome as an improved capacity to help her children with their homework: *"I think mainly it's to do with my daughters, and their schooling being able to look at what they're doing, spot mistakes, and . . . you know basically help them out. . . ."* Social benefits, in meeting new people in and through the courses, were also highlighted by some, including Melanie Taylor and Kathleen Croft. Irrespective of where the skills and literacy practices are used, the indications are that they continue to develop with use in different contexts and in ways that are likely to be as beneficial inside the workplace as outside it. The wider benefits of learning, combined with personal confidence gains, run through employees' accounts of the particular motivations and priorities that caused them to sign up initially. Although workplace learning may have immediate or obvious relevance to job skills, the long-term impact of learning on adults' capabilities—and thus on productivity as well as on life satisfaction—will often depend upon out-of-work activities as much as on workplace participation.

SHIFTING ATTITUDES TO LEARNING

Underlying the diversity of motivations to learn found in the personal accounts, a more fundamental process, which we term "shifting attitudes to learning," can be discerned in many of the sixty-four cases studied in depth. Again, the eleven cases focused on in this paper exemplify this and can be considered in relation to the benchmark indicators (see Table 10.1). Originally developed by a team of researchers from Bristol University, the Effective Lifelong Learning Inventory (ELLI) is a standardized instrument aimed at identifying the elements of an individual's capacity for lifelong learning or "learning power."[9]

Kathleen Croft (who shows above average [for the sample] scores on growth orientation/challenge and creativity at Time 1) had an approach to learning that emphasized the value of knowledge for its own sake whilst also embracing the potential for learning to have career benefits (though not necessarily in the short term or according to a well-defined program): *"Well I just feel that people should continue learning . . . you shouldn't give up, I don't care what age you are, you shouldn't give up learning, and keeping the brain clicking and because of that I just feel okay I'm here but I could be learning something, improving myself all the time so I think on that basis, I get to feel a bit guilty if I'm not doing something."*

Kathleen Croft described her (B grade in GCSE English) as "fantastic" and was on the point of attending a company awards ceremony at the time of the first in-depth interview. The confidence she had gained through her experience on the workplace courses encouraged her to embark on further learning: *"Well it's done a lot because right now I feel it's given me the impetus, to go further and do more, that's really made me feel right, you can do this and try and do something else. . . ."*

As a result of undertaking the course, she was considering taking a GCSE maths: *" . . . it's a brave step, I'm not good with figures, but it appeals to me. I do like maths and English."* She also had some long-term plans to study Spanish, psychology, or commerce. She was now prepared to take some risks in and through learning. The course had expanded her horizons, and this extended to others in her immediate circle of friends and family. She is now encouraging others to embark on learning: *"In fact I think I've encouraged one of my friends to try and do something as well . . . even my brother I'm pushing him a bit . . . they're thinking well you know if she can do it I can go for it as well. . . .*"

Victoria, whose initial ELLI scores were also above the average for growth orientation and creativity, also emphasizes that the courses have left her "wanting to do more" in many different ways, and believing that she can indeed achieve more. To questions about what she sees herself doing in the future, this is how she responds: "*. . . I would love to be a counsellor, you know counselling people, I think I've got my heart in that. I would love to do that. . . . Yes. That would be my dream.*"

For Trevor, a more vulnerable and less learning-oriented person at the outset according to the ELLI scores, the workplace course has both stimulated and responded to Trevor's shifting approach to learning in the few years since leaving school: "*I suppose when I was a kid I was always the clown of the class as well, so . . . now I'm more grown up I can actually sit there and like learn something, where as before I was too, I suppose too active . . .*" The course "pushed me a bit more" and gave him "more confidence to go for it." He too says that he would like to do another course, although he is not sure what the company will offer.

Melanie Taylor, who initially scored relatively highly on imagination/creativity, identified a major shift in her attitude to learning over the last few years: "*No, I just wish I'd done it years ago now. . . . I probably just thought that . . . I'd never be able to do it.*'" Faiza Anwar, whose ELLI scores were close to the average at the outset, had been motivated by "going back to the basics" of grammar and punctuation and stimulated by course activities which she talked of with enthusiasm: "*And then you'd take information out of like newspapers and gather that together and sort of come up with what exactly the editor or journalist was trying to tell you.*" Her desire to continue learning was put into practice immediately—she proceeded to undertake online ICT courses in the company's LearnDirect center. In relation to future study, Faiza Anwar's response reveals interests that extend well beyond the day-to-day roles of work and family life: "*I would like to carry out some further courses in English and literacy because I'm into poetry you see.*" She enjoys writing poetry about "*every day life and emotions and just whatever you know, when you've got a lot on your mind and just put pen to paper. That's what I . . . I tend to write a lot so I just sort of put it in to a bit of a rhythm and make something out of it.*" Her attitude to learning has shifted substantially over recent years, overcoming initial fears that originated from her experience at school: "*because they treat you obviously like an adult because you are an adult so they don't sort of shame you in to answering anything, or embarrass you or anything like that and I think at school they tended to do that and it put me off a little. . . .*"

These findings are consistent with the follow-up survey findings, in which 75 percent of the participants reported that they now felt differently about education. How shifting attitudes are reflected in take-up of new courses is also being tracked. The majority of participating employees expressed an intention at Time 1 to take further courses in the future. By Time 2, 37 percent had already taken further workplace courses, and 18 percent had taken new nonworkplace courses. Time 3 data will reveal how far this translation of intention into actual participation will go within the time frame of the project.

CONCLUSIONS

Our findings show how adult employees' experience of learning is rooted as much in biographical experience as in the contexts of workplace activity and culture. Many of the learners' experience of workplace learning can be described as "compensatory" in nature insofar as they are frequently seeking to "make up" for negative educational experiences in the past. The cases studies provide differing examples of how workplace courses can respond to employees' shifting attitudes to learning.

The combined data show that:

- workplace programs are successful in enrolling adults with very little or no previous experience of formal postschool learning.
- the most important general outcome of course participation, a year on, is an increase in learners' confidence: most noticeably in work, but also outside work.
- the outcomes that learners expect from courses are, more often than not, different from the outcomes they actually report afterwards.

We have also explored employees' perceptions of whether they are coping or struggling with their 'basic skills' in the workplace. The research benefits from two different perspectives on the literacy levels of employees: the assessment scores which offer an independent measure of literacy as a set of skills and the learners' own perceptions of whether they are coping or struggling with their existing skills. Indeed, several employees have indicated they have struggled to carry out aspects of their job as a result of poor literacy skills. Yet we have also identified many examples of employees who have coped sufficiently in the workplace with their existing skills, in which case the literacy component of the course is often viewed as a chance to "brush up" on their skills.

Workplace courses can play an important part in engaging individuals who would otherwise be intimidated by studying in a college. During the course of in-depth interviews, we pursue the various advantages and disadvantages of undertaking a course in the workplace as opposed to a college. The overriding majority of learners have emphasized such factors as accessibility, familiarity, and convenience as being key advantages of workplace learning. Other factors of major significance include the differing environments for learning in different organizations (including diverse strategies for promotion).

Current research on adult literacy in the UK has tended to be shaped by differing epistemological approaches that can broadly be associated with quantitative and qualitative methods. Those who advocate the importance of researching literacy as a "social practice"—emphasizing the social context of literacy usage—tend to employ qualitative methods framed by more relativist, hermeneutical perspectives. The concept of literacy itself is dissolved into a plurality of literacies shaped by differing contexts (e.g. Papen, 2005; Street, 1995). These perspectives tend to contrast with UK government reliance on quantitative data on literacy that implicitly assumes an "autonomous" rather than "ideological" model of literacy (Street, 2001) insofar as literacy is perceived as a clearly defined set of technical skills, the absence of which can impact negatively on an individual's economic and social opportunities. In advocating a "pragmatist" approach to mixed methods, we view differing research perspectives on literacy that tend to coalesce around either quantitative or qualitative research methodologies as potentially useful avenues for exploring varying facets of literacy rather than as being the intrinsic components of sharply divided and irreconcilable epistemological traditions. The employment of mixed methods allows for a more detailed exploration of how literacy is actually employed in differing contexts. In doing so, we are in a position to take issue with assumptions underpinning SfL discourses about the existence of large skill deficiencies in literacy and official numeracy that inevitably impact negatively on performance at work. Our more nuanced approach illustrates the diverse range of techniques that are employed in literacy practices whilst highlighting those cases where skills deficiencies exist.

The evidence yielded by mixed-methods approaches suggests that it may be more productive to think beyond the polarized conceptualizations of literacy and numeracy learning as either technical skills development or the expansion of social practices, towards a more ecological understanding of the phenomena observed. A "social ecology" of learning in the field of adult basic skills leads us to consider the relationships between the affordances of the workplace (or those features of the workplace environment that invite us to engage and learn), the types of knowledge afforded by literacy and numeracy learning (including knowing how and "knowing that you can"), and the agency or intention to act of the individual employee, reflected in their diverse motivations.

These are triangular relationships and mutually interdependent sets of interactions. There are affordances for learning in all workplace environments. Some are more accessible and visible than others. The intention of employees to act in particular ways in pursuit of their goals and interests, whether in their jobs or personal lives, makes the affordances for learning more visible to them. The know-how associated with literacy practices, such as report writing or finding better ways of expressing oneself, and the confidence of "knowing that you can" often develop as the person engages with the opportunity. The process of making the affordances for learning more visible itself can generate some employees' will to act and use those affordances, and new knowledge results. In the shifting attitudes to learning, the changing levels of know-how and the confidence that comes from "knowing that you can" both stimulate action and the seeking out of affordances within and beyond the workplace in the form of further opportunities.

These reflexive relationships, as the cases considered in this chapter illustrate, point the way forward in developing a better understanding of literacy learning as part of a wider social ecology of adult learning in workplace environments.

NOTES

1. The other major issue for the design relates to the existence (or rather nonexistence) of teaching content related to learners' workplaces or occupations. There is some strong research evidence (cited in our original proposal) that vocationally and job-relevant material is learned more effectively and faster by adults than general content. Moreover, government guidance explicitly recommends that workplace basic skills tuition be preceded by and built on a careful analysis of workplace needs and practices. However, in actuality, such detailed needs analyses are rarely conducted and teaching material is generalist. (In the case of LearnDirect-based schemes, this is invariably true.) We have therefore concluded that it would not be appropriate to use occupational-specific tests to measure skill changes and will administer only general tests (but designed explicitly for adult learners).
2. For example, accounts of shifting attitudes/compensation for poor previous educational experience may be compared with quantitative data on qualifications and inventory scores.
3. "Go" is a new set of tests of literacy skills, designed specifically for adults, and with parallel forms, whose development has been supported by our cofunder, the National Research & Development Centre for Adult Literacy and Numeracy.
4. 3 percent of the 2003 UK employed workforce is defined as being in the ESOL category (Labour Force Survey). Of the "ESOL" employees surveyed, 30 percent were white, 50 percent Asian, and more than two-thirds were male. They had been with current employer somewhat less than overall sample average. The ESOL subsample included some who were highly educated (about one-third were at Level 3 or above), but there were also many with no qualifications.
5. This is slightly lower than the rate we projected when first designing the study, but most of the dropouts are ESOL learners who had moved away and could not be contacted. This will not be a source of significant error for the planned statistical analyses involving correlation and analysis of variance.
6. Certificate of Secondary Education, a school-leaving qualification subsequently incorporated into the lower tiers of the General Certificate of Secondary Education, GCSE, in England.
7. Certificate of Secondary Education, phased out in Britain in the 1970s.
8. As defined by Noon and Blyton, 2002.
9. We used a modified version of the original instrument in order to enquire about forty-four items (covering seven broad dimensions of learning) using a

four-point Likert-type scale, ranging from *not at all like me* to *very much like me*. Factor analysis indicated the existence of three underlying factors that could explain the variance in the data in a meaningful way: 1) *growth orientation/challenge seeking*, 2) *dependence/fragility*, 3) *imagination/creativity*. Generally speaking, we would expect effective learners to score highly on *growth orientation/challenge seeking* and *imagination/creativity* and low on *dependence/fragility* and vice versa for ineffective learners. The analysis of ELLIs at Time 2 will allow us to trace changes in these underlying factors. In particular, we are in a position to pursue the hypothesis that learners' mean scores on *growth orientation/challenge seeking* and *imagination/creativity* will increase after attending a workplace training program and, conversely, that their mean score on *dependence/fragility* will decrease.

REFERENCES

Billett, S. (2004). *Work, change and workers*. Dordrecht, Netherlands: Springer.
Brewer, J., & Hunter, A. (1989). *Multimethod research: A synthesis of styles*. Newbury Park, CA: Sage.
Comings, J., & Soricone, L. (2005). Massachusetts: A case study of improvement and growth of adult education services. *Review of Adult Learning and Literacy*, 5, 85–124.
Evans, K ., Hodkinson, H., Rainbird, H., et al. (2006). *Improving workplace learning*. London: Routledge.
Labour Force Survey. (2003). London: Office for National Statistics.
Livingstone, D. W. (1999). Lifelong learning and underemployment in the knowledge society: A North American perspective. *Comparative Education*, 35(2), 162–86.
Noon, M., and Blyton, P. (2002). *The realities of work*. London: Palgrave.
Papen, U. (2005). *Adult literacy as social practice: More than skills*. London: Routledge.
Scheeres, H (2004). The textualised workplace. *Reflect*, 1, 22.
Street, B. V. (1995). Social literacies: Critical approaches to literacy in development, ethnography and education. Harlow, UK: Longman.
Street, B. V. (ed.) (2001). *Literacy and development: Ethnographic perspectives*. London: Routledge.
Tashakkori, A., & Teddlie, C. (1998). *Mixed methodology: Combining qualitative and quantitative approaches*. London: Sage.

11 Improving the Literacy and Numeracy of Young Offenders

Jane Hurry, Laura Brazier, and Anita Wilson

INTRODUCTION

Young people in the criminal justice system tend to have lower than average attainment in literacy and numeracy, a finding consistently reported internationally (Andrews, 1995; Farrington, 1996; Hawkins, Herrenkohl, Farrington, Brewer, Catalano, Harachi, & Cothern, 2000; Rutter, Giller, & Hagell, 1998). Recent surveys in the United Kingdom find that in the custodial setting, 51 percent of young offenders were below Level 1 in literacy and 52 percent in numeracy (ECOTEC, 2001). For young offenders supervised in the community, 57 percent were below Level 1 in literacy and 63 percent in numeracy (Hurry, Brazier, & Moriarty, 2005). Longitudinal studies have documented the negative pathways associated with weak basic skills. In particular, it makes it more difficult to find consistent employment and heightens the chances of becoming socially marginalized (Bynner, 2004; Parsons & Bynner, 1999). Such marginalization is likely to be a decisive factor in whether or not these young people desist from crime in adulthood. Sampson and Laub (1993), on the importance of social bonds, wrote "the stronger the adult ties to work and family, the less crime and deviance among [former] delinquents." On the basis of longitudinal data, Schoon (2003) emphasizes the importance of the transitional period around the age of sixteen, when important decisions about future careers are made.

This line of evidence suggests that improving the literacy and numeracy skills of young people in the criminal justice system will improve their chances of employment and reduce their chances of reoffending in adulthood. Indeed, based on this kind of information, education/training is identified as one of the promising approaches to reducing reoffending (Lipsey, 1995; McGuire, 1995; Sherman, Gottfredson, Mackenzie, Eck, Reuter, & Bushway, 1997). In the United Kingdom this has been rigorously translated into policy, with a requirement that "at least 90 percent of young offenders are in suitable full time education, training and employment" (Offenders' Learning and Skills Service [OLASS], 2004). There has always been an intention that young people are gainfully employed, but this is difficult to achieve. Also, the prioritization of basic skills in the United Kingdom following the Moser Report (DfES, 1999) has influenced education and training within the youth justice system. There is a particular commitment to improving literacy and numeracy standards with an expectation that young people will improve by one skill level or more within six months (OLASS, 2004). Policy initiatives have been given teeth, with funding for education providers contingent on student learning gains. As the research discussed in this chapter began, the Youth Justice Board (the UK Government agency with responsibility for the youth justice system) had just developed a new basic skills strategy, PLUS, specifically to address the literacy and numeracy, which included learning and enrichment materials for students and professional support for teachers (YJB, 2004).

Logical though all this is, some issues remain untested. The associations observed between educational attainment, qualifications, and offending cannot be interpreted as cast-iron evidence that improving educational attainment will reduce offending, that is, that low educational attainment is a cause of offending. There are a range of plausible alternative explanations for such an association. Most troublesome in this area is that a young person has underlying difficulties, either social or individual, which lead to doing badly at school, to offending, and to being unemployed. Longitudinal studies are helpful here because they can show whether or not problems with educational attainment may predate psychological problems, offending, and difficulties with employment. They can also control for "confounding" variables, such as socioeconomic status and parenting. However, where there is a complex web of relationships, it is difficult to predict the effect of intervention, and those "confounding" variables may have consequences for the type of intervention that might work. In the end, the only way to be sure is to try it out.

To pursue the implications of existing longitudinal data, an experimental design is helpful in answering two key questions:

- Can educational interventions improve the literacy and numeracy skills of young people in the criminal justice system?
- Do improvements in literacy and numeracy lead to higher employment and lower rates of offending?

Here we address the first of these questions. Educational intervention studies targeting young offenders are rare, and there is a shortage of evidence both on the best methods of improving the literacy and numeracy of young people, previously hard to engage in mainstream education, and on the impact of such intervention on recidivism (Hayward, Stephenson, & Blyth, 2004; Stephenson, 2007). Because of our focus on literacy and numeracy skills, and in line with UK initiatives described previously, we seek to test the impact of either increasing discrete provision (i.e., embedded in a subject based course) or improving the quality of that provision. The curriculum for the discrete provision of literacy and numeracy within this context is substantially defined by the Adult Literacy and Numeracy Core Curricula (DfES, 2001) and offers a fairly conventional coverage of skills. There is a debate about the desirability of such a focus with young people who tend to have negative attitudes towards schooling. Alternative provision of vocational training or employment addresses a range of skills, but in terms of literacy and numeracy, learning is embedded within the tasks of the workplace. However, literacy and numeracy are almost invariably addressed implicitly and we hypothesize that such an implicit focus will produce smaller learning gains than a more explicit approach. There remains the threat to the effectiveness of discrete provision that it will be rejected by students, or that they will be turned off and fail to attend. It is therefore necessary not only to measure literacy and numeracy gains but to speak to students about their opinions on education and training and to observe classroom dynamics.

METHODS AND RESEARCH DESIGN

An experimental design with pre- and posttest is a powerful way to assess the effectiveness of discrete basic skills provision on literacy and numeracy skills. However, there are a number of barriers to implementing such a design with young people within the criminal justice system. Random assignment to learning condition typically presents difficulties. In community settings, their histories of nonengagement with academic programs may make sustained teaching difficult to deliver. In custodial settings it is difficult to experimentally manipulate provision. All those under eighteen receive a full-time, mandatory programs of education and training; therefore, comparisons would involve either a) comparing quality of provision or b) comparing under eighteens with over eighteens not in education or training. In both settings attrition is likely to be an issue due to high rates of dropout and mobility within or out of the system.

The design adopted therefore needed to be flexible enough to enable a number of different types of comparison. The research was conducted in four sites, two in the community (each site comprising a number of projects) and two in custody. In each of the four research sites, samples were divided into "treatment" and "control" groups as follows:

Community site, South Wales (Com SW—all offenders). Learners were divided by project into those projects offering vocational training but little or no discrete basic skills provision (n = 6 projects—the control group) and those offering discrete basic skills as part of the education/training package (n = 6 projects—the treatment group).

Community site, London (Com L—disaffected youth, some offenders). All students were on a pre-employment scheme in one of two projects. The scheme included both vocational training and discrete basic skills classes. Students were divided into two groups: 1) pre- and 2) post- staff training in embedded basic skills provision. Group 1 was the control group, Group 2 the treatment group.

Custodial site, North England (Cust NE), was divided into two groups, one receiving compulsory education (judged good by inspectors—the treatment group), the other, marginally older, receiving no education (the control group). Education in all custodial contexts in England has a substantial formal element covering literacy and numeracy and using the PLUS and the Adult Core Curricula.

Custodial site, Central England (Cust CE). All students were in education or training. They were divided into two groups, those: 1) pre- and 2) post- basic skills staff training and reorganization. Group 1, the pretraining group, were either offered education or vocational training, and the education provision was judged in need of improvement by inspectors. Group 2, the posttraining group, all attended education (four participants worked on industrial cleaning full-time and were therefore reassigned to Group 1). Group 1 was the control group, Group 2 the treatment group.

This identification of treatment and control groups offered four sets of comparisons. Two of the comparisons were based on a quasi-experimental design with some provision offering more discrete basic skills provision and some less. Two of the comparisons were based on comparing provision before and after staff training and reorganization. Changes in provision aimed to increase the contextualization of discrete basic skills provision and an explicit understanding of the embedding of basic skills within the vocational elements of the courses.

In addition to these four comparisons, two further sets of comparisons were possible: a comparison of students receiving discrete basic skills provision with those not receiving such provision; a comparison of students in custody attending educational provision deemed good by inspectors (Cust NE) with those attending provision deemed in need of improvement (Cust CE).

SAMPLE

The community sites were selected as being experienced and successful providers of post-16 education for young offenders or disaffected young people. The custodial sites were selected to offer a comparison of good education provision and provision in need of improvement.

Students attending the provision in these sites were approached to participate in the research if they satisfied two criteria: 1) they had literacy or numeracy scores of Level 1 or below, and 2) they were scheduled to be in the provision for three months or more. Two hundred seventy students satisfied the selection criteria and were assessed at the beginning of their course (community) or on entry to the Youth Offender Institution (YOI) (custody). Of these 270, 147 (54 percent) were reassessed, and only this reassessed group are considered here. The attrition rate was substantially a consequence of young people leaving provision before follow-up. In the case of those in custody, this was normally due to problems in getting tests administered before release and was not due to the normal kind of 'drop-out' associated with disengaged students. In the community, attrition was a combination of difficulties in getting tests administered by project staff and students moving on. There was, however, no difference in the initial levels of literacy and numeracy between those followed up (mean literacy and numeracy levels = 2.7) and those not followed up (mean literacy and numeracy levels = 2.6).

Participants ranged in age from sixteen to nineteen years (mean age 17.4 years). All but three were male (the custodial sites were only for males). Seventy-one percent were white British, 13 percent were black Caribbean, and 8 percent black other. All were convicted offenders except those in the Com L group of whom only 35 percent reported being convicted and were attending the provision because they had failed to find their niche in education and training.

DATA COLLECTION

Literacy and numeracy were assessed using the Basic Skills Agency (BSA) Initial Assessment (2002). At pretest they were assessed on Version 1 of this assessment, at posttest on the parallel forms of Versions 2 or 3. Levels and raw scores were both coded. The levels map to the National Standards for Adult Literacy and Numeracy (Qualifications & Curriculum Authority & DfES, 2000); Entry levels 1, 2, and 3, Level 1 and Level 2. Progression from one level to the next in a school context represents approximately two years' progress. Level 1 is the average attained at the end of primary schooling. For the purposes of analysis, these levels have been converted to a five-point scale ranging from one (Entry level 1) to four (Level 1). Raw scores on the literacy assessment range from zero to seventy-two and on the numeracy assessment from zero to fifty. Students were posttested on average twenty weeks after initial assessment. The average time between assessments for the Com L control group was slightly longer at twenty-six weeks due to changes in the provider's program length and assessment practices.

Students were interviewed on two occasions concerning a range of dimensions, including their experience of school, their attitudes towards education and training in their current provision, and their future aspirations in terms of education, training, and employment.

A range of education/training provision was observed over a period of at least four days at each research site.

Data were collected from all sites concerning students' attendance overall and on discrete basic skills classes (this was defined as literacy and numeracy only; IT classes, social skills classes, etc., were not classified here as basic skills). For the community sites information was available on the amount of time individual students received discrete basic skills provision. In the custodial sites, information was only available for the amount of education provision overall, and in Cust NE, only at aggregate level. Approximately half the education classes were literacy and numeracy classes, the remainder being concerned with IT, art, social and life skills, drama, and so on.

RESULTS

Young People's Disposition to Learn

Consistent with other research on offenders, many of these young people came to literacy and numeracy classes with a fairly negative education history. As Table 11.1 shows, less than half had completed compulsory schooling, even fewer had gained any qualifications at school, and over a third rated their enjoyment of school as "very bad" or "awful."

However, they did (slightly) prefer education and training they had received during the research period, either in the community projects or in custody. Asked to represent their enjoyment of provision on a 7-point scale, where 1 was great and 7 was awful, their mean score for school was 4.5 (standard deviation [SD] = 1.9). Their mean score for current education/training was 3.8 (SD = 1.5), which was a slight but significant improvement on their enjoyment of school ($p < .01$). In many cases, current provision was mainly vocational or half and half, vocational and education, and the vocational element was particularly popular. Twenty-six students in Cust CE, who were attending education classes for half the day and vocational training for the other half of the day, were asked to rate their enjoyment of education and vocational training separately. They significantly preferred vocational training to education with mean enjoyment scores of 2.1 (SD = 1.9) and 3.9 (SD = 1.7) respectively ($p < .01$).

Table 11.1 Young People's School History (N = 149)

<table>
<tr><th></th><th>Completed school</th><th>Stopped attending</th><th>Excluded</th><th>Custodial sentence</th></tr>
<tr><td>Compulsory schooling</td><td>44%</td><td>25%</td><td>27%</td><td>4%</td></tr>
<tr><td></td><td colspan="2">Achieved qualifications</td><td colspan="2">No qualifications</td></tr>
<tr><td>Qualifications at school</td><td colspan="2">35%</td><td colspan="2">65%</td></tr>
<tr><td></td><td>Great or very good</td><td colspan="2">OK</td><td>Very bad or awful</td></tr>
<tr><td>Enjoyment of school</td><td>14%</td><td colspan="2">51%</td><td>35%</td></tr>
</table>

This is consistent with young people's views on their future education. The majority didn't want to do academic courses for their own sake. When asked what courses they would like to do in the future, they almost invariably mentioned vocational courses such as plumbing, cooking, mechanics, and bricklaying, though a small minority were interested in AS levels or art and crafts.

Those who were positive about attending future courses addressing literacy and numeracy (38 percent) thought that it might improve their skills and be useful in terms of getting a job.

> "It would be useful if it wasn't too easy. The YMCA was too easy, like infants school—apostrophies, spelling etc. I would enjoy it because it might help me get a job" (community).

Those who were negative (38 percent) commented on disliking such courses and being more focused on getting work.

> "I wouldn't want to do it. I'd do it if I had to do it for a job, like to be a warehouse person. I don't like writing and I don't think it would be useful for the kind of job I want to do" (community).
>
> "I'm not very keen on the idea. I want to do a work related course like for building and decorating. Something practical" (custody).
>
> "I won't do courses because there is nothing wrong with my reading and writing. I'm not interested in further qualifications" (custody).

LITERACY AND NUMERACY PROVISION

In the community, both the London-based projects and half of those in South Wales offered a mixture of vocational and a more academic focus. Students typically attended vocational training, in training establishments or on the job, for four days a week and discrete literacy and numeracy for one day a week. In the remaining projects in South Wales, provision was almost exclusively vocational with minimal basic skills provision.

In custody, students also usually had a mixture of basic skills and either vocational courses or further educational options such as art, drama, social and life skills, IT, physical education, or music. A minority of students were enrolled in vocational activities full-time.

In both contexts students entered and left classes on a "roll-on, roll-off" basis, and there were typically around six students attending class (though there were often eight on roll). Within basic skills classes, students were not only at different points in their work but also varied considerably in their competencies. There was often an attempt to group students by ability, but at best this could only be fairly crude, typically working with two groups, those at Entry Levels 1–3 and those at Levels 1–2. Much of the teaching was therefore individualized, with students working their way through worksheets or workbooks at the appropriate level. Some whole-class teaching was observed, but in the main, teachers or learning support assistants worked with students individually.

The content of the work was rather relentlessly focused on "skills," very often reflecting the levels of the students, covering material that they would have encountered in primary school (e.g., prefixes and suffixes, basic fractions). Indeed, students complained that it was like being in primary school.

> "I am in prison, not in f***ing primary school" (custody).

> "Are you f***ing joking, giving me f***ing tracing" (community).

The elements of worksheets which involved activities, such as discussion or data collection, tended to be omitted (particularly in custody). Whilst some students were observed to work diligently in these discrete basic skills sessions, a lot of restlessness, task avoidance, resentment, and frustration was also apparent.

This was in contrast to behavior observed in other formal educational lessons, such as social skills, art, and drama, where the same pupils were at worst quietly disengaged and at best involved in active participation, discussion, and recording. Similarly, there was less disruption in workshops and at best clear engagement in completing required tasks. Classroom-based vocational sessions lent themselves to contextualized tasks. For example, students enrolled on a Childcare NVQ (National Vocational Qualification) were asked to plan the basic equipment for a nursery within a budget of £1,000. In these tasks students would have to make mathematical calculations, some of them fairly straightforward, but they were not associated with a demeaning request to do "primary maths," but were seen as the sorts of skills required for future aspirations.

IMPLEMENTING CHANGE

In the original research design it was planned to offer additional discrete basic skills provision in six of the twelve community-based projects in South Wales, which would have enabled comparisons to be made between students' progress before and after the introduction of the new and more intensive provision. However, this had to be abandoned, falling back on the naturally occurring contrasts in discrete basic skills provision in the twelve projects described earlier. The failure of the intervention is nonetheless very informative. Essentially, it failed because many students refused to accept a program requiring them to spend two days per week on discrete basic skills. They wanted to spend more time engaged in either vocational training or work. This is consistent with young people's comments and our observations of their behavior in the different types of provision. An additional and unanticipated problem was that although these community projects had a very good record of keeping students attending regularly for an average of fourteen weeks, students did not attend the same place for the whole of this time. Attendance tended to be punctuated with arguments with a particular provider, or a change of mind or the lack of availability of a desired option, leading to students changing to another course.

Based on this experience, our observations, and interviews with students, we worked with staff in the other community site (Com L) and the custodial site, where provision was deemed in need of improvement (Cust CE), to increase the degree to which basic skills provision was contextualized and embedded. Substantial changes were observed, particularly in the custodial site, where education provision was radically overhauled. Here the final result was a highly contextualized discrete basic skills program and basic skills embedded in the vocational training which formed half of most students' program. However, the process of change took much longer than anticipated, over eighteen months in the Young Offenders Institution, and the "treatment" groups of learners were assessed in the middle of the process of change. This seriously undermines the extent to which we can confidently say there was a difference in the basic skills provision in the treatment and control groups in these two comparisons. In the following section, where we look at experimental effects on literacy and numeracy, we are essentially only able to compare little or no discrete provision with discrete provision (in both Com SW and Cust NE) and poor discrete provision with good discrete provision (Cust CE control with Cust NE treatment).

EXPERIMENTAL COMPARISON BETWEEN BASIC SKILLS CONDITIONS

Students in the four treatment groups received fairly similar amounts of discrete basic skills provision, on average about one day per week (6.9 hours, Table 11.2). It was not possible to collect official registers for education attended by the learners at Cust NE. However, all respondents reported that they had attended education regularly. Education ran for fifteen hours weekly, approximately half of which (7.5 hours) was devoted to basic skills, and this is the figure estimated in Table 11.2, though it will be a slight overestimation as it does not take account of any absences due to illness, visitors, and so on. As expected, the students in control projects in South Wales (Com SW) spent a very limited amount of their time on discrete basic skills. The Cust NE control group were not offered any education, and the Cust CE control group also received fewer hours

Table 11.2 Average Number of Hours of Discrete Basic Skills Received Weekly

Research condition		*Hours of basic skills weekly*	
		mean	*(SD)*
Com SW	Control (n = 12)	1.25	(2.3)
	Treatment (n = 13)	6.7	(7.0)
Com L	Control (n = 14)	6.1	(5.3)
	Treatment (n = 23)	6.2	(5.3)
Cust ME	Control (n = 29)	3.9	(3.8)
	Treatment (n = 16)	7.0	(2.6)
Cust NE	Control (n = 17)	0	(0)
	Treatment (n = 23)	7.5	(0)
Total	147	4.9	(4.6)

of basic skills on average because nearly half (n = 14) were enrolled on a full-time vocational course.

LITERACY

There was significant improvement between pre and posttest literacy levels overall, amounting to about a third of a level (roughly the expectation for eight months in school) (Wilcoxon's $Z = 4.43$, $p < .001$). However, inspection of Table 11.3 shows the gains made by the treatment groups to be only marginally better than those made by their controls, and in one case, slightly poorer. Effects were tested using a fixed entry regression, with posttest score as the dependent variable and controlling for pretest level before looking at group differences. For literacy, in the four mini-experiments/quasi-experiments, in no case did the experimental group do significantly better than the control group. This was true even in the custodial site, where those in education were compared with those with no access to education. In addition to these four within-site comparisons, differences between the two custodial sites were also explored. The students in the "good" Cust NE group made 8.7 points progress in the BSA literacy assessment from pre- to posttest, as opposed to 4.9 points progress in the control group at Cust CE (those students who had experienced a less adequate provision), but this did not reach statistical significance. Finally, comparisons were also made between those who had attended at least six hours of discrete basic skills provision weekly with those who had not received any such provision. The basic skills group made slightly greater progress but this was again not statistically significant.

Table 11.3 Progress in Literacy by Experimental Condition

Research condition		*Literacy assessment*			
		*BSA level**		*BSA score (max score 72)*	
		Pretest mean (SD)	*Posttest mean (SD)*	*Pretest mean (SD)*	*Posttest mean (SD)*
Com SW	Control (n = 12)	2.6 (1.0)	2.8 (.9)	51.1 (13.4)	55.0 (13.4)
	Treatment (n = 13)	2.85 (.7)	3.0 (.8)	54.4 (10.9)	58.4 (12.0)
Com L	Control (n = 14)	2.6 (.9)	2.9 (.7)	51.8 (14.3)	54.4 (12.3)
	Treatment (n = 23)	3.1 (.7)	3.1 (.8)	59.1 (8.9)	61.3 (8.6)
Cust ME	Control (n = 29)	2.9 (.9)	3.2 (.8)	54.7 (12.1)	59.6 (11.3)
	Treatment (n = 16)	2.6 (.9)	3.1 (.7)	52.3 (13.3)	57.3 (10.9)
Cust NE	Control (n = 17)	2.4 (1.3)	2.8 (1.4)	53.2 (17.9)	59.5 (17.0)
	Treatment (n = 23)	2.5 (1.2)	2.9 (1.1)	47.3 (21.7)	56 (17.0)
No discrete basic skills v minimum 6 hrs wkly	No ed (n = 44)	2.7 (1.0)	3.0 (.9)	52.4 (15.8)	57.8 (12.0)
	6 hrs plus (n = 67)	2.7 (1.0)	3.1 (.9)	52.7 (16.2)	58.3 (14.1)
Total	147	2.7 (1.0)	3.0 (1.0)	53.4 (14.6)	58.1 (12.4)

*Entry Level 1 = 1; EL 2 = 2; EL 3 = 3; L 1 = 4.

NUMERACY

As for literacy, overall there were significant improvements in numeracy between pre- and posttest, amounting to about a quarter of a level ($p < .001$, Table 11.4). In one of the four mini-experiments/quasi-experiments the treatment group did significantly better than its control group. The Welsh community group receiving discrete basic skills made more progress in numeracy than the control group who were at work or receiving vocational training only ($\beta = .25$, $p < .05$). The effect size (Cohen's d, Cohen, 1988) of .47 is small, verging on medium sized according to Cohen's classificatory scheme. Similarly, in the other comparison, where we could be reasonably confident that there were differences in basic skills provision between treatment and control groups (Cust NE), students attending numeracy classes made more progress than those who did not (.6 of a level as opposed to .1 of a level). This difference was not statistically significant; however, the treatment group did make statistically significant progress between pre- and posttest ($p < .01$) whereas the control group made an insignificant amount of progress. Comparing Cust NE with the control group in Cust CE, there was no statistically significant difference, but students in the less secure Cust CE provision made less and statistically nonsignificant progress over the time period (.1 of a level). There was no significant difference in the progress made by students receiving at least six hours weekly of discrete basic skills compared to those receiving none of this provision.

Table 11.4 Progress in Numeracy by Experimental Condition

Research condition		*Numeracy assessment*			
		*BSA level**		*BSA score (max score 50)*	
		Pretest mean (SD)	*Posttest mean (SD)*	*Pretest mean (SD)*	*Posttest mean (SD)*
Com/SW	Control (n = 11)	2.4 (1.0)	2.55 (.8)	32.4 (9.3)	33.4 (8.8)
	Treatment (n = 13)	2.6 (.7)	3.0 (6)	35.3 (7.6)	39.1 (6.2)
Com/L	Control (n = 14)	2.4 (1.0)	2.7 (1.0)	32.4 (10.0)	35.3 (9.8)
	Treatment (n = 24)	2.7 (.7)	3.0 (.9)	36.2 (8.3)	38.9 (8.2)
Cust/ME	Control (n = 28)	2.9 (.7)	3.0 (.8)	37.0 (8.7)	38.5 (7.8)
	Treatment (n = 14)	2.9 (.5)	2.8 (.6)	37.2 (7.1)	38.9 (5.9)
Cust/NE	Control (n = 16)	2.9 (1.0)	3.0 (1.3)	36.9 (11.4)	38.4 (11.8)
	Treatment (n = 24)	2.2 (1.0)	2.8 (1.1)	30.7 (13.1)	35.3 (13.4)
No discrete basic skills v minimum 6 hrs wkly	No ed (n = 44)	2.7 (.9)	2.9 (1.0)	35.2 (10.2)	36.9 (10.0)
	6 hrs plus (n = 67)	2.6 (.9)	2.8 (.9)	34.7 (10.5)	37.2 (10.1)
Total	144	2.65 (.8)	2.9 (.9)	34.9 (9.8)	37.4 (9.4)

*Entry Level 1 = 1; EL 2 = 2; EL 3 = 3; L 1 = 4.

DISCUSSION

Young people of sixteen to eighteen are at a critical stage in their working lives, described by Bloomer and Hodkinson (2002) as the "most volatile stage in human growth and development" (p. 151). For those caught up in the criminal justice system, there are real dangers that they will fail to gain the necessary literacy and numeracy skills and qualifications, with consequences both for themselves and society. Two-thirds of the young people we interviewed had left school before school-leaving age, without any qualifications. We know from the 1970 British Cohort Study that this augurs badly for their future employment (Parsons & Bynner, 1999). It would therefore seem of critical importance to try to encourage them to develop their basic skills.

The young people in our study, all in education or training, made significant gains in literacy and numeracy of, on average, a quarter or a third of a level over a period of five months. Students in full-time, mainstream education would be expected to make this amount of progress in eight months. Our participants, with a history of disengagement from education and training, might be expected to make less progress than the average teenager but they appear to have done rather well. It would seem that keeping them in education or training can offer them real benefits. This is in itself challenging as these young people are often out of education and employment. A recent study by the Youth Justice Board reported that at any given time only 35 to 45 percent of those in the youth justice system are in full-time education, training, or employment (YJB, 2006).

Because improvements were observed in control and treatment groups, this does raise doubts as to whether the overall improvements signify learning gains or some artifact of the research, such as familiarity with the test or depressed scores on entry to new provision or a selection effect due to sample attrition. Our conclusion is that the gains are probably real. As different versions of the test were used at pre- and posttest, familiarity with the test is an implausible explanation for improved scores. The punishing attrition, with a loss of 46 percent of the sample, admits the possibility that only the better learners were followed up, those with little interest in learning being lost to the study. However, there were no differences in the initial levels of literacy and numeracy between those followed up and those not followed up. Also, we know that, particularly in the custodial sample, we frequently failed to reassess students because of institutional factors, such as early release, or difficulty in accessing students' test results, rather than because of student dropout. In previous research, using a similar sample and measures, we found similar sized learning gains in literacy and numeracy (Hurry et al., 2005). Here, those who attended education and training provision for longer than fourteen weeks made greater progress in literacy and numeracy than those who left earlier. If improvement was some artifact of reassessment or attrition, these results are difficult to explain. If, then, these learning gains are real, how good are they? On average, learners in the present study received an estimated 143 hours of education/training in the five months over which progress was measured. For literacy, Comings and colleagues (Comings & Soricone, 2005; Comings, Sum, & Uvin, 2000) suggests that 150 hours of education should lead to about one grade gain in literacy. A BSA level equates to about two grades. So, bearing in mind that the 143 hours was by no means devoted solely to literacy and numeracy, this seems a good result.

The purpose of the experiments was to explore the impact of more or better discrete basic skills provision. There were no significant differences in learning gains between treatment and control groups in literacy and only some in numeracy. The reasons for this need to be considered carefully. Firstly, there were problems with the implementation of the treatments. In the two comparisons where we worked with staff to maximize the impact of their basic skills provision, we measured the "improved" group too early. The provision in the custodial site (Cust CE) changed quite radically over a period of one and a half years (Hurry et al., 2005), and this gives confidence that intervention is possible, but that there are no quick fixes. In the community site, where we worked with staff to increase contextualizing and embedding of basic skills (Com L), the intervention only provided four days of staff training and a couple of support visits per trainer. From conversations with tutors it seems likely that whilst this enriched their teaching, in the absence of a systemic approach to change, involving their own management systems, substantial changes in embedding did not happen.

We therefore need to rely on those comparisons between presence or absence of discrete basic skills provision (from sites Com W and Cust NE) and the more adequate provision of one custodial site (Cust NE) with the less adequate initial provision of the other (Cust CE). None of the comparisons find any significant difference in literacy gains between groups. Although those in the control groups received less discrete basic skills provision, they received more vocational training, which in itself probably produced improvements in students' basic skills. However, students in these treatment groups did make slightly greater progress than the controls and sample sizes were small. Much statistical power and a highly reliable assessment instrument would be needed to detect statistically significant differences, and we have to conclude that the case here is still unproven. The story for numeracy seems to be a little different. Here, those receiving discrete basic skills lessons on the whole made more progress than those receiving little or none, though in only one comparison did this reach statistical significance. This suggests that for numeracy, formal, decontextualized teaching may be important, and we would be interested if this finding could be replicated. In primary school contexts it has been observed that educational factors (as opposed to home factors, etc.) are more important for mathematics than reading (Reynolds & Muijs, 1999). Our own evidence is not secure enough to make any strong statements but it does pose a question.

The implementation of the quasi-experimental design was problematic. However, it was also fruitful. In the process of trying to mount the sort of intervention suggested by the evidence from surveys and longitudinal studies, a number of important things emerged. In particular, the reasons why some students lack qualifications and expertise in basic skills are associated with their attitudes towards learning and their future aspirations. These associations have implications for the kinds of education they will tolerate. Given the choice of whether or not to attend discrete basic skills provision (as in the community contexts), many will reject basic skills programs. This in turn undermines the viability of such programs. Where there is no choice, many will find alternative ways of avoiding engagement. For these reasons, the design of educational interventions needs to take account not only of what needs to be learnt but also in what ways this can be presented in a palatable form.

It also highlights the fact that randomized controlled trials, although scientifically rigorous, are not always possible to mount where participants are unenthusiastic about the "treatment." Alternative research methods must be adopted in these circumstances, qualitative studies, longitudinal studies, and quasi-experimental designs. Natural comparisons offer a viable solution, and we would argue that the scientific community should be more open to this methodology, whilst requiring large sample sizes and suitable information on participants' characteristics at baseline.

The present study has a number of limitations, but findings suggest that educational intervention can improve the basic skills of young people within the criminal justice system, or disaffected with education. There is no real evidence to suggest that discrete basic skills provision offers an advantage over vocational training in terms of literacy progress, but may be more effective in terms of numeracy progress, and this needs to be followed up, perhaps making use of naturally occurring comparisons. The critical issue may be to ensure that young people are involved in some form of educational or vocational activity. There do remain unanswered questions concerning the effectiveness of discrete basic skills provision. The basic skills lessons observed were formal, decontextualized, worksheet, and skills based. Students reported and were observed as being more engaged in more practical learning environments where they had greater opportunities for activity or interaction. The effectiveness of discrete basic skills provision which is contextualized to a greater degree and involves more active learning remains to be tested.

REFERENCES

Andrews, D. (1995). The psychology of criminal conduct and effective treatment. In J. McGuire (ed.), *What works: Reducing reoffending.* Chichester, UK: John Wiley & Sons.

Bloomer, M., & Hodkinson, P. (2002). Learning careers and cultural capital: Adding a social and longitudinal dimension to our understanding of learning. In R. Nata (ed.), *Progress in education* (p. 5). Hauppauge, NY: Nova Science.

Bynner, J. (2004). Literacy, numeracy and employability: Evidence from the British birth cohort studies. *Literacy and numeracy studies*, 13, 31–48.

Cohen, J. (1988). *Statistical power analysis for the behavioral sciences* (2nd ed.). Hillside, NJ: Lawrence Earlbaum Associates.

Comings, J., & Soricone, L. (2005). Massachusetts: A case study of improvement and growth of adult education services. *Review of adult learning and literacy*, 5, 85–124.

Comings, J., Sum, A., & Uvin, J. (2000). *New skills for a new economy: Adult education's key role in sustaining economic growth and expanding opportunity*. Boston: Institute for a New Commonwealth.

Department for Education and Employment. (DfEE) (1999). *A fresh start: Improving literacy and numeracy*. London: DfEE.

DfES. (2001). *Adult literacy core curriculum*. London: DfES.

DfES. (2001). *Adult numeracy core curriculum*. London: DfES.

ECOTEC. (2001). *Education, training and employment*. London: Youth Justice Board.

Farrington, D. (1996). *Understanding and preventing youth crime*. York, UK: Joseph Rowntree.

Hawkins, J. D., Herrenkohl, T. I., Farrington, D. P., et al. (2000). *Predictors of youth violence*. Washington, DC: U.S. Department of Justice.

Hayward, G., Stephenson, M, & Blyth, M. (2004). Exploring effective educational interventions for young people who offend. In R. Burnett and C. Roberts (Eds.), *What works in probation and youth justice*. Cullompton, UK: Willan.

Hurry, J. Brazier, L., & Moriarty, V.(2005). Improving the literacy and numeracy skills of young people who offend: Can it be done and what are the consequences? *Literacy and Numeracy Studies*, 14(2), 61–74.

Lipsey, M. (1995). What do we learn from 400 research studies on the effectiveness of treatments with juvenile delinquents. In J. McGuire (ed.), *What works: Reducing re-offending—guidelines from research and practice*. London: John Wiley & Sons.

McGuire, J. (1995). *What works: Reducing re-offending—guidelines from research and practice*. London: John Wiley & Sons.

OLASS. (2004). *The offender's learning journey: Learning and skills provision for juvenile offenders in England*. London: DfES. Retrieved 12 June 2007 at http://www.dfes.gov.uk/offenderlearning/uploads/documents/05%200111_Juvenile_OLJ%20v04.doc.

Parsons, S., & Bynner, J. (1999). *Influences on adult basic skills: Factors affecting the development of literacy and numeracy from birth to 37*. London: Basic Skills Agency.

Reynolds, D., & Muijs, D. (1999). Numeracy matters: Contemporary policy issues in the teaching of mathematics. In I. Thompson (ed.), *Issues in teaching numeracy in primary schools*. Buckingham, UK: Open University Press.

Rutter, M., Giller, H., & Hagell, A. (1998). *Antisocial Behaviour by Young People*. Cambridge: Cambridge University Press.

Sampson, R., & Laub, J. (1993). *Crime in the making: Pathways and turning points throughout life*. Cambridge, MA: Harvard University Press.

Schoon, I. (2003). *Teenage aspirations for education and work and long-term outcomes: Evidence from the 1958 National Child Development Study and the 1970 British Cohort Study*. Paper presented to the ESRC seminars on "How to motivate (demotivated) 14–16 year old learners, with particular reference to work related education and training"; seminar held 16 May 2003 at CEP, London. Retrieved 13 June 2007 at http://cep.lse.ac.uk/events/seminars/motivation/schoon.pdf .

Sherman, L.W., Gottfredson, D.,Mackenzie, D. L., et al. (1997). *Preventing crime: What works, what doesn't, what's promising: Report to the United States Congress*. Washington, DC: National Institute of Justice.

Stephenson, M. (2007). *Young people and offending: Education, youth justice and social inclusion*. Cullompton, UK: Willan.

Youth Justice Board. (2004). *Annual review 2003–4: Building confidence*. London: YJB.

Youth Justice Board. (2006). *Barriers to engagement in education, training and employment*. London: YJB.

Part IV

Social and Economic Outcomes in Context

12 Using a Longitudinal Approach with State Administrative Records to Evaluate Adult Education Programs in the United States

Stephen Rose[1]

The primary federal program in the United States supporting the development of basic literacy skills and English-language proficiency among adults is the Adult Education and Family Literacy Act (AEFLA) Grant to States. In fiscal year 2004, this program distributed $584 million to designated state agencies, serving 2.7 million learners. Adult Education (AE) grants support various kinds of programs, mainly divided into:

Adult Secondary Education (ASE) prepares students who have skills that are close to the level needed to pass the General Educational Development (GED) or external high school diploma. These students typically are under twenty-five years old and dropped out of school before receiving their high school diploma.

Adult Basic Education (ABE) serves students with relatively low skills who need considerable instruction before being ready to pass the high school equivalency exam or GED. These students tend to be older students who have economic, educational, or family reasons to enter an AE program.

English Literacy or English as a Second Language (EL or ESL) provides instruction for adults residing in the United States with limited English proficiency. ESL students also tend to be older and predominantly immigrants.

AE takes place in a variety of instructional settings. Sponsoring organizations range from local library branches and educational districts to community colleges and correctional institutions. Class sizes and pedagogical approaches vary. Instructors have varied levels and types of training and may work full- or part-time.

Classes are organized to make it easy for students to attend. In most programs, there is an open enrolment, meaning that students can enter at any point in the year as soon as there is an opening. Although programs try to group by skill level, this is not always possible and forces teachers to mix whole group, small group, and individual instruction techniques. A further challenge is that students can miss several weeks because of a work or family crisis; again, the teacher is forced to cater to requirements of the student.

In *A Blueprint for Preparing America's Future*, the U.S. Department of Education (ED) lays out the principles on which AEFLA could be strengthened when it is reauthorized: creating accountability for results, funding what works, expanding options and choices, and reducing bureaucracy and increasing flexibility. As with the No Child Left Behind Act, these principles reflect the thinking that educational practices must be tied to demonstrable results. On the basis of clear performance measures, successful service providers can be recognized and rewarded, while less successful providers can be given models and methods for improvement as well as clear program goals. In this way, accountability leads to enhanced performance as the average level of success improves and the bar for what is acceptable moves higher.

Accountability can only be as successful as the quality of the data that measure performance and the capacity of organizational and information systems to manage and utilize that data.[2] In general, data systems are structured to meet three ends.

1. At the simplest level, reporting systems answer the "how-many" questions (e.g., how many students are served, how many hours do they attend, how many students have learning gains, how many students are of each demographic type).
2. At the second level, reporting systems perform an accountability function by determining performance levels for providers. By having measures that identify the number of students with successful outcomes, key benchmarks can be established to determine who is performing adequately. For those who are performing below average, they can be given clear goals to demonstrate improvement over time.
3. At the highest level, reporting systems provide data for program evaluation to help determine "what works." In this instance, program descriptors and demographic characteristics are analyzed in relation to student outcomes to identify factors that closely correlate with success, thus providing a guide to implement new policies and hypotheses for the next wave of program evaluation. Over time, these analyses feed back to the second purpose, by helping states and providers learn what to do to *improve* their performance levels.

Until recently, many states did not maintain statewide records of AE student attendance, progress, and achievements, relying instead on local program providers to maintain student records. The Massachusetts Department of Education (2001) surveyed state AE programs in 1999 and found that up to a third of states did not maintain individual student records. Performance accountability provisions in the 1998 Workforce Investment Act Title II mandated states to begin to provide annual reports of AE student progress.

The National Reporting System (NRS), adopted in 2000, requires that states submit summary information on student attendance and achievement. Annual NRS reports must include information on student demographic characteristics, average attendance, and educational and economic goal attainment. Although NRS reports can be made without statewide data systems on individual students, more and more states have developed centralized Management Information Systems (MISs) to prepare NRS reports. Often local sites can enter data directly into these MISs using the Internet. The Department of Education has also expended much effort on training to improve NRS data quality.

As part of the department's effort to help states implement effective MISs, a study was commissioned to identify three states with the most advanced MISs. The primary goal was to determine the procedures and elements used by these states with the best practices in order to provide guidance for other states in improving their MISs. Secondarily, it was hoped that high-quality administrative data could be used to identify best practices in encouraging student persistence and performance. Finally, a third component was added to link adult education data to economic outcome information through the use state unemployment insurance wage records.

In addition to helping states with less-developed MISs improve their data collection, there was also the hope that thorough state administrative records could be used to answer the "what works" question. This was particularly important to the department because Beder (1999) had identified 115 outcome and impact studies that had been conducted since the late 1960s on the "changes in learners that occur as result of their participation in adult literacy education" and found that twenty-three studies that met minimum scientific rigor. Although the twenty-three credible studies tended to support similar conclusions about the effectiveness of AE programs, Beder cautioned against drawing strong conclusions because of limited sample sizes, inconsistencies in variable definition and units of measurement, and problems in the methodologies measuring outcome changes.[3] State AE administrative records, with large numbers of cases and consistently defined and measured outcomes, seemed to offer a rich source of existing data on AE student experiences.

This paper reports on this study of three states with a reputation for highest quality data (chosen on the basis of suggestions from experts in the field and interviews with twenty state administrators). The study collected four years of individual student information from each state. Most of the focus, however, was on a single cohort with unique student identifiers permitting a transcript-style record to be created for all four years. The NRS, by contrast, focuses on results from a single program year, which will be shown later to create certain distortions.

FINDINGS

Data Coverage

In terms of data coverage issues, all of these states (referred to as States 1, 2, and 3) had accurate records of student demographic background information and classroom participation. Therefore, learning outcomes could be aligned with student backgrounds and participation in AE classes. Unfortunately,

Table 12.1 Available Data Elements: Background Characteristics of Students

	State 1	*State 2*	*State 3*	*State 4*
Gender[a]	✓	✓	✓	✓
Age[a]	✓	✓	✓	✓
Ethnicity[a]	✓	✓	✓	✓
White	✓	✓	✓	✓
Black	✓	✓	✓	✓
Hispanic	✓	✓	✓	✓
American Indian	✓	✓	✓	
Asian	✓	✓	✓	✓
Country of origin/original language	✓	✓		✓
Immigrant	✓	✓	✓	✓
Veteran	✓		✓	
Marital status	✓			
Single parent status[b]		✓	✓	✓
Presence of children/dependents	✓	✓	✓	✓
Number of children/dependents	✓	✓	✓	✓
Children in school/age of children	✓		✓	
Disability status[a]	✓	✓	✓	✓
Homeless/resides in shelter	✓	✓	✓	✓
In nursing home or other institution		✓	✓	✓
Public assistance status[a]	✓	✓	✓	✓
Received food stamps			✓	
Received subsidized housing			✓	
Received Medicaid			✓	
Previous years of schooling		U.S. only		
Previous adult education experience	✓			
Labor force status[a]	✓	✓	✓	✓
Mandated enrollment	✓			

Source: The Pilot Analysis of Student Attendance, Instruction, Student Achievement, and Economic Outcomes (2004), conducted by ORC Macro under contract to the U.S. Department of Education, Office of the Under Secretary, Policy and Program Studies Service.

[a]Core NRS measure.

[b]Supplemental NRS measure.

there was very little information about the classes themselves (e.g., support services, form of instruction, qualifications of the teachers, hours received per week of instruction).

Table 12.1 lists the kind of variables that were available in each of four states. The fourth state was dropped from the final study because it created new student identifiers each year, making it impossible to create a longitudinal record for that state. Surprisingly little information is available about family characteristics. All states but one record whether or not someone is a single parent, while States 1 and 3 record the number of preschool children. Only State 2 has a question about marital status.

Assessments

The most difficult challenge faced by AE programs was measuring student performance. Without grades or a graduation milestone (relatively few students will earn a GED or high school diploma), learning gains were determined by an assessment of student competencies at intake and exit with the appropriate testing instruments. Unfortunately, students entered the program at different times during the year and often left without prior notification to the teacher. As a consequence, a fairly large proportion of students had no basis on which to measure their progress.

Another fundamental issue concerns measuring educational gains, which is the core product of AE programs. There are many problems in how this is organized in various states. First, there is no consistent practice in how student progress is assessed. At intake, students are assessed in many different areas (e.g., reading, speaking, mathematics, etc.), yet their progress is only measured in one subject. This is usually determined by the area in which the students scored lowest at their intake evaluation (which is the recommended but not required procedure). While some states have information on all evaluations for each student in every subject, other states just report the scores for the subject for which student progress is measured. Consequently, measures of student progress in literacy, reading, and numeracy are rarely available in all three areas.

Another complication is that the testing instrument can vary within the same state. In one state of the study all a single testing instrument was used. In another state, three different instruments were used depending on whether the student was destined for ASE, ABE, or ESL. Finally, in the third state, local agencies were permitted to choose which instrument they used and there were a wide range of products used.

In order to accommodate various testing instruments, the Office of Vocational and Adult Education (OVAE) established six educational functioning levels for both English speakers in ABE and ASE program and for non-English speakers in ESL programs. Educational gain was then defined as having a posttest with the same instrument that demonstrated a level increase.

Although this seems to be a clean and easy way to measure results, there are difficulties with this approach. First, literacy (or any other educational) skills are not necessarily neatly ordered into different levels but are a continuum of abilities. Therefore, any measurement system that applies levels has width and boundary issues: Anyone who scores close to the boundary needs just a little increase to "demonstrate" a learning gain, whereas those who score far away from the boundary point can have sizable test score gains and not have a reported level gain (this is particularly true among beginning ESL students who start with very low English skills). For example, a student who scored 560 on the Test of Adult Basic Education (TABE) total reading exam initially and 570 on the TABE posttest would be classified as having a level gain. By contrast, if another student's scores went from 520 to 560, she would not be shown with a level gain. So, the student with a considerably higher gain in scale score would be shown as not doing as well as the student with a lower gain in scale score.

Over the four years of data to which we had access, the assessment records in the three pilot states improved dramatically over the studied period (see Table 12.2). In two states, the share of students with a reported pretest and posttest rose dramatically between program years (PYs) 2000 and 2003. In

Table 12.2 State Assessment Practices, 2000 to 2003

				Percent of student records with two matched tests		
	Year	*Percent of student records with no tests or inconsistent scores*	*Percent of student records with one test (assessment only)*	*(National test used)*	*(Local test used)*	*Total*
State 1	2000	5%	47%	17%	31%	100%
	2001	3%	50%	25%	22%	100%
	2002[a]	1%	66%	13%	19%	100%
	2003	7%	36%	54%	2%	100%
State 2	2000	41%	37%	20%	2%	100%
	2001	16%	52%	31%	1%	100%
	2002	18%	53%	28%	1%	100%
	2003	4%	45%	51%	0%	100%
State 3	2000	44%	33%	23%	0%	100%
	2001	23%	31%	46%	0%	100%
	2002	7%	35%	59%	0%	100%
	2003	5%	37%	57%	0%	100%

the other state, most assessments prior to 2003 were conducted with local instruments of unknown quality and accuracy. In 2003, a new policy was instituted, requiring local providers to use one of only three possible assessment instruments. By 2003, the share of students with proper assessments at intake and entry using a scientifically calibrated assessment was 54 percent in States 1 and 2, and 57 percent in State 3. Since another 10 percent of students earned a general education development (GED) certificate or alternative high school diploma, approximately two-thirds of students in these states exited the program with some kind of final evaluation in PY 2003.

Attendance

Table 12.3 demonstrates the prevalence of multiyear attendance: between 72 and 78 percent of AE enrollees in PY 2000 only attended in that year alone. This means that approximately one-quarter of students attended in multiple years even though very few students attended classes in all four years for which we have data. As for the simple question how many students received at least eighty hours of instruction in State 2, there are two answers: while 22 percent of students in State 2 had more than eighty hours in PY 2000 alone, 43 percent of students who started classes in 2000 received at least eighty hours of attendance over the course of their entire participation in an AE program.

Another way to show the difference of the longitudinal versus the single-year approach is to look at total attendance hours. As can be seen in Figure 12.1, the multiple-year perspective shows that many more students have

Table 12.3 Multiyear Attendance Patterns

Percent of PY 2000 cohort that attended in:	*State 1*	*State 2*	*State 3*
PY 2001	25%	20%	14%
PY 2002	10%	12%	10%
PY 2002	5%	8%	8%
Percent of PY 2000 cohort that attended in:			
First year only	74%	72%	78%
First and second year only	16%	15%	9%
First, second, and third year only	5%	4%	3%
All four years	2%	0%	2%
Other multiyear combinations	4%	10%	9%
	100%	100%	100%

Note: this table displays the share of students enrolled in PY 2000 who attended classes in the subsequent three years. At least 22 % percent (State 3) of students attended in more than one year.

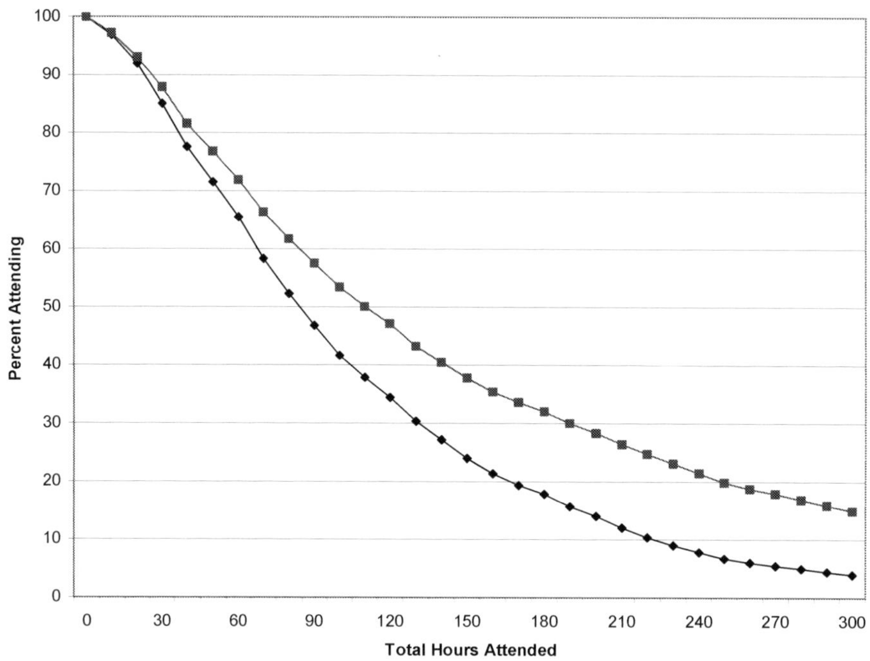

Figure 12.1 Survival curve of ESL students (State 1). Top curve: Four years of attendance data. Bottom curve: One year of attendance data.

higher levels of attendance. For example, 14 percent of these students had attended at least two hundred hours in a single year while the comparable figure was 28 percent reaching this attendance level if their history of participation over many years is tracked.

Although programs are supposed to retest their students at regular intervals, administrative records show this to not be always the case. An unanswerable question is whether the students with many instructional hours but no posttest just missed the day of the assessment or whether teachers did not test students that they thought were not ready for a posttest.

A final advantage of longitudinal data is that it can determine whether when a student starts a program affects her level of attendance. This may be important because many programs have summer breaks and lose a lot of students once they get out of the routine of coming to class. In some states, enrollments are limited in certain periods to maintain class continuity. Other states and programs have open enrollment and allow students to join a class at any time. Therefore, it is useful to know if students who begin in the months before summer have the same attendance as those who begin in other months.

In Figure 12.2, the relationship between starting quarter and total attendance is depicted. As can be seen, in State 1, 58 percent of students who

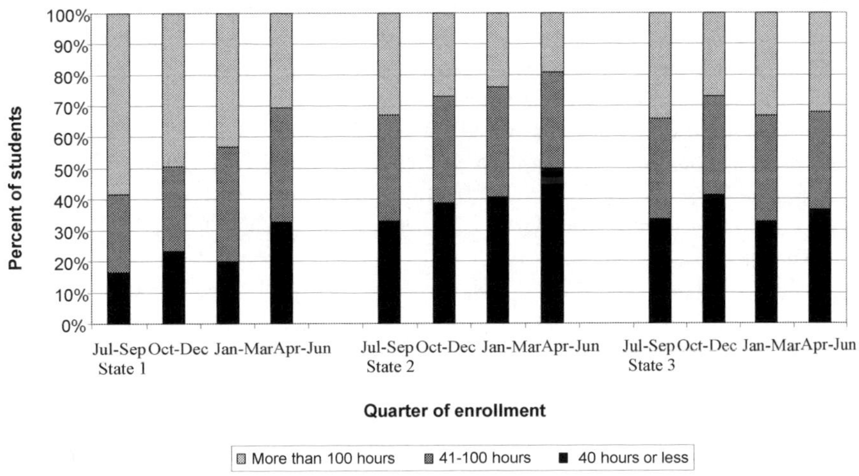

Figure 12.2 Attendance by quarter of enrollment (PY 2000 cohort).

started in the first quarter of the program year received more than one hundred hours of instruction while only 31 percent who started out in the last quarter of the program year were able to stay in class this long. In State 2, there is a dramatic difference in share of students with relevant few class hours (forty or fewer hours): 33 percent of those who started in the first quarter had this low level of attendance as opposed to 50 percent of those starting in the last quarter. It is only in State 3 that starting quarter does not have a strong impact on total attendance hours.

Aligning Attendance and Achievement

Obviously, the multiyear transcript approach allows for more instructional time and hence permits a longer window for posttesting leading to a higher share of students having a match pre- and posttest. However, it also creates a situation in which some students may have multiple assessments after their intake assessment. Unless there are multiple level gains, the extra tests have no practical effect.

Even if the learning outcomes measure were cleaner, AE programs suffer from another difficulty in aligning attendance with positive learning gains (a key component in any comparison of relative performance of different programs). As opposed to the normal approach where each student has a complete transcript of his/her overall participation over many years, the NRS only requires very gross yearly progress measures—e.g., hours of attendance, records of results on assessments, and number of students receiving a GED or HS diploma. Other than initial score on placement assessments at time of entry into the program, no data from previous years are included.

Consequently, the single-year approach can provide many summary measures about the number and type of students served, but it is not necessarily the most accurate way to follow student progress. For example, consider a student who attends sixty hours in one program year and another thirty hours in the following year. At the end of the second year, she is posttested and has a level gain from her initial placement level of the previous year. In the official state records, she will be listed in year 1 with sixty hours of attendance and no exit assessment; in year 2, she will appear in the records with thirty hours of attendance and a gain of one NRS level. This official yearly record will yield misleading information in both years.

Another example of the problem of aligning learning gains with attendance is evidence in the data about receipt of a GED. For example, in State 2, 50 percent of students who earned a GED in PY 2000 attended fewer than forty hours in that year. However, of students who enrolled in PY 2000 and who eventually received a GED in any year between PY 2000 and PY 2003, only 31 percent had fewer than forty attendance hours. Figure 12.3 shows the division of students who enrolled in PY 2000 that participated in AE programs over multiple years before receiving their GED.[4] In State 2, 30 percent of those who received a GED did so in a year after their initial enrollment; in State 3, the comparable number was 36 percent. This demonstrates that contrary to some conventional wisdom, many GED-aspiring students require many hours of instruction before passing this hurdle.

Where did this mistaken wisdom come from? Figure 12.4 shows the distribution of hours of instruction that students received depending on the data collection strategy. Using the single year of PY 2000, one-half of students had at most forty hours of instruction. But this figure includes those who were continuing on from other years. Looking at the true cohort who began in PY 2000 and eventually earned a GED, just 31 percent accomplished this in less than forty hours. At the other end of the attendance spectrum, it was relatively uncommon (22 percent) for students reporting

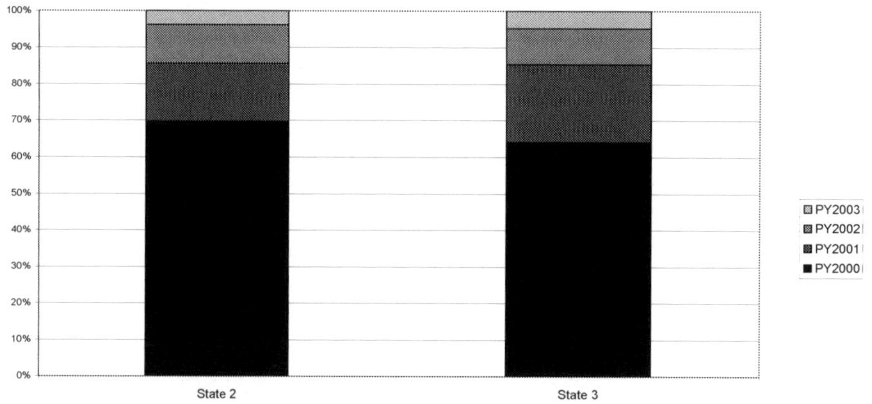

Figure 12.3 Year of GED attainment among PY 2000-cohort GED recipients.

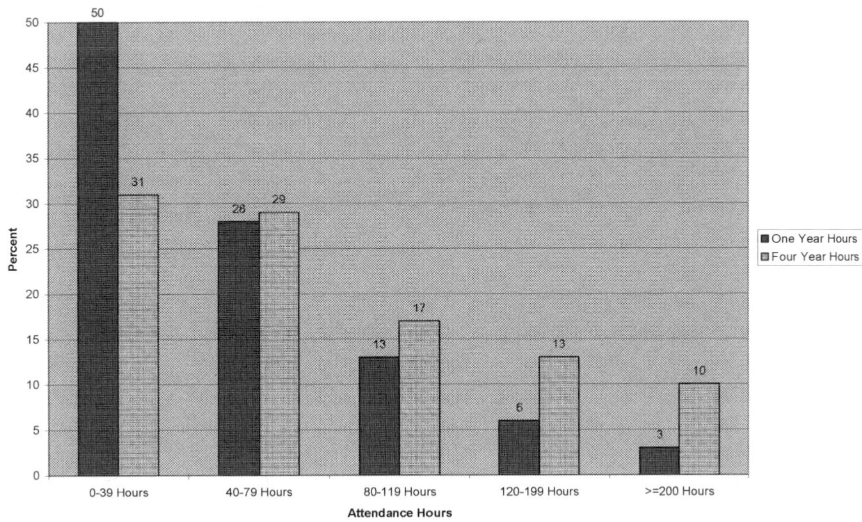

Figure 12.4 GED recipients' attendance hours as shown in their one-year and four-year records (State 2).

receiving their GED in PY 2000 to require more than eighty hours of instruction. By contrast, using the longitudinal record approach, we find that 40 percent of students beginning in PY 2000 needed over eighty hours of instruction before they could successfully pass the GED.

These numbers highlight clearly the difference between a report based on attendance in a single year versus a longitudinal, transcript-style approach.[5] Approximately 20 percent of the distribution is moved from low to high hours. One-year records make it appear that one-half of GED recipients require at most forty hours and only 22 percent have more than eighty hours. Yet, the real experiences of students differ from this.

Economic Outcomes

One of the major motivations of AE programs is to improve the labor market performance of the students who are served. In order to answer how effective programs are in meeting this goal, we would need a comparison group of workers with similar backgrounds that did not enroll in any program (see Reder, this volume). Without this comparison group, the best we can do is track the labor market experience of AE students before and after enrollment to see how much progress is made in terms of employment and earnings gains. Since we can't determine whether these gains are "normal" for workers as they age, the best we can do is to compare different groups of AE students to determine if those with greater attendance and learning gains have the highest labor market gains.

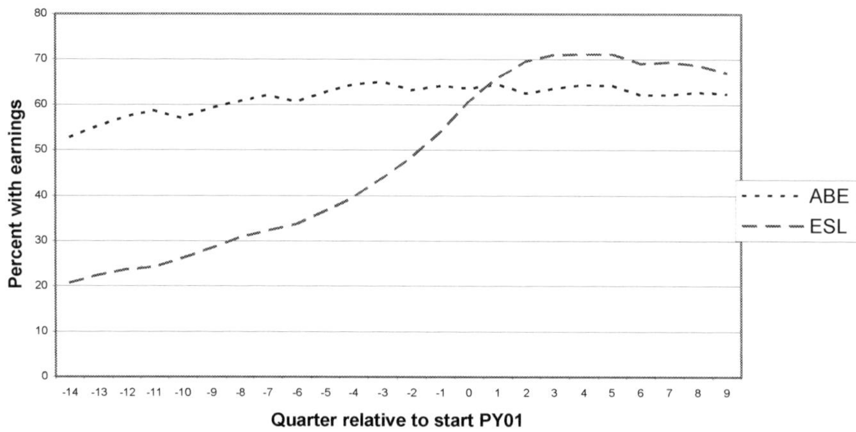

Figure 12.5 Employment rates relative to time of program enrollment (State 2).

Six years of quarterly unemployment insurance records were collected for each student who had a valid SSN. As Figure 12.5 shows, ESL students had tremendously rising rates of employment in the years prior to their enrollment. In State 2 the employment rate of ESOL students rose from just over 21 percent fourteen quarters prior to entry to more than 71 percent in the first quarters after entry. In State 3, the pattern is similar but less extreme (from 25 to 66 percent).[6] In both States, ABE/ASE students also increased their employment rates in the quarters before entering an AE program (from about 50 to 64 percent).

In both states and both programs, employment rates level off or slightly decline, starting about the fourth or fifth quarter after entry. This pattern

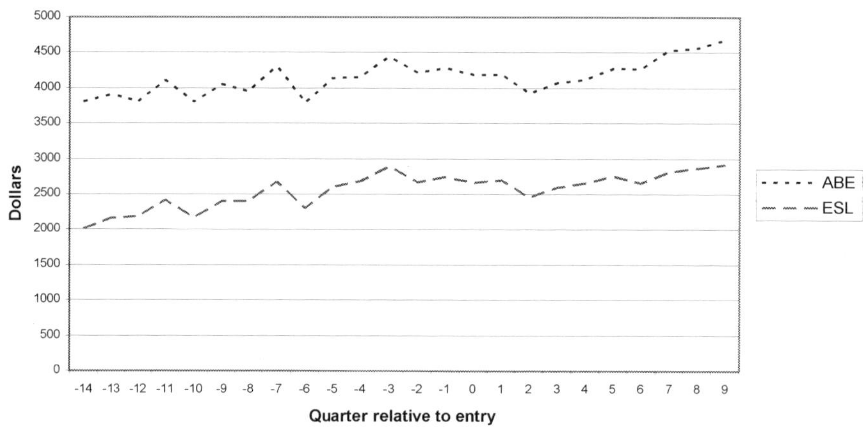

Figure 12.6 Earnings relative to time of program entry (State 2).

indicates that AE participation is not associated with being out of the labor force. Since being employed did not seem to affect hours of attendance, it appears that employed AE students are able to combine work and learning. The decline in employment rates, found during the last quarters of the data, may be due to some learners leaving the state. Because of the limitations of this data source, all of these people would appear as not working even if they were employed elsewhere.

Figure 12.6 shows the average earnings of students when they are employed over this same time period. The trend lines for ESL and ABE students are parallel, with ESL students in both states earning considerably less than their ABE counterparts. The common trend is that earnings levels rise during the first quarters for which we have data and then level off or decline around the time of program entry; for the first few quarters after entry, earnings are flat and then rise again and seem on a fairly steep positive trajectory in the last quarters for which we have data.[7]

In order to measure how participation (as measured as attendance hours) affected earnings, Table 12.4 presents a simplified version of the results. As is evident for both ESL and ABE/ASE students in both states, the net positive change grew with hours of attendan ce. The most dramatic change occurred among ESL students in State 3, where the net positive value for those with forty or fewer hours was 26 percentage points while it was 50 percentage points among those with one hundred or more hours.

Table 12.4 Employment and Earnings, Before and After Participation

		Job Loss or Lower Earnings	*No Change in Earnings*	*Job Gain or Higher Earnings*	*Total*	*Net Positive Change*
ABE/ASE Students						
State 2	LT 40 Hours	32	24	44	100	12
	40–100 hours	30	25	45	100	15
	GT 100 hours	30	25	45	100	16
State 3	LT 40 Hours	29	25	47	100	18
	40–100 hours	27	23	50	100	23
	GT 100 hours	25	24	52	100	27
ESL Students						
State 2	LT 40 Hours	28	15	57	100	30
	40–-100 hours	23	13	63	100	40
	GT 100 hours	23	13	64	100	42
State 3	LT 40 Hours	30	15	55	100	26
	40–-100 hours	27	12	61	100	34
	GT 100 hours	19	12	69	100	50

CONCLUSION

Of all of the parts of U.S. educational system, adult education is the least developed and least funded. Nationally, the annual average cost per student is approximately $400, and AE programs often have just one to three full-time employees per provider. Most of the teachers work part-time and are paid accordingly. All of these factors make it difficult to have high-quality data collection and reporting systems.

In establishing the NRS guidelines and associated training grants, OVAE has established minimum standards for data collection. As part of the initiative to reward achievement, the department plans on using these reports to provide bonus awards to states that are meeting its performance targets. Furthermore, states are encouraged to reward providers who meet and exceed their performance targets.

These incentives can be very powerful in increasing the performance of all providers as long as the data reported are accurate and not subject to manipulation. There is reason to believe that this is not true. For example, in the fourth state in which data were collected, a higher percentage of the students had learning gains with an average total attendance that was significantly less than the three states in the study. In addition to not having student identifiers that could link multiyear records, there were very few posttests as part of the official record. This state used a "portfolio" assessment approach in which teachers evaluated the range of student performance—e.g., quizzes, classroom exercises, verbal ability, etc.

It does not seem plausible that one state with much less student attendance would outperform in terms of learning gains another state in which students attended more than twice the number of hours. The lack of confidence in these results is bolstered by the fact that the gains in this state are measured by teacher judgment rather than a valid posttest. For this outcome to be valid, the teachers in this state would have been remarkably more productive than teachers in other states even though their credentials were no different than those in other states.

This pattern is repeated within states. An added component of the project was to look at different sites and determine what kind of methodology could identify high performers. One problem with high-stakes assessments is that providers may try to avoid very poor students who are unlikely to show learning gains as defined by an increase of one level—ESL students can actually have high learning gains if they started with very low English skills and moved up below the cut point to move from Level 1 to Level 2. Depending on how the formulas of success are computing, there may be an incentive to avoid posttesting students if the share of successful students among those who took posttests is one of the metrics.

To date, states have used simple formulas to assess sites. In this study, a more sophisticated approach was used to look at point-score increases (not level gains) using longitudinal data that were adjusted for student mix and share of students with posttests. When we shared our list of top six performing sites with the AE leaders of each state, there were few surprises. Nonetheless, our list did not match the sites that had been rewarded with state performance grants. Furthermore, in one case, the state staff insisted that one of the programs identified by our methodology was not known for running a good program or for keeping good records.

These divergences between states and between sites highlight the importance of good MIS systems and accurate measures of performance. Because so little money is spent per student, it is very difficult for states to require elaborate administrative records. With the limited amount of data, lack of consistency of testing procedures, and no state creating longitudinal transcript records, the quality of these data are compromised and can only be used with care to make broad evaluations of states and programs.

NOTES

1. This study was conducted when the author was at ORC Macro and was commissioned by the Program Policy Studies Services of the U.S. Department of Education (Adult Education Secondary Data Analysis: Pilot Analysis of Student Attendance, Instruction, and Student Achievement). Neither ORC Macro nor the Department of Education is responsible for the content of this paper.
2. In any high-stakes evaluation, there are pressures to manipulate the results to make them look better. Many times, this practice is quite subtle and does not constitute intentional deception. Therefore, careful training and oversight are necessary to ensure that the data coming from the local providers are accurate.
3. See Medina (1999) for a review of all 115 studies.
4. Approximately 15 percent of learners in each of these states received a GED.
5. Similar data are available for the other states.
6. Although we have no data, we surmise that many of these students are new immigrants who were not in the country until near the time of their enrollment.
7. Approximately 15 percent of learners in each of these states received a GED. Although we have no data, we surmise that many of these students are new immigrants who were not in the country until near the time of their enrollment.

REFERENCES

Beder, H. (1999). *The outcomes and impacts of adult literacy education in the United States.*

Boston: The National Center for the Study of Adult Literacy and Learning, NCSALL Reports #6. Available at: http://gseweb.harvard.edu/~ncsall/research/report6.pdf.

Massachusetts Department of Education. (2001). *The "State of the Art" for ABE Performance Accountability: Profiles of What's Happening in Each of the Fifty States.* Available at http://doe.mass.edu/acls/pawg.html.

Medina, P. (1999). *Appendix to the outcomes and impacts of adult literacy education in the United States.* Boston: The National Center for the Study of Adult Literacy and Learning, NCSALL Reports #7. Available at http://gseweb.harvard.edu/~ncsall/research/report7.pdf.

13 The Tennessee Longitudinal Study of Adult Literacy Program Participants

Mary Beth Bingman

The work presented in this volume looks at lives and changes in lives over time. By looking in a variety of ways at moments in people's lives, the authors have tried to understand who these people are and how they have changed. We all are drawing conclusions from variously assembled fragments. The Tennessee Longitudinal Study of participants in adult literacy[1] programs may be more fragmentary than most.

This chapter discusses a study that started and stopped and started again. Data collection began in 1990 and the first report from which this chapter draws extensively was published in 1993 (Merrifield, Smith, Rea, & Shriver). We collected final data in 1998 and published the final report in 2000 (Bingman & Ebert). An "occasional paper" based on the study and focusing on policy implications was also published in 2000 (Bingman, Ebert, & Bell). Now we revisit the study in this chapter, considering again what we learned about people's lives, the uses and usefulness of our approach, and some implications for practice and policy.

In some of the studies discussed in this volume we learn about particular people; in others we look across a large sample of people. Designed as a mixed method study, the Tennessee project attempted to do both. Our initial design was for a longitudinal study collecting baseline data on three cohorts of new enrollees in selected adult education programs with at least three annual follow-up interviews conducted with these participants to assess changes in their quality of life. We planned qualitative community, program, and individual case studies to add context to the longitudinal data. However, after a change in administration at the state adult education agency, funding for the study was withdrawn in the third year of data collection and many planned components were not completed. Subsequent support from the federally funded National Center for the Study of Adult Learning and Literacy (NCSALL) enabled us to analyze some of the longitudinal data (Bingman, Ebert, & Smith, 1999) and to carry out a small life-history study.

In this chapter we first discuss the methodology and findings of the major component of the study, a series of structured interviews with 450 adult literacy students. We then focus on the qualitative component during which we conducted life-history interviews with ten of the original study participants. We present some of their voices as we discuss the findings of this part of the study. Finally, we look across what we learned from the various pieces of the study and some implications for policy, practice, and research.

RATIONALE FOR THE STUDY

The Tennessee Longitudinal Study conducted by the Center for Literacy Studies at the University of Tennessee, Knoxville, was designed to address the Tennessee adult education agency's interest in knowing the impacts of literacy education in the lives of participating adult students. The state agency makes grants to local providers of adult basic education and was interested in demonstrating the value of these programs. We, as researchers, were interested in the same issue, though from a somewhat different perspective. The state agency hoped to be able to use the study to demonstrate the value of Tennessee's investment in adult basic education. As researchers, we were interested in examining a wide range of possible outcomes of participation and in expanding our understanding of how increased literacy and numeracy skills and participating in education impact adults' lives. We were particularly interested in adding learners' perspectives to discussions on outcomes of adult education. We wanted to hear directly from adult learners their assessment of how their lives had or hadn't changed after they participated in adult education classes.

The Tennessee Longitudinal Study ultimately included three components: the original longitudinal study, a later analysis of part of the data from the longitudinal study, and the small qualitative study with ten of the original participants. All three components examined the outcomes of participation in adult literacy education, asking the question *What difference does literacy education make in the lives of individuals?*

In the longitudinal study we examined how participation in adult literacy programs changes adults' quality of life with a focus on four main areas:

- socioeconomic well-being (jobs, income)
- social well-being (family and community life)
- personal well-being (self-esteem, life satisfaction)
- physical well-being (health and access to health care.)

The research questions addressed in the later analysis of the longitudinal study data were: *What aspects of life change one year after enrollment in an adult basic education program?* In the final qualitative portion of the study our question was *How do adults describe the impacts of their participation in adult literacy programs in their lives?*

LONGITUDINAL STUDY RESEARCH DESIGN

The central component of this research was a panel study of participants in adult basic education programs in Tennessee. For each of three years, a cohort of adult literacy students was to be interviewed when they enrolled in an adult education program, establishing a baseline. They would then have a follow-up interview annually for three to five years.

The participants in this study enrolled in classes in federally funded adult basic education programs in the state of Tennessee sometime between 1991 and 1994. When these adults enrolled in adult basic education programs they, like all students in Tennessee then (and today under the National Reporting System), were asked to complete a standardized assessment. The assessment used at that time in Tennessee was the Adult Basic Learning Exam (ABLE). The adults who were asked to participate in our study scored at Level 1, defined as below grade level 5.9 on the reading component of the test. The 5.9 level is roughly equivalent to that expected of children age 10 or 11. This choice of test and definition of Level 1 were determined by the state adult education division. The decision to include in this study only students who tested at Level 1 on enrollment was based on the state's particular interest in the impact of instruction for adults at this "literacy" level.

The participants in the study were students in one of nine Tennessee adult education programs. Like most other Tennessee adult basic education programs, these were operated by local educational authorities under contract to the state agency. The selection of these particular programs was based on comparison of six demographic variables ("nonwhite," families living in poverty, population change, percentage of adult high school graduates, urban/rural, median years of education completed) deemed relevant to the study. (If we were conducting this study today we would be certain to include a substantial percentage of immigrant students studying English.) Counties were sorted as to urban and rural in the three major regions of the state. Tennessee stretches over 500 miles (800 kilometers) from east to west, and these regions differ economically and culturally as well as geographically, with East Tennessee being mountainous and West Tennessee similar to the Mississippi Delta. The counties in each region with most of the selected demographic variables falling within a half a standard deviation of the mean for their set (urban or rural in their region) were selected as possible sites. Then, based on expectations of sufficient potential enrollment, eight programs from these counties were chosen as research sites. A ninth program was added the second year of the project to better level the sample size for each region.

Local adult education program personnel conducted interviews with entering Level 1 students using a 116-question instrument. In addition to questions about demographics (age, gender, race, marital status, number of children), participants were asked about their education, employment,

Table 13.1 Tennessee Longitudinal Study Cohorts

	Cohort 1	*Cohort 2*	*Cohort 3*	*Total*
Baseline	133	149	168	450
Follow-up 1	70	64	65	199
Follow-up 2	45	39	84	
Follow-up 3	35	35		

social relationships, literacy and numeracy activities, self-esteem, health, and life satisfaction. In annual follow-up interviews, participants were asked about their experiences in adult education as well as about the factors addressed in the baseline interviews. These follow-up interviews were conducted by both adult education program personnel and other trained interviewers. Most were conducted in person, but some were by phone.

The baseline data for the first cohort were collected in 1991–92, the second in 1992–93, and the third in 1993–94. The last follow-up data were collected in 1995, the year funding was eliminated. Data were collected on 450 participants as indicated in Table 13.1.

The original plans for the Tennessee study anticipated comparing changes among individuals coming from different communities and programs and correlating changes with various factors in their lives. The analyses that were completed by the original research team included a description of the first two cohorts at baseline (Merrifield et al., 1993) and an analysis of the one-year follow-up data of Cohort 1 (Merrifield, Smith, Rea, & Crosse, 1994).

The Cohort 1 baseline interviews were analyzed first in frequency tables looking at the characteristics of the sample and various subgroups, and then subgroup differences in the areas of investigation were analyzed. The Cohort 1 participants ranged in age from seventeen to seventy-two, more than half were female, and more than half were African-American. While a few had graduated from secondary school, most had left school because they were not doing well, needed to work, or had married or had children. While they had low literacy skills as measured by a standardized test, they reported a range of everyday literacy activities. Only a third was employed, almost all at low-skilled jobs. As a group, their incomes were low.

These baseline data challenged two stereotypes commonly held about people with limited literacy skills, that they have low self-esteem and that they are not supportive of their children's educations. The participants completed the Rosenberg Self-Esteem Assessment (1965) as part of the baseline interview and showed a range of self-esteem. Their self-esteem scores seemed to be closely connected with aspects in their lives other than low literacy: employment, marriage/gender, personal and community satisfaction (Merrifield et al., 1993). Over half the participants had children under eighteen and most reported being involved in their children's schooling and expected their children would complete their secondary schooling. At least by self-report, they valued education.

In 1994 a second report on the study compared Cohorts 1 and 2 at baseline and analyzed the first follow-up interviews conducted with seventy of the Cohort 1 participants twelve to twenty months after their first interviews. The primary demographic difference in the two cohorts was an increase from 30 percent to 44 percent in participants from East Tennessee and a reduction from 62 percent to 45 percent of those from West Tennessee.

Of particular interest from the Cohort 1 follow-up interviews were findings about continued program participation. Thirty-five percent of participants were still active in ABE programs at the time of their follow-up interviews, a rate considerably higher than the national level of 12.5 percent (Young, Morgan, & Fleishman, 1993). The reason most gave for leaving the program (32 percent) was work schedule. All the people who gained a job during this period left the literacy programs, usually for a low-wage, low-skill job.

At this point, the study was interrupted when the state decided to no longer fund the project. While baseline interviews were conducted with Cohort 3 and follow-up interviews were conducted with Cohort 2 and again with some of Cohort 1, these data were not analyzed or reported.

The last follow-up interviews were conducted in 1995. When we returned to this study in 1997, we considered how to analyze the data we had. Because we had second follow-up interviews with fewer than 20 percent of the 450 participants (eighty-four), we decided that only the first follow-up interviews provided a sufficiently large sample for purposes of analysis. We combined the three cohorts and analyzed the data from the baseline and one-year follow-up interviews available from 199 of 450 original participants. Because three slightly different versions of the baseline interviews and the follow-up interviews had been used in the original research, we decided to use only the items that were consistent across the surveys. We looked at changes in variables from: Work (four items); Self-esteem (ten-item version of the Rosenberg Self-Esteem Assessment); Community (fourteen items); Literacy Practices (nine items); Family Life/Children (six items); and Life Satisfaction (thirteen-item scale).

After one year, participants (combined cohorts, $n = 199$) in the study reported positive change on at least one item in each of the categories examined by this study: employment, self-esteem, community, and children's education. These changes included:

- a higher rate of employment (from 32 percent to 48 percent).
- increased self-esteem (from 3.52 to 3.66 on a five-point scale, $p < .01$).
- increased involvement in community organizations (religious, parent-teacher association, social/sports) ($p < .05$).
- positive changes and numeracy practices examined (paying bills, working with numbers on the job, not needing to memorize because of limited reading ability) ($p < .05$).

- increase in number of people who thought a book was a good gift for a child ($p < .05$).
- an increased overall satisfaction with their financial situation in Cohorts 2 and 3 ($p < .05$).

There were, however, no significant changes in community awareness or in how people felt about their community. There was not a significant increase in reading reported, nor a significant increase in involvement in children's education. There were few significant changes in life satisfaction.

When we compared our findings in the one-year follow-up of Cohort 1 with the follow-up of the combined cohorts, we found some differences. In both, employment had increased and those who gained employment were more likely to leave the program. There was modest increase in self-esteem in both groups. The positive changes in the combined cohorts in community involvement, literacy and numeracy practices, and overall satisfaction with their financial situations were not found in Cohort 1 when considered separately.

From the two sets of interviews we found a variety of changes over the course of a year. There were positive changes in at least one item in each of the broad areas examined: employment, self-esteem, community involvement, and children's education. But we failed to find significant changes in other items in each area. We can conclude that participation in adult education programs seemed to have had some impact on participants and that these impacts affect various life aspects. But we were not able to gain a clear picture of what these changes meant in people's lives. The life histories helped fill in this picture.

LIFE-HISTORY COMPONENT

In the beginning of this chapter we referred to fragments of lives. With the survey questionnaires we heard people's responses to questions we posed at a few points in time. We ended up with aggregate pictures of situations and perceptions across nearly two hundred people, with some "subaggregates"—employed or not, still in the program or not, and so on. But in this process we lost the individual people. We used a life-history approach to bring some individual lives and voices to this study.

Life-history interviewing results in a set of narratives from the participants generated as they respond to general questions with their own focus. Mishler (1986) described this approach as encompassing " . . . much more of a person's life than a narrowly specified particular situation experienced by all . . . respondents. Nonetheless, . . . attention is centered, or focused, primarily on a general 'situation' that they have in common" (1986: 99). In our study, the situation was participants' experiences related to literacy practices before, during, and after their enrollment in literacy programs.

Analysis of life-history narratives " . . . produces paradigmatic typologies or categories" (Polkinghorne, 1995: 5).

The ten adult learners in this life-history study were a subset of the original sample of 450 participants in the longitudinal study. They were chosen from the 139 participants who had participated in a follow-up interview in 1995, the last year of data collection in the original study. We anticipated that these people would be easiest to locate. They were from all three cohorts of the original study.

The follow-up interviews in the original longitudinal study were conducted with participants whether or not they remained in the program and without regard to how many hours they had participated. For the life-history interviews, we wanted participants who had persisted in adult education programs long enough to expect some impact on their literacy and numeracy skills. We narrowed our potential sample to the thirty-six of the 139 who had taken part in at least eighty hours of adult literacy class. Of these thirty-six, we located twenty-three who were willing to take part in this phase of the study. From this set of twenty-three we chose ten people who as a group roughly matched the demographics of Adult Basic Education students in Tennessee in terms of gender, age, and race, were both urban and rural, and were from the three regions of the Tennessee.

These ten people were interviewed in their homes or another place of their choosing. The audiotaped interviews were structured by very general questions asking participants to talk about their adult education experiences, their family and work lives, their childhoods and earlier schooling, and the changes in their lives that they attributed to adult education participation. Second interviews were conducted with eight of the participants, clarifying points and addressing issues from the first interviews.

Space does not allow us to present each person as we came to know them through their stories of their lives. We will present some basic information about them in the table following and then tell more about their understanding of their lives and changes in their lives in our discussion of what we learned. (All names are pseudonyms.) We include race here because of the effect that racial segregation had on the education of African-American participants. The hours of participation in adult education are from time of interviews, were often interrupted, and are estimations in most instances.

We analyzed the data from the life-history interviews using an inductive iterative process. We noted both particular stories and common themes and categories. The broad categories that cut across all interviews were, to some extent, determined by the questions asked, for example, about work, adult education, early schooling, and family. We also identified themes that emerged from the interviews that cut across these categories, for example, value placed on education, the impact of poverty and race on education, literacy practices, and sense of self.

From these life stories, some much richer in detail than others, we learned about and from these lives and about the changes the participants

Table 13.2 Participants in the Tennessee Life-History Study

Name	*Gender*	*Race*	*Birth Year*	*Employment*	*School Year Completed*	*Hours of Adult Education*
Bert	M	B	1970	Some past employment, mostly temporary jobs	9	122+
Elizabeth	F	B	1933	Retired after l ong-term job in restaurant.	7	148+
Harry	M	W	1924	Retired after long-term employment in manufacturing	8	247 GED
June	F	W	1966	Little past employment. Has job as kitchen aide	9	101+
Kris	F	W	1968	Past employment in restaurant and chicken plant	9	95+
Laura	F	B	1952	Past employment in service jobs and manufacturing	12	127+
Marvin	M	B	1945	Retired from long-term job as mechanic and truck driver due to poor health	5	98
Ruth	F	W	1955	Long term employment in food processing	8	137+
Suzanne	F	W	1966	Intermittent employment ın manufacturing	10	82 GED
Will	M	W	1945	Long-term employment in manufacturing	8	214

perceived. We learned about the complexity and the ordinariness of adults who were also literacy students. We learned about the impacts of racism and poverty on their earlier educations. As in the longitudinal interviews, we found challenges to assumptions about self-esteem and intergenerational illiteracy. We heard how people with limited reading skills were able to navigate both text and long-distance truck routes. We also heard about the changes that participants attributed to their participation in literacy programs, changes in how they used literacy and in their sense of self.

Adult students bring a history of schooling to their adult education experiences. While these experiences are in some ways those of particular children with particular abilities and interests, they are shaped by factors that are well beyond the individual child. In the lives of these Tennessee adults we clearly see the impacts of poverty and legalized racism. Laura, Elizabeth, and Marvin all were children while schools in southern states of United States were legally segregated by race. Schools serving African-American children were typically underresourced and the school year in rural areas was often scheduled to allow children to be available to work in the fields.

> *I was going to school but I didn't get a chance to go every day. Because back then if you were able to pick cotton and chop, . . . whenever cotton chopping time come, school would shut down, we would have to go to the field. . . . The tiny little time that I did have in school, I forgot it all when my father died.*—Marvin

> *Most of the books we got then was books passed out to us from the White school. And that was better than nothing, not having books at all in the school.*—Laura

Marvin left school when his father died in order to help support himself and his mother. He was ten. Laura graduated from secondary school, but described herself as having been "slow" in school and suspects she has a learning disability. She had received no special help and had limited reading and writing skills. Harry and Will went to "White" schools, but both left school in the eighth grade to go to work. They were not able to complete their secondary education.

The older adults among these ten had their schooling impacted by poverty exacerbated by racism. Many of the others described leaving school for reasons that might be linked to poverty—feeling out of place in a new school when their families moved to find work; their mother working "three, four jobs" when an addicted father did not have a job. We might have expected to find what has been called a "cycle of illiteracy" (Darling & Hayes, 1989) in their families, that their children would have limited education because their parents did. Instead they seem to have been quite successful in supporting the education of their children. At the time when we conducted our interviews, the participants had among them a total of twenty-six children. Eighteen of these twenty-six were aged eighteen or older, old enough to have completed secondary school. All but one of the eighteen had graduated from secondary school and eight had some postsecondary education. We can't generalize from these eighteen, but these life histories suggest that the factors that affect school leaving may be systemic as much as individual and should be addressed systemically rather than only focusing on individual solutions.

In their interviews, participants identified changes in their lives that they attributed to their participation in adult literacy programs. We grouped the changes into two broad areas, literacy practices and sense of self. Nine of the ten described particular skills they had acquired from their participation in adult education and eight talked about changes in literacy and numeracy practices. These are summarized in the following table:

Table 13.3 Changes in Literacy and Numeracy Skills and Practices (Direct Quotes in Italics)

Name	*Skill Change*	*Practice Change*
Bert	*Read quicker* *Learned a lot of new words*	More reading Going to library
Elizabeth	Breaking words into syllables Compound words Understanding words better *How to do math better and better*	*Counting my pennies better* Reading Bible better Reading (official) mail Making out a money order Reading Sunday school book Reading newspaper occasionally
Harry	Algebra and geometry for GED test Format for essays	
June	*Read a little bit faster* Using sounds to spell words How to use the dictionary	Paying attention to news on TV Reading more to youngest child Opening and using a checking account Using the dictionary Reading newspaper
Kris	No changes	*It didn't make no difference.*
Laura	Reading maps Read even better Using sounds in spelling	Use maps to find new apartment Reading the Bible better Reading "little" books
Marvin	*Learned how to read some* Multiplication tables	Reading to grandchild Using instruction to program remote control for television Reading mail Reading newspaper
Ruth	Elements of standard written format— paragraphs, margins Fractions	Reading instructions Writing letters in standard format Relating better to the newspaper
Suzanne	Measurement How to write an essay	Using measurement at work Reading more to youngest child
Will	Spelling Math	Filling out job reports Reading books Reading the Bible more Reading the newspaper regularly

This group read from a wide variety of texts from children's books to court papers. They described changes in literacy and numeracy practices that had impact in various aspects of their lives. Many noted changes in conducting the everyday "business" of life—sending money orders, reading mail from the housing authority, reading and using instructions, using maps. June, for example, opened a checking account.

> *I got it in May. Doing good on that. I ain't bounced no checks. Make my deposits that I need to make to cover what bills I have to pay and always make sure that monthly charges [are] took off, and make sure I got the money in there to cover the checks that I do write.*

There were changes in literacy and numeracy practices at work for Bert, Suzanne, and Will, who described filling out "tickets" on work accomplished and now, he joked, "I don't have to count on my fingers and toes as much as I did."

Five participants described new uses of literacy to engage with the wider world, reading the newspaper regularly and having an increased knowledge of and wider interest in the news. For Marvin, this reading the newspaper impacted his interactions with friends:

> *I'm not . . . don't get me wrong now, I am not real good at reading still, but I've learned a lot about reading. It makes me want to read now, try to read better. Because there's a lot of interesting things that you miss when you can't read. Like newspapers, you can't sit down and read it, you miss everything. You try to sit down with other people and get in on the conversation and you don't know what they're talking about. You didn't read the newspaper. Because I've got a lot of friends right now that we sit down and talk, and hey, I done looked in the paper and see the same thing that he sees, so we can discuss this matter. You know, "Hey, so-and-so, and so and so"—"Yes, I see that, too."*

Some of the changes came from new reading skill; others came from greater general knowledge. Laura developed an interest in the early African-American poet Phyllis Wheatley. June described a new interest in international news.

> *The news talks about what's going on overseas and stuff. Before, I thought, well, that's just news, something to report. Then after I read the book and learned how this become, I understood more. I was, "OK, now this is how this ended up at." I'd be walking in there to get something [near the television] and something overseas happen. I'd stop and come back. "Wait a minute now, I've got to catch this." Grow more, more interested now. I understand more now.*

Perhaps the change in literacy practice most strongly noted in this study was a change in how participants perceived reading and themselves as readers. Most participants reported reading as something they now do regularly, not just when needed for a particular task. Suzanne, Kris, Bert, and Harry said they read for pleasure before attending adult education, but for others this habit of reading was a change. June and Laura talked about practicing reading in texts they found interesting, "dream books," horoscopes. Marvin and Will said they now "pick up books" and read. They mentioned continuing difficulty with reading, but they read.

The other change that most participants attributed to their participation in adult literacy programs was in their sense of self. When we analyzed how these ten people described their lives and described themselves, we noted a strong sense of self-efficacy in eight of them. (The other two both had serious health problems.) They talked of themselves as able, competent people who had accomplished a great deal in their lives, often in the face of difficult circumstances. They had held jobs, raised children, been involved in church; they had survived and more. They did not seem to suffer from the "low self-esteem" often ascribed to adult literacy students.

At the same time, they did attribute positive changes in their sense of self to their participation in adult education. Three of them spoke of their sense of shame about their limited literacy skills and how that changed when they realized that they were not alone in their lack of skill or, as Will described it, "I mean everybody was in the same boat when you got there." They also felt a sense of accomplishment in what they had learned. Suzanne and Harry passed the GED (General Educational Development) test and were granted a secondary certificate. Suzanne described what that meant to her:

> *Four times and I finally, finally done it! And it was all kinds of certificates. I got them on my wall and I keep looking at them and think, "Well I did that." I believe now that I have more determination in me now that I ever had.*

For some their participation in adult education led to a change in voice, to a stronger sense of having something to say and being willing to say it. Marvin and Harry both spoke publicly about the importance of literacy. Ruth had worked many years in a plant that processes chickens. She said she had been shy and "afraid to speak my opinion," and that that had changed.

> *Also about speaking up, I can do this better now. Like recently, at my work I was scheduled to have a vacation for Christmas. And then my supervisor comes to me and says that this other woman will get Christmas week off, not me. And I was already scheduled and she even had less seniority. So I spoke up. I said, "No, it isn't right. I want my vacation." And I got it. And before I was so shy.*

WHAT DID WE LEARN?

When we look across the results from the longitudinal study interviews and the life histories, we can see some answers to our research questions: *What difference does literacy education make in the lives of individuals? What aspects of life change one year after enrollment in an adult basic education program? How do adults describe the impacts of their participation in adult literacy programs in their lives?*

The (one-year) longitudinal data indicate that literacy and numeracy education may contribute to increased employment, but is not likely to lead to substantial change in income, at least in the short term. Levels of employment in the combined cohorts increased, but the Cohort 1 follow-up found most of the newly employed were in low-paying jobs with limited benefits. For the life-history participants, lack of a credential seemed more important than lack of skills. Ruth could not become a supervisor in the chicken plant until she had the GED credential and Will lost a new job as a mechanic when his employer learned he did not have a secondary credential. Kris and Bert both wanted to enter computer training but needed a GED to qualify.

We found some reported improvements in social well-being, family and community life after enrollment in adult literacy programs. The combined cohort showed significant increases in involvement in community organizations including church, parent-school organizations, and social or sports groups. Bert, June, and Marvin talked about increased comfort in social situations. While we did not find an increase in involvement in children's schooling beyond increased involvement in parent–school organizations in the combined cohort, Marvin, June, and Suzanne reported reading more to their children. The parents in this study had already supported their children's education in a variety of ways.

The new literacy and numeracy practices that participants reported facilitated some of the changes in their social life and may have also contributed to increased self-esteem. The self-esteem of the combined cohorts made a small but significant increase as measured by the Rosenberg scale. Nearly all the life-history participants spoke of ways they "felt better about themselves," because of participation in adult education or because of the skills they gained there.

The participants in the Tennessee study reported positive changes in a variety of life domains after their enrollment in adult education. These changes were not dramatic, did not change their lives in immediate substantial ways, but did seem to be seen as positive changes, as improvements. Lives of adults, including adults with limited literacy and numeracy skills, are complex, include many challenges, and are affected by a wide range of factors internal and external. The outcomes of participation in literacy program participation in learners' lives are also diverse and often complex. Participation in adult educations seems to increase options that adults have or perceive, and that may be value enough.

What are the implications of the Tennessee study? First, our research illustrates some of the challenges that need to be taken into account in doing longitudinal studies in adult education and the impact of decisions about funding. A primary methodological issue we faced is evident from Table 13.1. We were not able to maintain our participants over time. When we compared demographic data of participants who completed the baseline interview and those who were interviewed one year later, we did not find significant differences in gender, marital status, previous schooling, or employment between the two groups. However, the follow-up group was significantly older, more likely to be white, and more likely to be rural than the baseline group. We do not know how these differences affected outcomes. Even in the life-history study, we lost two participants between the first and second interviews. This may be a particular challenge working in adult education, but it also reflects the challenges of working across a large geographic area to conduct face-to-face interviews. The LSAL study (Reder, this volume) has had success maintaining contact with participants by starting in a smaller geographic area and by investing considerable resources in maintaining contact with participants.

The connections of our study with the Tennessee adult education agency had major implications for the research. First, their decision to stop the funding three years into the project had a major impact on what we were able to do. After the director who had made the initial decision to fund the study resigned, the remaining staff determined that the study or at least the results as found in the first two reports told them nothing that they did not already know. They did not see any value from the study to their day-to-day work and felt the funds might be better spent on provision of services. In retrospect, we might have done a better job of defining the value of longitudinal studies to practitioners.

The state's concerns about including all the regions of the state in the study, essentially a political decision although there are geographic and cultural differences, complicated the research and seemed not to have mattered much in the end, though we could not know that in advance. We had also decided to use local program staff to conduct interviews, and in some instances they made decisions about whom to interview and when; those decisions probably impacted the quality of the data. This was a factor in who received follow-up interviews for at least the first cohort. We might have had better control if we had used our own interviewers, although this would have required substantially more resources.

Our lack of information about the programs in which adults participated seriously limited what we could learn about impacts of program practice on learner outcomes. The planned case studies of the programs would have helped fill this gap. If we were doing this research today we would have access to a much more sophisticated state data collection and reporting system that could provide a better understanding of various program factors.

The Tennessee study was planned as a mixed-methods study. A few initial case studies of programs and communities were conducted by graduate students but did not really contribute to the overall study. We were, however, able to conduct the set of life-history interviews discussed in this chapter, and the stories of those particular people help bring the longitudinal data to life. We can in some sense "know" a few of the people who were part of the larger study. Their life stories help us understand the experiences of older adults from the South and the limits placed on their schooling as a result of their poverty and race. Adult educators often refer to self-esteem as either an accomplishment or a reason for their students' lack of accomplishment. The ways the adults we interviewed described their own efforts and accomplishments, as well as their discomfort with what they were not able to do, complicates this notion of self-esteem. In a mixed-methods study the quantitative data can give power to the conclusions and qualitative data add the depth of context and the complexity of life as lived.

Our study—the richness and complexity of the lives we learned a bit about and the range of changes we saw—also has implications for determining program accountability in adult education. This research indicates that common measures of literacy program outcomes such as student test scores and employment gains may miss much of the impact that participation in literacy education has in people's lives. These narrow measures may show to what extent programs are effective in helping some students make the changes they want in their lives. For many other students these measures tell us little or nothing about what these students hoped to achieve or the effectiveness of programs in helping them reach their goals.

NOTES

1. In this chapter I use the word *literacy* as it was used in the United States at the time of the study, for example, referring to literacy students and literacy programs. Numeracy was an aspect of what the programs offered and what the students studied, and literacy in those contexts should be assumed to include numeracy in most instances. When discussing our findings I refer to literacy and numeracy as is relevant.

REFERENCES

Bingman, M. B., & Ebert, O. (2000). *"I've come a long way": Learner-identified outcomes of participation in adult literacy programs.* NCSALL Report 13. Cambridge, MA: National Center for the Study of Adult Learning and Literacy.

Bingman, M. B., Ebert, O., & Smith, M. (1999). *Longitudinal study of adult literacy participants in Tennessee.* NCSALL Report 11. Cambridge, MA: National Center for the Study of Adult Learning and Literacy.

Bingman, M. B., Ebert, O., & Bell, B. (2000). *Outcomes of participation in adult basic education: The importance of learners' perspectives.* NCSALL Occasional Paper #1. Cambridge, MA: National Center for the Study of Adult Learning and Literacy.

Darling, S., & Hayes, E. (1989). *Breaking the cycle of illiteracy: The Kenan family literacy model program*. Wilmington, NC: University of North Carolina.

Merrifield, J., Smith, M., Rea, K., et al. (1993). *Longitudinal study of adult literacy participants in Tennessee: Year One report*. Knoxville, TN: Center for Literacy Studies.

Merrifield, J., Smith, M., Rea, K., et al. (1994). *Longitudinal study of adult literacy participants in Tennessee: Year Two report*. Knoxville, TN: Center for Literacy Studies.

Mishler, E. G. (1986). *Research interviewing*. Cambridge, MA: Harvard University Press.

Polkinghorne, D. (1995). Narrative configuration in qualitative analysis. In J. A. Hatch & R. Wisniewski (eds.), *Life history and narrative* (pp. 5–23). Bristol, PA: Falmer Press.

Rosenberg, M. (1965). *Society and the adolescent self-image*. Princeton, NJ: Princeton University Press.

Young, M., Morgan, M., & Fleishman , H. (1993). *National evaluation of adult education* programs, second interim report: Profiles of client characteristics. Arlington, VA: Development Associates, Inc

14 Outcomes of Literacy Improvement

A Longitudinal View

Barbara A. McDonald and Patricia A. Scollay

Literacy holds a pivotal and unique place among the necessary skills for successful living in our society; yet we know little about adults who have low literacy skills. In fact, it is difficult for those who are literate to imagine what life would be like without knowing how to read and write. Nevertheless, many Americans do have trouble reading, making it difficult for them to fill out forms, read instructions, and apply for and get good jobs.

There has certainly been recent interest in the literacy levels of adults in the United States (Beder, 1999; Grubb & Kalman, 1994; Sheehan-Holt & Smith, 2000; Sticht, 2002). Concomitantly, there is a growing interest in adult education (Askov, Johnston, Petty, & Young, 2003; Comings, Reder, & Sum, 2001; Drago-Severson, 2004; Kegan, Broderick, Drago-Severson, Helsing, Popp, & Portnow, 2001; Reder & Strawn, 2001; Sticht, 2005; Strawn, 2003). Nonetheless, our understanding of both the social context (LeVine, LeVine, & Schnell, 2001) and the process itself remains incomplete (Kansanen, 2003).

One reason we understand so little about adults who don't know how to read is that people in this situation do not make it widely known. There is a stigma associated with low literacy. Brandt (2001) asserts that while knowing how to read enhances economic and political opportunities, not knowing decreases them. And literacy is more than a set of processes to take in and communicate information or get a job. Formal literacy skills are related to the distribution of social power (Gee, 1999). In fact, Gee says that people use discourse to display membership in particular groups, especially if they are proud of themselves. He calls literacy abilities "identity kits" describing them as ways in which we identify ourselves to others like us. Children spend years in school learning formal literacy skills and the appropriate style of discourse to accompany them.

Such learning is a complex process, viewed by Hedegaard and Lompscher (1999) as fundamental to psychic development. Thus, school provides children with an environment not only for learning reading and math but also for acquiring social knowledge and skills. They suggest that learning is the leading activity of personality development. Of course, since children's literacy acquisition is partly derived from social interactions with peers in the classroom, the acquisition or improvement of literacy for an adult may be quite different.

Rogers (2002) discusses the difference between discourse practices used in formal settings such as school (secondary discourse) and discourse practices used in the home and community (primary discourse). There is widespread belief that only the school-based discourse is valuable, and this perception is often held by those who lack it (Heath, 1983; McCarthey, 2000). Indeed, of seventeen adult learners from library literacy programs throughout California interviewed for a documentary film in 2004, all expressed a lack of confidence, a need for the basics learned in school, and a fear of the school environment. These feelings persisted even though some of them were financially quite successful: one had gone on to complete his BA and two were successfully attending college (Rameriz & Talan, 2006).

From a societal standpoint, the issue of low adult literacy levels is often addressed preventively by K–12 school reform. However, the obvious connection between adult literacy and children's test scores is sometimes missed. Family literacy research has attempted to rectify this by showing that as a parent becomes more literate the school performance of the children improves (Brooks-Gunn, Berlin, & Fulgini, 2000; Brown, 1989; Debruin-Parecki, Paris, & Siedenburg, 1997; Sticht & McDonald, 1989).

The importance of parental involvement in children's learning is well documented (Christian, Morrison, & Bryant, 1998; Leseman & de Jong, 1998; Senechal & Lefevre, 2002). Parental involvement is highly correlated with children's academic success (Eccles & Harold, 1996; Epstein, 2001; Hoover-Dempsey, Bassler, & Burrow, 1995). For parents who have low literacy skills, full participation in their children's school activities can be challenging and even intimidating. In addition, since literacy skills are social as well as academic, many activities are affected by parental skills and abilities.

Based on the literature we would certainly expect that adults who improve their literacy skills would experience positive change in many areas of their lives. In fact, there is considerable anecdotal evidence about these changes. Based on responses to open-ended questions in another study, Terry (2006) documented that the learners showed "improvement in six major areas: (1) general attitudes toward life, (2) levels of interpersonal awareness, (3) self-esteem, (4) academic confidence, (5) learning goals, and (6) employment plans."

The problem of adult literacy has been addressed in many ways. Because of the evidence that parental literacy is crucial to children's developing literacy skills, some programs focus on "family literacy" as a whole. They attempt to improve the literacy skills of the parents and in many cases train the parents to work with their children. The Families for Literacy Program of the California State Libraries is one of those. Local branch libraries implement the statewide Families for Literacy Program in which a trained tutor is paired with a learner, allowing each learner to proceed at his or her own pace. Adults who qualify for this program must have a preschool-age child. Although the individual lessons may cover various things, the primary focus is on reading and writing and not numeracy or other basic skills.

Our research was specifically designed to look at life change in adults and we drew our participants from that program. Our project was a three-year longitudinal interview study of adults who were learning or improving basic literacy skills. To engage in a literacy program as an adult is a big undertaking that contributes to greater life satisfaction, new jobs, and better outcomes for the children. We wanted to know how literacy improvement connects to these broader outcomes. While there is a considerable amount of correlational evidence that increasing educational attainments of adults will have positive effects on their lives and their children's literacy, the institutions and social contexts that accompany the changes are still unclear (LeVine, LeVine, & Schnell, 2001). In addition, the instructional process itself and how it actually contributes to such changes remains a mystery (Kansanen, 2003).

Our study provided an opportunity to get a deeper view of the learners and the impact their literacy experiences had on their lives. We chose to combine qualitative and quantitative methods, using interview protocols with open-ended questions and interviewing a large number of participants. This allowed us to hear about the learning process from the learners' and the tutors' viewpoints and also to describe the population statistically. The open-ended interview questions were designed to provide data about the key themes of the study. We asked about the learners' children, about the tutoring sessions, about both learners' and tutors' social networks, about literacy practices, jobs, and hopes for the future.

Throughout the study we, as researchers, were pulled by opposing forces. On the one hand we wanted to adequately represent the individuals we interviewed. On the other hand, as scientists, we understand that conclusions cannot be based only on anecdotes and stories, and we know that patterns in data found by analyzing the larger group can provide valuable insight into how adult literacy learning changes people.

RESEARCH QUESTIONS

The following research themes were key to our understanding of our participants:

- First, the description of these adults: who they were demographically, socially, and psychologically.
- Second, the description of the tutoring sessions.
- Third, documenting personal changes for the learners, particularly changes in identity and social networks.
- Fourth, describing the impact of the parents' participation in the literacy program on the children.

The following table presents examples of some of the research themes with interview questions used to gather information.

Table 14.1 Some Research Themes and Corresponding Interview Questions

Research Theme	*Question(s) Asked in Interview*
Literacy Tutoring Activities	To Learner: "What are you working on in your tutoring session?" To Tutor: "What are you working on in your tutoring session?" "How do you know if your methods are working?"
Personal Change: Due to the Program	To Learner: "What has been most helpful to you about this literacy program?" To both: "Has being in this program changed the way you see yourself or the world?" "Is the way you see yourself now different from the way you saw yourself in the past?"
Personal Change: Identity	To both: "How would you describe yourself to someone else?"
Personal Change: Social Networks	Method of Antonucci (1986) using social circles: "Who would you put in the first circle closest to you, that you could not imagine life without?" and second and third circles distinguished by how close people are
Impact on the Children	"How is your child doing in school?" "What do you do to help your child with learning or reading?"

STUDY SAMPLE

Participants were selected from all libraries participating in a statewide family literacy program: the Families for Literacy Program of the California State Library. Approximately 10 percent of participants of each local program were interviewed the first year. The aim was to sample in such a way as to represent the diversity of the library's adult literacy learners. In addition, we were interested specifically in life change that resulted from program participation. Therefore, we asked the program coordinators at each library to make recommendations of people who would be willing to participate. Although we recognize that this biases our sample, we felt that this was an acceptable bias because we were more interested in questions of how people changed while they participated in the program than in assessing whether the program was effective overall. In other words, we felt that to study the impact of developing literacy we needed participants who were successful in the literacy program.

In a three-year longitudinal study we interviewed 132 learners the first year and had a 14.5 percent attrition rate per year. We also interviewed each learner's tutor. In the first year we interviewed 127 tutors and had a 15.5 percent attrition rate per year. Attrition in our study was due primarily to being unable to locate the participant in successive years. Rarely did a participant refuse to be interviewed after the first year. However, when we tried to call for follow-up interviews it was sometimes the case that the contact information we had was no longer correct and that the library had also lost contact with the participant.

Attrition from the library literacy program was a different issue. We continued to interview learners even if they were no longer part of the Families for Literacy Program. Of these learners interviewed, 61.2 percent were still in the literacy program in the second year and 55.2 percent remained in the program by the third year. Analyses were conducted to determine what if any variables might distinguish those who persisted in the program from those who dropped out. The two groups were remarkably similar demographically; they had similar goals and similar relationships with their tutors. There were two major reasons people left the program. Either they had life events that prevented their continuation or they moved on to higher education or better employment.

RESEARCH METHODOLOGY

This was a qualitative investigation using the narrative method, and in particular using a psychological framework (Denzin & Lincoln, 2000). Our study was different from most qualitative studies, however, in the large number of participants we used. When it came to coding the data, we used the constant comparative method of analysis, which is part of Glaser and

Strauss's Grounded Theory approach (Glaser & Strauss, 1967). We used the data analysis method without claiming to build theory. The method did help us select theories to assist with explanation of our findings. Understanding of our participants and the issues they face was made much clearer by using the method.

Qualitative data are difficult to analyze because of the wide range of possible answers to all of the questions. This makes it difficult to understand the patterns in the data. This is not such a problem with questionnaires to which participants select from a group of four or five alternatives, making it easy to see how many people selected each possible answer. This allows quantitative analysis of data and the use of statistics to ensure that the patterns observed are not due to chance.

We needed a methodology that would incorporate the strengths of both quantitative and qualitative approaches. A qualitative approach turned out to be essential in understanding literacy learning from the standpoint of the individuals involved. Initially we developed categories for coding the data based on previous research; however, these did not adequately represent what the participants actually said. What worked better for our data was to read all of the responses and search for themes: the qualitative approach. We then used the themes that emerged as the categories for coding: the quantitative.

Two other major benefits accrued from the method. First, we had two distinct groups of interviewees: the learners and the tutors. These groups were involved in the same activity but with different roles, motivations, and demographic profiles. Second, we categorized our learners into various subgroups based on variables like years in program, language spoken, number of children, and so on, in order to see patterns in the data more clearly. We could then analyze the data comparatively, looking for differences between learners and tutors, or between learners of particular subgroups.

FINDINGS

Our project findings are reported in detail in a technical report (McDonald & Scollay, 2005).[1] For purposes of this paper we will present a summary of our main research findings for each of the four research questions.

First, the description of these adults

Predominantly our learners were women (89 percent) with a mean age of 35.41 years. They represented the diverse population of California with 65 percent identifying themselves as Latino, 10 percent as African-American, and 5 percent as Caucasian. As a requirement of the literacy program, learners had at least one child under five years of age. They reported a mean number of 2.64 children, almost all under ten years of age. Because of these older children, we were able to find out how parents were helping their children in school.

While there is narrow variability on some dimensions, for example, there are only eighteen men in this sample, there is broad variability on some of the demographics within the sample. Of 132 learners we interviewed the first year, thirty-seven were native English speakers while seventy-seven were native Spanish speakers and twenty-one more were native speakers of Russian, Chinese, Cantonese, Japanese, Korean, Armenian, Punjabi, Portuguese, Arabic, Urdu, Hindi, and Hmong. Of the seventy-seven who were native Spanish speakers, forty-four preferred to be interviewed in Spanish and the other thirty-three were interviewed in English. Those native speakers of all other languages were interviewed in English. For all the nonnative English speakers, one of the main reasons for coming to the library for literacy training was to learn the new language.

To assess the learners' goals we asked why they joined the program and where they saw themselves in five years. Three major types of motivation were given as reasons for entering the program—improving literacy, improving English, and helping their children—each accounted for about one-quarter of the responses. Learners often had multiple responses, and about half of the participants mentioned helping their children as one of their responses. However, their five-year plans focused on employment and education rather than family changes.

The tutors were much less diverse, largely middle-aged women. They ranged from twenty-three to eighty-nine years old; the average age was 53.46 years, and 83 percent were women. The overwhelming majority (92 percent) were native English speakers, although 6 percent were Spanish speakers. They were largely white (72 percent) although 13 percent were Hispanic and 5 percent were African-American. Most (63 percent) lived with a spouse. Although they averaged 1.6 children, only four of the 170 children were under seventeen years of age; most of the tutors had empty nests.

Second, the description of the tutoring sessions

We asked both learners and tutors what they worked on in the sessions but relied on those of the tutor for analysis. We found that the tutors gave very elaborate and detailed answers to this question but that the learners gave much more general answers to this question. For example, the tutor might give extensive lists of activities in the sessions while the learner might have only said "we practice reading."

The tutors made a significant investment in the program. They had been tutoring for between one month and eighteen years, with an average of just over three years (37.3 months). They averaged 2.24 learners over that period, and some tutored more than one learner at a time. They interacted with their learners on a regular basis, averaging 1.5 times a week. In addition, many of them talked about the time needed to prepare for their tutoring sessions. This was a long-term commitment of several hours a week.

Tutors are given training in a variety of techniques and suggested materials and had many different approaches to working with their learners. Overall, thirty-six techniques were described ($M = 3.90$ methods per tutor). There were many indications that the tutors tailored the lessons for the individual learner. When discussing the techniques they used, 72 percent of the tutors made very specific personal statements about the learners' needs or desires. Of the 115 tutors to give detailed descriptions about their strategies, 47 percent revealed the in-depth knowledge they had of their particular learners.

Only 26 percent used an established program and they tailored the lessons to the learner. Many tutors used reading material of particular interest to the learner such as driver license manuals or even comic strips. However, 42 percent did use books provided by the library literacy program. Whatever methods the tutor chose, she had well-defined ways for assessing their effectiveness. The tutors had given considerable thought to assessment and readily described their methods, which ranged from obvious signs like performance on tests to more subtle cues like noticing whether the learner found time to work on an assignment.

We also found that the tutor not only provided literacy instruction but also assisted with a multitude of life and social issues concerning the learners' children, their job searches, their desires to participate in the community, and their housing situations. In these cases, literacy materials related to these real-life needs were used: paperwork, applications, newspapers, government forms, recipes, and the like. Certainly one of the reasons that tutoring is an effective way to increase literacy is that a personal relationship develops between the tutor and the learner. The tutor/learner relationship is highly valued by both parties. It allows the tutor to personalize the lessons in terms of both level and content and the relationship provides social support for the learning process and the accompanying life change.

Third, documenting personal changes for the learners, particularly changes in identity and social networks

Changing the definition of learning

A change that we hadn't anticipated, but that we feel is very important to the lives of the learners and their children, was the change in their perception about the learning process itself. Both tutors and learners were asked the question "Could you describe an important learning experience you remember from the past?" Tutors have so many positive learning experiences that a common response is that there are too many to mention. Learners, on the other hand, reported that they were learning to read. Many of them said they had no memorable learning experience until this program. This suggests that they defined learning experiences in a somewhat narrow way, limiting them to academic. The tutor's rich background of positive learning was passed on to the learner, who came to see learning in a new light. By the next year, learners mentioned more life lessons from their past, perhaps revealing a shift in their perspectives. There was no indication that they had come to devalue school-based discourse, only that they had come to value primary discourse as well.

Identity

We asked the learners and tutors two questions to look at identity and change. One question was "How have you changed as a result of the program?" When it came to whether or not learners felt they had changed, they overwhelmingly (96 percent) agreed that they had. By different, they said they meant "better," "happier," and "more responsible." These positive feelings toward themselves as a result of the program continued into Years 2 and 3 of the project.

There was also one main question about their identities: "How would you describe yourself to someone else?" Participants' answers were followed with gentle probes to encourage them to explore their ideas more fully. Often the learners made several statements. The most common response (20 percent) was that they were hardworking and persistent. Many described themselves as being a "people person" or as "friendly" (18 percent), as "outgoing" (14 percent), and as being "a good person" (13 percent). They also described themselves as "family-centered" (10 percent), as being a "student" (9 percent) and as being helpful (7 percent).

In addition to looking at types of statements learners made, we also categorized their answers based on the work of McAdams and colleagues (Mansfield & McAdams, 1996). They suggested that a person's self-concept or identity is both internal and external. It is composed of an internal sense of ability to effect the world, as might be expressed in agency statements. But it is external also with ties to important relationships through roles played and emotions experienced, which a caring statement would express. The self-descriptions of learners and tutors were coded as trait description (nonelaborated or elaborated) and, using the coding scheme of Mansfield and McAdams, as an agency statement or a caring statement. Agency statements could be about self-mastery, achievement/responsibility, self-empowerment, or status. Caring statements could be about love, dialogue, helping, or unity. Table 14.2 presents these data.

Table 14.2 Learners' and Tutors' Descriptions of Self

Category	*Learners*	*Tutors*
Trait Description	1.36	3.65
Agency Statements	2.33	.71
Caring Statements	1.26	1.19

Overall, tutors and learners made approximately the same number of overall statements about themselves (learners = 4.94 and tutors = 5.94), but what they said about themselves differed. Tutors made more trait descriptions of themselves than did learners. Furthermore, tutors made many more elaborated trait statements. They gave examples, qualified their words, told stories about themselves. Learners did not make as many statements and they were not very detailed. When it came to caring statements, learners and tutors made similar numbers of comments. Clearly they all saw themselves as individuals who were needed by others and who wanted to help.

However, the learners made many more agency statements than the tutors. Learners made mastery statements about being hard workers who could accomplish goals; they made empowerment statements about their abilities to succeed (often saying that their tutors had told them so). Tutors' agency statements were more about responsibility, and there were not nearly so many of these as might be expected given how accomplished, busy, and hardworking our tutors were. Although tutors could have made the same types of statements about themselves, they didn't; perhaps they didn't feel the need to say these things about themselves. They did, however, frequently make agency statements about their learners, as a way of providing positive feedback.

As well, achievement plays a large role in self-concept and self-esteem. The self-concept is a work in progress. As one masters important tasks, one becomes more confident of ability to undertake new tasks. According to Holland, Lachiotte, Skinner, and Cain (1998), people use self-descriptions to tell others who they are, but they also use their self-statements to help guide their actions. They try to live up to their own self-descriptions.

Each year the learners continued to stress the positive changes in their lives as a result of participation in the program. By Year 3, the learners' descriptions of themselves had changed dramatically in two important ways. First, they began to use more trait descriptions and, second, they elaborated those statements. In fact, their self-descriptions became more like the tutors' self-descriptions. Using a subset of twelve native English speakers and their tutors, we were able to demonstrate statistically that the learners' self-descriptions were more linguistically complex in Year 3 than Year 1. They used more words and longer sentences in Year 3 than they had in Year 1, and they used more elaborated statements. A similar change was not observed in the tutors.

Social Networks

In order to better understand the social context of literacy learning, measures of the participants' social network were taken using the social convoy method of Antonucci (1986). This method allows each person to name individuals who belong in concentric circles around them, such that those

Table 14.3 Mean Size of Social Circles for Year 1

	Inner Circle	*Middle Circle*	*Outer Circle*	*Total*
Learner	4.96	3.99	2.47	10.07
Tutor	6.42	6.13	6.06	18.61

Table 14.4 Changes in the Learners' Social Circles

	Inner Circle	*Middle Circle*	*Outer Circle*	*Total*
Year 1	5.23	4.39	2.61	12.23
Year 2	7.48	9.86	9.46	26.70
Year 3	8.59	7.90	9.01	25.50

in the inner circle are perceived to be closer socially than those in outer circles. Table 14.3 presents the mean size of the social circles of learners and tutors for Year 1. Clearly, the social networks of the learners are not only smaller than those of the tutors but they are particularly smaller in the middle and outer circles.

In addition, the members of the inner circle many times lived some distance from the learner, further reducing the functional amount of the social support. The children were also very important in the learners' social networks. A large number of learners, 86 percent, included their children in this circle. What is unusual about this was that for many learners the children were the only people in this circle who lived close by.

Looking at annual differences in the mean sizes of the learners' social circles shows interesting changes. Not only do their total circles get larger after the first year but the distribution of the members changes dramatically. These means are shown in Table 14.4.

Over two-thirds of the learners (68 percent) had larger social convoys the second year. Enlargement was especially dramatic in middle and outer circles, often including people met at the library. In Year 3 many of the learners had new jobs or had moved into higher educational settings. Their new social contacts were present in the circles. The larger circles in all cases represented a broadening of social contacts rather than increasing contact with neighbors or family.

Fourth, describing the impact of the parents' participation in the literacy program on the children

Over half of the learners gave helping their children as a major reason for joining the adult literacy program. Even in the first year they felt confident to assess their children's progress at school. Many of the children

(60 percent) were doing well in school; the others were labeled as below average, as having problems, or as having special needs.

However, these parents felt able to help their children with homework: 25 percent reported they helped them with literacy skills, 23 percent said they supervised them, and 14 percent reported that they used outside resources to help them. Only 14 percent said they were unable to help. Detailed comparisons between the learners and their tutors showed that adult learners used many of the same, often highly individualized, techniques their tutors used with them to help their children.

When asked how their school-age children were doing in school the second year, 73 percent reported that they were in the same successful categories used before. This was a dramatic increase. By this time most of the parents also reported taking their children to the library and attending children's literacy programs. Because the library was the location for the literacy tutoring in most cases, the learners became comfortable coming to the library for other events geared toward their children. All of these activities promoted and supported secondary discourse and are the sort of activities highly educated parents typically select for their children.

The tutors influenced the children; 81 percent of them said they knew them personally. Many (50 percent) said they helped the parent to help their children. Some (20 percent) actually tutored the children, helped them directly, or had direct contact with the children's schools. Another 22 percent said they expected the children to be helped indirectly.

The third year interviews revealed a shift in the parents' statements toward much more comfort in talking about their children. A big difference in the interviews in the third year is in the amount of elaboration parents used when talking about their kids. Whereas in the first year it seemed difficult for the parents to come up with stories about their children, in the third year they had many stories. There was more concern about how the children were doing in school or in their preschool learning activities and talked more specifically about which aspects of literacy were or were not going well. The change in these responses must be related to both the improved ability to use complex language for descriptions and the fact that the children were doing better.

SUMMARY OF RESEARCH FINDINGS

- First, the description of these adults:

The learners were an ethnically diverse group of parents of preschoolers. They were mostly women in their thirties, and the majority were native Spanish speakers.

- Second, the description of the tutoring sessions:

The tutoring sessions were highly individualized for the learner. Methodology and materials were diverse but geared to the learner's goals and motivation.

- Third, documenting personal changes for the learners, particularly changes in identity and social networks:

Learners broadened their definitions of learning to include life experiences as well as learning that might occur in a traditional classroom. This added value to their own experiences without devaluing secondary discourse.

There were dramatic changes in self-descriptions. Learners described themselves as being more confident, more capable, and happier. The type of self-descriptions changed. Initially learners used more agency statements while tutors used more trait statements and more elaboration in their self-descriptions. In Year 3 the learners were using more trait statements and more elaboration.

There was indirect evidence of increased literacy. Their language became more complex. They used more sentences, longer sentences, and richer descriptions.

Learners' social networks became larger, expanding particularly in the outer and middle circles. This often included members whom they had met at the library or in their new workplace or school.

- Fourth, describing the impact of the parents' participation in the literacy program on the children.

The percentage of children doing well in school increased dramatically, about 12 percent. Children are benefiting both directly and indirectly from their parents' participation in the adult literacy program. Their parents are actively promoting literacy behavior and academic success at home, are interacting more effectively with the school and the community, are more able to access necessary support in the community, and in many cases the child has direct help from the tutor.

INTERPRETING THE DATA

Clearly we have documented dramatic changes in the lives of the adult literacy learners that we studied. The question arises why were we able to document change that at best had been described anecdotally and the occurrence of which was questionable. We believe that the answer to that is twofold. First, the Families for Literacy program's one-on-one tutoring arrangement is particularly strong in fostering personal growth along with improvements in literacy skills; the changes were real. Second, our methodology was particularly sensitive to these changes and therefore able to detect them. Let us discuss these one at a time.

One of the most important conclusions we have drawn from our research findings is that identity is an important component of learning and can be seen in the context of the activity of learning. Identity shifts and changes through the activity of learning. Sfard and Prusak (2005) affirmed the importance of self-referential remarks as clues to the effects of learning on personal change. Thus, the literacy activity itself was the mechanism for increased knowledge and perspective and a catalyst for important life changes. Within the literacy tutoring sessions there were mini-activities that acted to create a climate conducive to personal change as well as literacy improvement. We found that the library designed a program to help the learners with whatever they wanted to learn, featuring cooperation at many levels.

The library literacy program is set up with the learning implementation done by a volunteer tutor and the nature of the tutoring relationship was one of the keys to the change learners experienced. One-on-one tutoring is done on an individualized basis with specific and relevant feedback. The learners are helped to set explicit goals and lessons are designed to help them meet those goals.

The literacy lessons took place in the context of a relationship between two people who both had important roles to play in that activity. Tutors planned lessons and modeled literacy behaviors. Learners had very personal contact with a person higher in literacy and could see how the tutor was thinking and hear the types of comments the tutor made. Tutors were very positive and caring toward the learners. This positive environment encouraged the learner to stay engaged in literacy tasks over months and sometimes years and reinforced literacy behaviors without punishing imperfect performance.

The activity of learning was conducted in a different way using different "rules" than are typically used in school. The learners and tutors set the learning goals together instead of teachers deciding what students will learn. Techniques were used but if they did not suit the learner, a change was made. Since specific feedback was given, learners learned to evaluate themselves from a literacy perspective. This probably helped them understand their children's literacy efforts better.

Critically important, the sessions were private. There were no comparisons made to others, so there was less concern about being judged. This essentially decoupled the literacy lesson from the institution "school," which for many adult learners had negative connotations and memories and for others seemed too childish. This was also powerful because it decoupled learning from the school setting, freeing the learners from the need to engage in school-based secondary discourse which they didn't know. This allowed them to value their own learning (and discourse) more while still understanding the value of secondary discourse. They could encourage it in their children and still feel positive about their own different kind of literacy.

The combination of being successful at improving literacy skills, being in a reinforcing environment, and being encouraged to redefine learning must provide a very fertile ground for personal change. It is not surprising that the participants report improved self-esteem, broadening social networks, and higher goals.

The fact that we could document it lies in the unique combination of qualitative and quantitative methodologies we used. Because of our difficulty arriving at coding categories from the literature, we believe that as highly literate, educated women we had little intrinsic understanding of the adult literacy learners in this study. The use of open-ended questions allowed the participants to educate us about how they viewed the world and what was important to them. Using their categories to analyze the data was critical to "hearing their voices."

On the other hand, their individual stories were as unique as they were compelling. Though we often complained about the overwhelming amount of data, it was the large sample that allowed us to make definitive statements about the general patterns of the learners' experiences. With a small sample we could not have know whether what we observed was the result of the unique experience of a few individuals or a general pattern. We are comfortable in asserting that our data document what others have suggested based on anecdotal evidence: this adult literacy learning program fostered life change in its learners. Learning literacy was accompanied by improvements in self-image, social support, and their children's performance in school.

NOTES

1. If you want a copy, please contact Barbara McDonald at bmcdonal@mail.sdsu.edu.

REFERENCES

Antonucci, T. C. (1986). Hierarchical mapping technique. *Generations*, 10, 4, 10–12.

Askov, E. N., Johnston, J., Petty, L. I., et al. (2003). Expanding access to adult literacy with online distance education. *NCSALL Occasional Paper.*

Beder H. (1999). *The outcomes and impacts of adult literacy education in the United States*. Cambridge, MA: The National Center for the Study of Adult Learning and Literacy.

Brandt, D. (2001). *Literacy in American Lives*. Cambridge, MA: Cambridge University Press.

Brooks-Gunn, J., Berlin, L. J., & Fuligni, A. S. (2000). Early childhood intervention programs. What about the family? In J. P. Shonkoff, & S. Meisels (eds.), *Handbook of early childhood intervention*. New York: Cambridge University Press.

Brown, P. (1989). *Involving parents in the education of their children.* Urbana, IL: ERIC Clearinghouse on Elementary and Early childhood Education (ERIC Document Reproduction Service Number: ED 308 988).

Christian, K., Morrison, F. J., & Bryant, F. B. (1998). Predicting kindergarten academic skills: Interactions among child care, maternal education, and family literacy environments. *Early Childhood Research Quarterly*, 13, 501–21.

Comings, J., Reder, S., & Sum, A. (2001). Building a level playing field: The need to improve the national and state adult education and literacy systems. *NCSALL Occasional Paper.*

Debruin-Parecki, A., Paris, S. G., & Siedenburg, J. (1997). Family literacy: Examining practice and issues of effectiveness. *Journal of Adolescent and Adult Literacy*, 40, 596–605.

Denzin, N. K., & Lincoln, Y. S. (2000). Introduction: The discipline and practice of qualitative research. In N. K. Denzin & Y. S. Lincoln (eds.), *Handbook of qualitative research* (pp. 1–28). Thousand Oaks, CA: Sage.

Drago-Severson, E. (2004). *Becoming adult learners: Principles and practices for effective development.* New York, NY: Teachers College Press.

Eccles, J. S., & Harold, R. D. (1996). Family involvement in children's and adolescents' schooling. In A. Booth & J. F. Dunn (eds.), *Family school links: How do they affect educational outcomes?* (pp. 3–34). Mahwah, NJ: Lawrence Erlbaum Associates.

Epstein, S. (2001). Commentary: The self and emotions. In H. Bosma & E. Kunnen (eds.), *Identity and emotion: Development through self-organization.* Cambridge, MA: Cambridge University Press.

Gee, J. (1999). *An introduction to discourse analysis: Theory and method.* New York: Routledge.

Glaser, B. G., & Strauss, A. L. (1967). *The discovery of grounded theory.* Chicago: Aldine.

Grubb W. N., & Kalman J. (1994). Relearning to earn: The role of remediation in vocational education and job training. *American Journal of Education,* 103(1), 54–93.

Heath, S. B. (1983). *Ways with words.* Cambridge, MA: Harvard University Press.

Hedegaard, M., & Lompscher, J. (1999). *Learning activity and development.* Langelandsgade: Aarhus University Press.

Holland, D., Lachiocotte, W., Jr., Skinner, D., et al. (1998). *Identity and agency in cultural worlds.* Cambridge, MA: Harvard University Press.

Hoover-Dempsey, K., Bassler, O. C., & Burrow, R. (1995). Parents' reported involvement in students' homework: Strategies and practices. *The Elementary School Journal*, 95, 435–50.

Kansanen, P. (2003). Studying—the realistic bridge between instruction and learning. An attempt to a conceptual whole of the teaching studying-learning process. *Educational Studies,* 29, 2, 221–32.

Kegan, R., Broderick, M., Drago-Severson, et al. (2001). Toward a new pluralism in ABE/ESOL classrooms: Teaching to multiple "cultures of mind." *NSCALL Report 19a.*

Leseman, P. M., & de Jong, P. F. (1998). Home literacy: Opportunity, instruction, cooperation and social emotional quality predicting early reading achievement. *Reading Research Quarterly*, 33, 404–24.

LeVine, R., LeVine, S, & Schnell, B. (2001). Improve the women: Mass schooling, female literacy and worldwide social change. *Harvard Educational Review*, Spring.

Mansfield, E., & McAdams, D. (1996). Generativity and themes of agency and communion in adult autobiography. *Personality and Social Psychology Bulletin*, 22(7), 721–23.

McCarthey, S. J. (2000). Home-school connections: A review of the literature. *Journal of Educational Research*, 93, 3, 145–54.

McDonald, B. A., & Scallay, P. A. (2005). *The Final Technical Report of Project FULFILL*. San Diego, CA: Project FULFILL.

Rameriz, N. (Producer), & Talan, C. (Executive Producer). (2006). Literacy: Adult learners' perspectives [Videotape]. United States: Mortarotti-Rameriz. (Available from the California State Library)

Reder, S., & Strawn, C. (2001). Program participation and self-directed learning to improve basic skills. *Focus on Basics*, 4, Issue D.

Rogers, R. (2002). Between contexts: A critical discourse analysis of family literacy, discursive practices and literate subjectives. *Reading Research Quarterly*, 37, 3, 248–77.

Senechal, M., & LeFevre, J. (2002). Parental involvement in the development of children's reading skills: A five-year longitudinal study. *Child Development*, 73, 2, 445–60.

Sfard, A., & Prusak, A. (2005). Telling identities: In search of an analytical tool for investigating learning as a culturally shaped activity. *Educational Researcher*, 34, 4, 14–22.

Sheehan-Holt, J. K., & Smith, C. (2000). Does basic skills education affect adults' literacy proficiencies and reading practices? *Reading Research Quarterly*, 35, 2, 226–43.

Sticht, T. (2002). The rise of the adult education and literacy system in the United States. *Annual Review*, volume III.

Sticht, T. (2005). ABCs of investing in adult literacy education. *National Adult Literacy Database*.

Sticht, T., & McDonald, B. (1989). *Making the nation smarter: The intergenerational transfer of cognitive abilities*. Paper prepared for the John D. and Catherine T. MacArthur Foundation.

Strawn, C. (2003). The influences of social capital on lifelong learning among adults who did not finish high school. *NSCALL Occasional Paper*.

Terry, M. (2006). Making a difference in learners' lives: Results of a study based on adult literacy programs. *Adult Basic Education*, 16, 1, 3–19.

15 'The More You Learn the Better You Feel'

Research into Literacies, Learning and Identity in Scotland

Kathy Maclachlan, Lyn Tett, and Stuart Hall

INTRODUCTION

This chapter explores the interconnections between learning, self-confidence, learner identities, and social capital. It draws from a two-phased study of over six hundred adult literacy and numeracy (ALN) learners in Scotland (comparable to adult basic education learners in the United States) that examined aspects of their learning experiences and the impact that learning had on their lives. The aim of the research was to evaluate the Scottish Adult Literacy and Numeracy (ALN) Strategy. Detailed findings are reported in Tett, Hall, Maclachlan, Thorpe, Edwards, and Garside (2006). The sample drew from over one hundred different institutions in nine areas in Scotland, representing both the geographical diversity and full range of ALN provision including further education colleges, community, work, and prison-based tuition. Unlike its English equivalent, adult literacies policy in Scotland is nested in broader community learning and development policies. Its "Adult Literacy and Numeracy in Scotland" report (ALNIS) (2001) defined literacies as:

> The ability to read, write and use numeracy, to handle information, to express ideas and opinions, to make decisions and solve problems, as family members, workers, citizens and lifelong learners (Scottish Executive, 2001: 7).

And significantly, it recognized that "Literacy and numeracy are skills for which sufficiency may *only* be judged within a specific social, cultural, economic or political context" (p. 7, emphasis added). Much learning is located in community settings where it is frequently integrated into and supportive of other activities, and policy stipulates that all learners should negotiate individual learning plans based on their personal aims or ambitions. Monitoring criteria are not premised on targets linked to the attainment of qualifications, but on, amongst other indices, learners' "distance traveled" in their learning journey and the impact that it has made in their personal, family, educational, community, and working lives (ALNIS, 2001: 30). This chapter reports on measured changes in the levels of confidence and social capital of the learners between the two phases of the research and discusses the complex interconnections between engagement in learning and the development of self-confidence, an increasingly positive learner identity, and enhanced social capital in this large sample of learners.

METHODOLOGY

Face-to-face interviews using a structured questionnaire were conducted with 613 learners in their place of tuition. Interviews took place between September 2003 and April 2004 and lasted around one hour. Around a third of the participants were from FE colleges, the remainder being from a mix of community settings, schools, voluntary service, private providers, and workplaces. Over half had been engaged in literacies learning for less than twenty weeks, and the vast majority was involved in group learning rather than one-to-one or flexible, drop-in provision. Open and closed questions were designed to ascertain demographic information; learners' pathways into ALN; learners' views of perceived barriers to entry; their experience of learning, teaching, and the curriculum; the guidance received at entry, during and at the end of the program; the effect of participation on their personal, family, public, education, and working lives; degree of satisfaction with the quality of the learning program; how provision could be improved; the "social capital" of the learner; and an assessment of "self-confidence." The instruments used for assessing these latter two aspects are detailed later in the chapter. The combination of qualitative and quantitative methods in the research enabled us to gain a deeper understanding of the processes and events that influenced the changes in learners' lives that the quantitative elements had identified.

After an interval of around one year, learners were reinterviewed between September 2004 and April 2005. Three hundred and ninety-three learners participated in round two, representing 64 percent of the original sample. The personal and program characteristics of the learners in both rounds were compared, and the overall profiles of the two were broadly similar, the only marginal differences being that in the second round, fewer were under twenty-one, studying in FE colleges, and in integrated provision. Figure 15.1 shows the distribution of learning locations in both rounds.

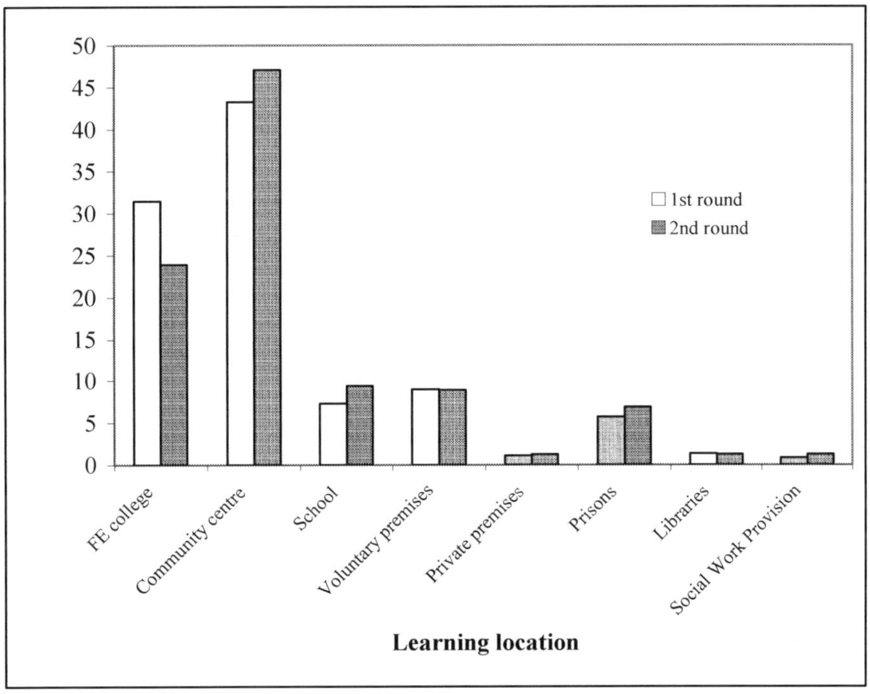

Figure 15.1 A comparison of the samples by learning location.

The focus of the second interview was on changes since the first, so generally the same questions were asked. This enabled changes in learners' social capital and self-confidence between the interviews to be assessed. In addition, their views on the barriers that might put people off joining a program; their experiences of learning, teaching, and the curriculum; guidance during the course and on exit; and the effect of participation on their personal, family, public, education, and working lives were explored. A new question asked for their general reflection on all aspects of the ALN experience. As in the first interview, the questions were both open and closed and a sample of two hundred learners (just over 50 percent) was drawn from people who had taken part in both rounds of interview. Cases were selected using a stratified random procedure to ensure that the sample included learners from all the areas involved in the research and the demographic profile of the qualitative sample closely matched that of the total sample.

In this chapter we detail the changes in social capital and self-confidence reported by learners between the first and second interviews and their effects on learner identity, but first we discuss the concepts of social capital and learner identity.

SOCIAL CAPITAL AND LEARNER IDENTITY

The generation of social capital is one of the claimed outcomes of educational intervention (Baron, Field., & Schuller, 2000; Field, 2005; Schuller, Brasset-Grundy, Green, Hammond, & Preston, 2004), but there are different ways of understanding the term. For example, Coleman (1994) discusses social capital and its links with "human capital" and argues that the former is important for generating the latter. Coleman essentially viewed social capital as facilitating collective aims—the power to do things—but these aims were mainly located within the family. On the other hand, Bourdieu (1990, 1997), writing from a conflict perspective where social processes result from organized actions required to produce goods necessary for daily living, characterized social capital as the power over others. In Putnam's work (1993, 2000) the emphasis is more on the combined effect of trust, networks, and norms of reciprocity in creating strong communities, so he regards social capital as the power to do things collectively. Both Coleman and Putnam use empirical data as well as theoretical accounts of social capital, which involves operationalizing the concept, and they see trust and reciprocity as both arising from the activities that create social capital and as contributing to social capital in their own right (see Croll, 2004: 403).

For our purposes we needed an operationalizable concept of social capital, so we adapted a definition mainly derived from Putnam and developed by adult educationalists (Baron et al., 2000; Schuller et al., 2004; Field, 2005). This definition was that "Social capital refers to the processes between people that establish networks, norms and social trust that facilitate co-ordination and co-operation for mutual benefit, leading to reciprocity and the achievement of mutual goals," because, as Croll suggests, "the central idea underlying social capital is that social relationships and the personal networks which they create are a resource which can be used to generate outcomes which are valued" (2004: 398). However, recent studies have highlighted the "dark side" of what were originally perceived as beneficially supportive arrangements because what were initially constructed as mutually reinforcing ties have also been recognized as potentially binding shackles in some circumstances for some groups in society (Baron et al., 2000; Croll, 2004; Schuller et al., 2004; Field, 2005).

Although there are difficulties in isolating social capital from other factors, nevertheless there appears to be a broad acceptance of an association between social capital and participation in structured learning, and generally this association is deemed to be a positive one because strong social networks seem to enhance participation, and learning in turn seems to increase social capital. The combined effect of trust, networks, norms, and reciprocity is also construed as creating strong communities and a sense of personal and social efficacy. The development of social capital involves the active and willing engagement of citizens within a participative community, that is, "links between like-minded people" (bonding social capital) and/or "the building of connections between heterogeneous groups" (bridging social capital; [1] Baron et al., 2000: 10). Both are highly normative in that they are constructed around commonalities that link people together. This is true even with bridging social capital, because it is the recognition of that which members of the groups have in common that enables the bridge to be constructed in the first place. However, where the bonding capital is high but the norms of the community are not associated with participation in learning, then it is unlikely to be easy for adults to bridge into a learning community because the normative pull of nonparticipation will be strong, and the pressure to conform, equally so.

Research also suggests that developing an identity as a learner is shaped by the complex interaction of a number of factors that relate to the social, because learning is essentially a social activity (Coffield, 1999). They include past learning experiences and the mediating effect of family influences upon them (Rees et al., 2000), as well as the norms and values of the social networks that individuals belong to (Gallacher & Crossan, 2000; Crowther, Hamilton, & Tett, 2001). These are the networks, associated norms, and levels of trust that are the basic building blocks of social capital (Putnam, 1993; Baron et al., 2000; Schuller et al., 2004). As Field suggests, therefore, "social capital is important for learning, and learning is important for social capital" (Field, 2005: 110).

Identity, or "sense of self" (Malcolm & Field, 2005), is not, however, only a product of the social. It is also determined by psychological (emotion, self-esteem, confidence) and positional (class, gender, ethnicity) factors that create who/what we are and aspire to be, and who/what we choose not to be (Preston, 2004). In addition, identity is both ascribed and created; in other words, it is the product of how people both define themselves and how they are defined by others (Schuller et al., 2004). So whilst situated experiences form the contexts in which identities are created and re-created, individual responses to these experiences, mediated by powerful others, determine their impact on identity formation. Stuart (1995) maintains that perceptions of our "intelligence" have a powerful formative influence on our learner identity, and that these perceptions are ascribed by professional educators, "powerful others," through schooling, during a person's formative years (1995: 199). They are subsequently internalized and become part of our created identity. So some adults grow to identify themselves as capable learners, whereas others do not. As active agents in our lives, however, we make choices that are influenced by strategies that Stuart (1995: 200) describes as the desire to amass esteem and avoid shame, and Preston (2004: 7) refers to as positionality, resilience, and affiliation . . We associate ourselves with like-minded people and avoid people and situations where we may be embarrassed because we see ourselves as "other," as not belonging. This in turn polarizes our being and doing; it reinforces our identity in the community of practice that we have chosen to belong to.

For adults ascribed as having "low intelligence" or as being "not capable" learners, which includes a disproportionately high number of ALN learners, the tendency will therefore be for them to associate with similar communities of people for whom engagement in structured learning is not integral to who they are. And so their identity as "nonlearners" becomes more and more entrenched, which in turn blocks the possibility of them creating an alternative image of themselves as capable and competent learners.

Identity is neither fixed nor static, however, and Vincent argues "for an understanding of identity as fluid, not fixed; multiple, not single; and transforming, not static" (2003: 5). She cites Hall's observation that

> Cultural identities come from somewhere, have histories. But . . . far from being eternally fixed in some essentialist past, they are subject to the continual play of history, culture and power . . . identities are the names we give to the different ways we are positioned by, and position ourselves within, the narratives of the past (1993: 394).

We change and can be agentic in determining how we change. In the same way that earlier experiences of learning affect our becoming, so too can experiences of learning in adulthood affect who we come to be later in our life. For those adults courageous enough to risk potential embarrassment and shame, the act of participation in a learning community, or, more accurately, successful participation in a learning community, alters their construction of their identity, of who they are. They not only learn to do; they learn to become. They come to recognize that their present learner identity is a subjective product of their past experiences and not a consequence of their personal inadequacies, and so begin to position themselves differently, to see learning as an integral part of who they are. What appears to operate is a series of complex interconnections between identity, confidence, and social capital that affect and are affected by learning experiences.

However, the precise nature of this interconnectedness is often fuzzy, always context dependent, and not easily isolated from a range of other determinants. This research, therefore, sought to explore potential links between social capital, ALN learning, and learner identity in those learners participating in the study. It was also mindful of the fact that a disproportionately high percentage of ALN learners in Scotland belong to communities that experience socioeconomic marginalization (Scottish Executive, 2001) where levels of participation in structured learning remain persistently low (Aldridge & Tuckett, 2004). However, these participants had engaged in learning, had begun to construct the bridge between different communities, and so it examined the impact that this had on their social capital.

MEASURING SOCIAL CAPITAL

Social capital provided a relevant framework for analysis, as it was hypothesized that learners might improve their social capital as a result of participation in programs. Strawn (2005: 551) has argued that the discourse of particular communities around education is an important component of social capital because it is a function of interpersonal interaction over time. Her research found that people who live in communities where education is seen as a means of advancement are more likely to participate in learning programs. So it was hypothesized that participating in ALN provision would build a discourse around education that would in turn lead to enhanced social capital. The research therefore asked learners the same questions in the first and second rounds to see if their views about themselves, their communities, and their networks had changed.

Particular social capital indices were selected because they represented some of the defining characteristics of the concept as identified in previous research (Baron et al., 2000; Bullen & Onyx, 1998; Campbell, Wood, & Kelly, 1999; Halman, 2001; Murtagh, 2002; OECD, 2001; Tuijnman & Boudard, 2001). They included: identification with and attitudes towards the neighborhood; social and civic engagement; feelings of safety and belonging; and social contacts and supportive networks. The questionnaire therefore sought to record such attitudes and activities and identify any associations between participation in ALN learning and these particular elements of social capital.

Learners were asked if they had taken part in a range of activities within and without their neighborhoods because research shows that activity

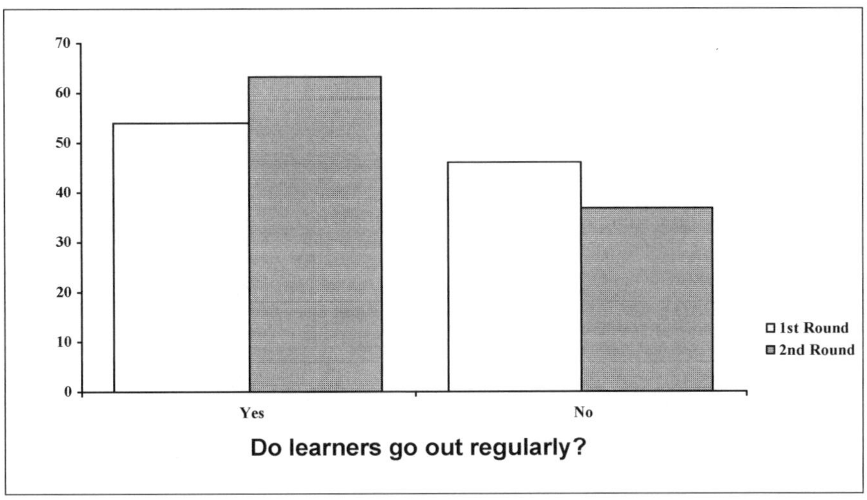

Figure 15.2 Do learners go out regularly?.

builds social capital through developing knowledge resources through opportunities for interactions with other members of the community (Falk & Kilpatrick, 2000; Putnam, 1993, 2000). Figure 15.2 shows the changes between the first and second rounds. These percentages have been calculated for participants who responded. Significance is set at the 1 percent level throughout unless otherwise stated.

There was an increase in the numbers of learners indicating they went out to pubs, clubs, and/or the cinema from (from 54 percent to 63 percent). The change was statistically significant for females with 61 percent reporting going out at their second interview compared to 50 percent at the first. Older learners showed a similar statistically significant increase in the numbers going out (from 49 percent to 60 percent). There was no significant change between interviews of learners going to meetings (from 35 percent to 36 percent) or engaging in voluntary work (from 29 percent to 30 percent). It would appear, therefore, that there had been an increase in social, but not civic, activity between the two rounds.

Learners were then asked if they wanted to become more involved in local activities. Figure 15.3 shows that there has been both an increase in people saying definitely no and also in those saying definitely yes. From the qualitative data reported on later, it appears that these changes are caused

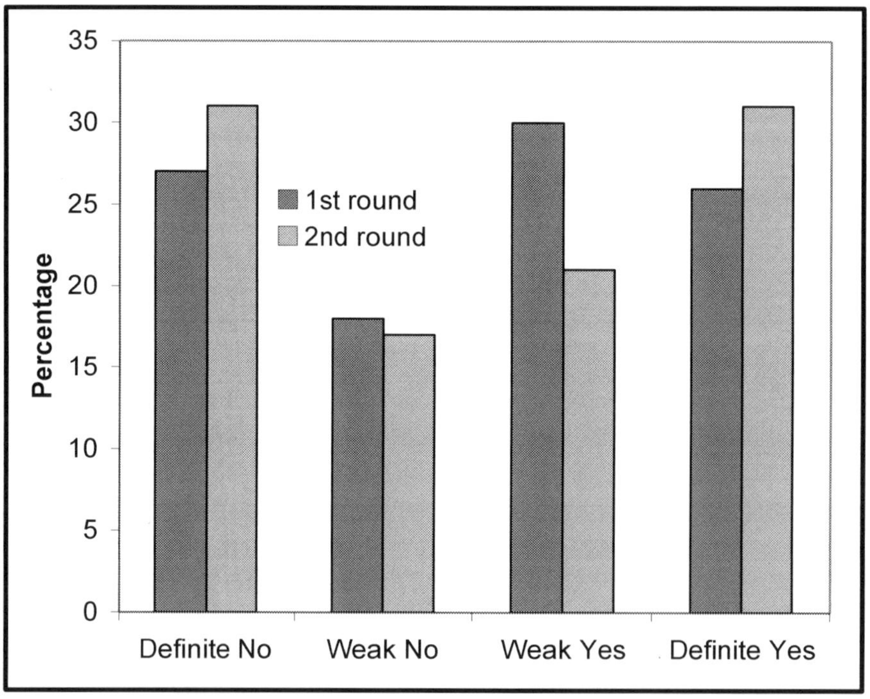

Figure 15.3 Do learners want to become more involved in local activities?

by learners becoming clearer about what they do and do not like to do with their lives.

Those learners who did not want to become more involved were asked why they did not. Reasons differed slightly between the first and second rounds as Figure 15.4 illustrates. There was a decrease in learners' concerns about their ability to cope with English or numerical ability and an increase in other interests. These, combined with the reporting of a lack of time to do things, suggest that the learners were increasingly involved in other chosen activities, and it is feasible that these increased levels of involvement were related to their reduced levels of concern about their ALN skills as their feelings about their capabilities were not as strong a barrier to social engagement as they had previously been.

The next questions were about personal safety, as again this is an area that research has shown impacts on social capital because feeling safe leads to a greater willingness to engage in networks and trust people in the neighborhood (Campbell et al., 1999); however, the changes were so minor that there is no measurable difference between the two rounds of interviews.

The final question was about the learners' contact with other people, as again research shows that being in social contact and getting help when needed lead to positive social capital, through building knowledge resources

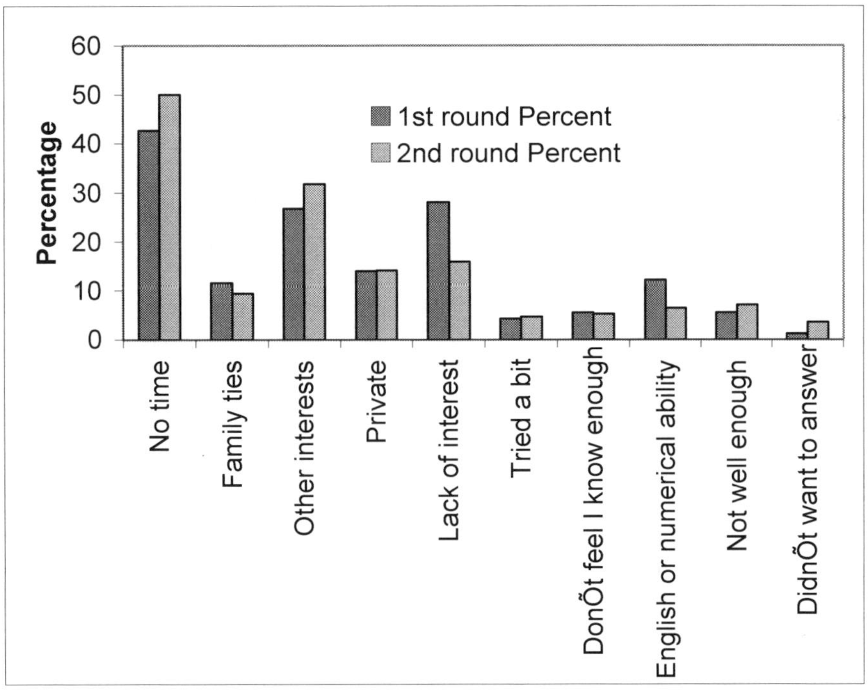

Figure 15.4 Why learners do not want to become more involved in local activities.

Table 15.1 Contact with Others—Comparison between First- and Second-Round Responses.

Learners who indicated.	*Percentages*				*N* =
	1st round		*2nd round*		
	Yes	*No*	*Yes*	*No*	
Visiting a friend or neighbor in the last 2 weeks	65	35	63	37	361
Often meeting friends or family when out shopping or walking in the neighborhood	79	21	82	18	362
Being able to get help from friends and others near where they live if they needed to	88	12	93	7	355

of who, when, and where to go to for advice or resources and through being willing to act for the benefit of the community and its members (Campbell et al., 1999; Falk & Kilpatrick, 2000). Engagement in learning can, in turn, lead to increased levels of social engagement, particularly for isolated and vulnerable adults (Field, 2005:108–9). Table 15.1 shows that there have been a number of changes between the first and second rounds.

There was a significant (5 percent level) increase in the number of learners reporting that they could get help from friends and neighbors if they needed to. Eighty-eight percent and 92 percent of learners at the first and second round of interviews, respectively, noted being able to get help from friends and others in their locality. There was also a significant (5 percent level) increase among older learners noting this.

SELF-CONFIDENCE

Findings from the research showed that there was a strong association between increased social capital and increased self-confidence as shown below. The literature shows that confidence in oneself in the social world and confidence in oneself as a learner interact in complex ways and are both linked to prior experiences of learning and to social capital (Field, 2005; McGivney, 2001; Schuller et al., 2004). Research with adult returnees to learning has shown that those who have failed previously in school gain in confidence, particularly from later successful learning experiences (Hammond, 2004: 42). Hammond's research found that confidence developed through learning was often accompanied by positive personal growth and an openness to new ideas. Since growth in self-confidence is the most widely documented "soft" outcome of learning, this research sought to measure changes in learners' levels of confidence over time.

There have been many debates about what is meant by self-confidence, so we reviewed the literature to decide on what would be operationalizable. There seemed to be considerable consensus with Erwin and Kelly's (1985: 395) definition as "assuredness in oneself and in one's capabilities"; Lawrence (1999 :92) as "confidence in abilities and confidence in personality"; and Owens (1993: 289) as "positive self-evaluation," so we looked at both individual and social situations. There are a number of existing instruments for measuring self-confidence, but the research team wanted a straightforward means of measuring change over time that would not be too intrusive into learners' lives and would be easily understood. We therefore devised a method that picked out relevant scenarios for the learners that were grounded in situations that they would face in their everyday lives. As the scale was primarily derived to measure change over time *within* the respondents rather than as a means of comparison with other groups, it was designed to pick up on elements of their lives in sufficient depth and variety to allow scores to be calculated. The scenarios asked how confident learners were when: meeting new people; making phone enquiries; joining a group of strangers; discussing things with officials; discussing things with a doctor; speaking up in a meeting; complaining about poor service; defending their position in an argument; agreeing within the family; and being interviewed. Responses to each scenario were allocated a score with 1 representing very uncomfortable and 4 very comfortable. Mean scores were computed for each of the 335 learners who completed all ten of the elements of the self-confidence instrument in both rounds of interview. The higher the number the more confident learners were.

Figure 15.5 shows a marked shift to the right, that is, more learners obtaining a higher overall confidence score in the second round than in the first. It also shows no learners in round two obtaining the very low confidence scores that they had initially. There appears therefore to be a marked correlation between engagement in learning and increased confidence.

The responses to individual scenarios were analyzed to see if there were statistically significant changes in the elements comprising the confidence scales between the first and second interviews. These are detailed next.

Learners were significantly (5 percent level) more likely to indicate confidence in the following at their second interview than at their first: a) making telephone enquiries (from 24 percent to 27 percent); b) joining in with a group of strangers (from 49 percent to 56 percent); c) speaking up in a group or meeting (from 54 percent to 62 percent); and d) being interviewed (from 82 percent to 87 percent). Analyzing the data by gender showed that females were significantly more comfortable making phone enquiries, speaking in a group, and being interviewed, and though more indicated increased comfort in joining a group of strangers, these figures were not significant. Older learners were significantly (5 percent level) more comfortable with joining groups of strangers, speaking at meetings,

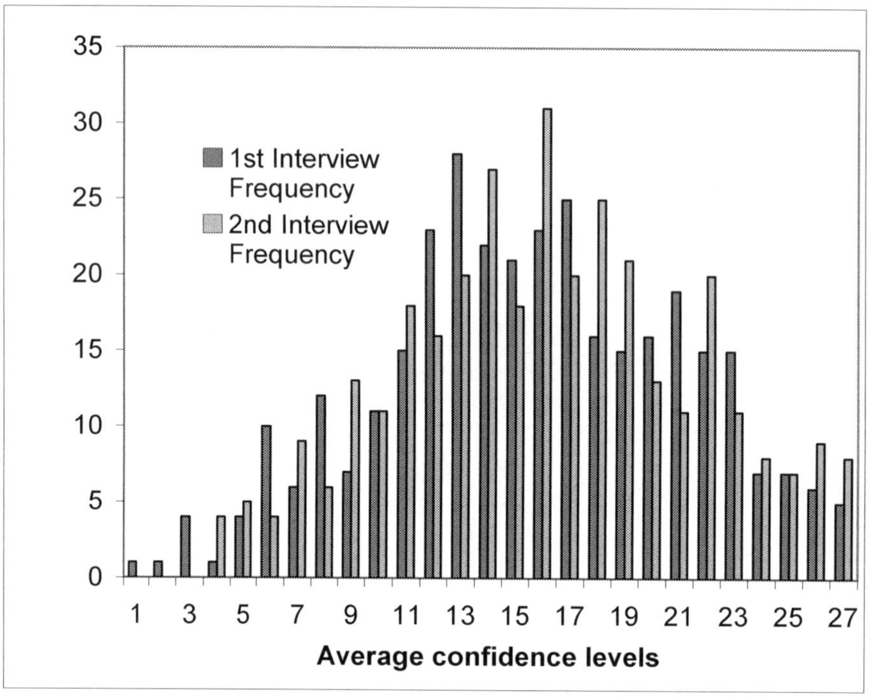

Figure 15.5 Average confidence levels.

and being interviewed, whilst younger learners showed a significant (5 percent level) increase in confidence in speaking in groups.

So by the second interview statistically significant numbers of learners, in particular women and older learners, had become more confident in these situations and their overall confidence scores had increased. In other words, their social and communicative abilities had increased during their learning episodes.

DISCUSSION

Field (2005: 19) suggests that interpersonal communications and connections are the core elements of social capital, and Norman and Hyland (2003: 269) have demonstrated the importance of social interaction as a factor in increasing confidence. Their study showed that "although the individual learner can affect his/her own level of confidence, tutors, peers, mentors and workplace supervisors can help increase the learners' confidence by providing support, encouragement and constructive feedback" (2003: 270). Schuller et al.'s (2004) studies of the wider benefits of learning provide strong evidence of the impact of learning on social meta-competencies that equip people with the confidence and ability to develop their social connections. Whilst we recognize the limitations of relying on descriptive bivariate results for theory testing, the evidence from the confidence scales together with the increases in social capital does indicate that the learners in this research experienced this same positive shift.

In order to explore this shift in more detail, qualitative data from two hundred learners were further analyzed. Confidence is a quite generalized concept, so the responses were analyzed to identify "confidence in what," that is, how the respondents anticipated that their increased confidence would manifest itself in various aspects of their lives. The data show that increased self-confidence acted as a key to opening up a wide range of other changes. The responses clustered into three broad groupings: "'psychological," "skills," and "activity related."

The psychological differences that learners identified included increased self-esteem, a growing sense of their potential, ability and achievements, more independence, being happier as a person, being more able to voice their own opinions, openly talking about their ALN difficulties, improved health, and an enhanced awareness and understanding of the world around them. The number of learners citing enhanced confidence and belief in their own abilities rose from 19 percent to 30 percent. For example:

> I believe in myself now that I can achieve things.
>
> The more you learn the better you feel about yourself.
>
> I'm not "crabbit" [bad tempered] anymore because . . . I'm not avoiding problems—I'm tackling them head on.
>
> It's boosted my confidence because I don't feel as thick as I did before.

The confidence to do things related to skills, some specific ALN skills, and others that had developed as a consequence of participants' enhanced ALN skills. One of the most evident differences in the learners between the two rounds of interviews was a sense of their achievement in learning. Thirty-two percent in the first interview referred to themselves as beginning to achieve things; however, 48 percent in the second interview talked of their pride in their learning achievements. For example:

> I'm more confident in speaking to others so I'm not scared to go to interviews now.
>
> It has helped me to use the computer and I need it for work. I can also interact with friends better because of the computer, because I know what they're talking about.
>
> I will now fill in forms which before I would have left to my husband.
>
> I'm more confident particularly in shopping because I can work out percentage reductions.

The third cluster of confidence indicators relates to a range of activities in facets of learners' lives that they now participate in, or do so with more confidence. Thirty-one percent of the learners (an increase in 11 percent) talked of things they now do because of their enhanced ALN skills. For example:

> I can help myself. I don't need to depend on others and have changed my mind to be very hopeful and helpful.
>
> I'm more confident approaching strangers for information.
>
> It's much easier to live, and I feel safer.
>
> If a conflict came up, I used to cry, but now I don't. At work I managed to say I didn't do something I was falsely accused of doing, and can stand up for myself.

Twelve percent (up 3 percent) said that they socialized more and were more comfortable doing so. They talked about meeting new friends, going out more, starting new leisure activities, and not being afraid of meeting new people. They had grown in confidence, and this was manifest in an increasingly positive sense of self and ability to tackle a whole range of things in their life, including learning.

Many learners' responses related to changes in the nature of familial relationships. These included relationships between parents and children, general relationships amongst family members, between partners, grandparents and their grandchildren, and other relatives. Although there was little quantitative difference in the volume of responses, there was a noticeable shift in emphasis in the comments. On the whole, learners in the second interview were less tentative about changes in their families and more specific about the precise nature of these changes. For example:

> There's no more fighting with my daughter when it's homework time because I can help her with it now.
>
> I'm helping the children with their homework, reading *Harry Potter* to my son and helping my daughter who has learning difficulties.
>
> I'm a bit more patient with my Dad when we go out. I learned from seeing the patience of the tutors.
>
> It saves my wife doing everything all the time and I don't want to have to rely on her.

The responses to both rounds of interviews show the considerable impact that ALN learning has had on relationships and activities within the family, and though the percentage responses in each aspect of familial change were not always greater in the second interview, cumulatively they represent a numerical increase in learners' observing positive differences in this area of life. The greatest changes related to improved relationships, primarily between parents and children, particularly parents' enhanced confidence and skills in supporting their children's education. This in turn gave them more in common to do together and to talk about, thus engendering an all-round better relationship between the two parties.

Learners remained positive about the link between their ALN involvement and employment in both interviews, though the proportion of those volunteering additional comments about work dropped from 64 percent to 51 percent, and fewer "perceived" their job prospects to have improved (from 55 percent to 28 percent). We hypothesize that this may in part be explained by learners gaining a more realistic sense of time frames in moving into employment and in part by the actual changes in working status that many had already experienced since their ALN involvement. However, the proportion of those reporting an increase in their confidence in themselves as workers had risen from 13 percent in the first interview to 26 percent in the second. For example:

> I am recently promoted, would not have been possible before, I would not even have thought of trying for it.
>
> I have more responsibility at work—it [ALN] has made work easier.
>
> I am more confident using the written word, write a lot more, which is required at work.
>
> I am more confident about filling in applications . . . my brother used to do it for me.

One hundred thirty-two learners (66 percent) from the first round of interviews spoke about the impact of ALN involvement on their educational aspirations. At this stage the overwhelming majority of informants were keen to continue with some form of education after their ALN course. Many of them indicated that the self-confidence gained through ALN learning had encouraged them to consider further study. One hundred forty-six learners (73 percent) commented on their education plans or progress at the second round of interviews; seventy-one of these indicated a willingness to engage in future study. For example:

> I'm much more positive about doing further courses.
>
> I intend to go back and finish the classroom assistant course.
>
> Would also like to learn basic accountancy skills to help with the Café project.

Learners had progressed to study a wide range of topics at various levels including computing/IT, communications, social and health care, and gardening. Fourteen said that they had no plans for further study. Of these, four reported that "personal circumstances" (employment, illness, family commitments) prevented them from continuing with education. Two suggested that their ALN experiences had been less positive and they were unlikely to continue in education. What does come across strongly from most of the learners who commented was the importance of ALN in building their confidence and providing them with positive educational experiences that encouraged them to undertake further study. In other words, their learner identities had grown.

Ninety-one learners (46 percent) commented on the impact of their involvement in ALN on their public lives during their second interview compared with eighty-nine learners (45 percent) in the first. Seventy-four indicated that their involvement had, or would result in, changes to their public lives, and many clearly related this to an increase in self-confidence. For example:

> I am more confident socially and able to attend local gyms.
>
> I am a Union representative at work, and doing this course has helped in this area. People at work now come to [me] for advice with problems.
>
> Walking outside, mixing more with the general public and going into a shop on my own.
>
> I am involved in voluntary work in the school. Confidence from the course helped with making this move.

Other commitments, including their studies, prevented nineteen learners from expanding their public lives. In a number of cases learners have begun to increase their public profile and develop new areas of involvement as a direct result of their ALN involvement.

Confidence, developing close family relationships, and social and civic engagement each comprise elements of social capital, and, as Field (2005) observes, they affect and are affected by learning. The evidence from this research suggests that the "virtuous circle" of social capital is operating and affecting those who have engaged in learning. It indicates that increased confidence and sense of self as learner are impacting on familial, social, and work relationships, which in turn add to the sum of learners' social capital, although the precise nature of the causal relationships between them cannot be easily or precisely determined. The data also show that learning produces multiple outcomes, some of which are anticipated, and others that are not, which collectively have had a positive impact on the participants' learner identity.

CONCLUSION

There is an extensive research literature that demonstrates the link between low ALN skills and economic and social status. Adults with low ALN skills are more likely to be unemployed, living on low incomes, experiencing poor health and early morbidity (Parsons & Bynner, 2002; Chisman & Campbell, 1990; Hammond, 2004; Raudenbush & Kasim, 2002; Schuller et al.,

2004; Willms, 2003). There is also a strong relationship between educational inequality, income inequality, and lack of social cohesion in terms of societal trust and community safety (Green & Preston, 2001; Green, Preston, & Sabates, 2003), especially in communities where "education is simply not part of [their] value system and behavior patterns" (McGivney, 2001: 25). Add to this the negative experiences of education endured by many ALN students throughout their schooling (Maclachlan & Tett, 2005) and it is not surprising that being a "nonlearner" is integral to the identity of many ALN students. Given these negative indicators, any positive changes in outcomes for learners as a result of ALN participation will contribute not only to a stronger learning identity but also to widening and increasing social, human, and economic capital.

The data show that learners were more likely to increase their contact with local people and go out regularly therefore indicating an increase in trust and more engagement in their local communities. Research has shown (e.g., Baron et al., 2000) that membership of networks and the ability to mobilize social capital provide access to employment opportunities and enhance people's ability to do the job effectively. Therefore, enhancing social capital through engagement in learning can increase economic and social activity, leading to wider benefits for the individual, their community, and society. This is particularly important in communities where people generally do not have educational aspirations because, as we have argued, it is extremely difficult for an individual living in such locations to behave differently. In other words, the effect of education in raising people's sights is experienced more widely as a positive influence on the cultural norms that encourage others to do the same (see Schuller et al., 2004: 191).

This research has indicated that on the whole, the learners had quite high levels of social capital at the start of their learning, and it is hypothesized that its "dark side" could have been operating in their lives, because many of them were located in areas where engagement in structured learning was not part of the norms of their networks, so they came to their learning tentative, apprehensive, with quite negative learning identities. The quantitative and qualitative evidence from the second round of interviews indicates that after their engagement with learning, these learners were generally feeling, behaving, and constructing themselves differently. Their confidence had increased, their learner identity had grown, and their levels of social capital had risen. However, its operational dynamics seem to have altered. Learning appears to have reduced the impact of its "dark side" and to have opened up a range of possibilities that were hitherto blocked for them, which in turn enabled them to accrue greater social and economic capital.

Learning and its benefits are dynamic in the sense that benefits gained in one domain, such as education, impact on functioning in other domains, such as family and community (see Bynner & Hammond, 2004:161). In combining the quantitative evidence with learners' own perceptions, the research has been able to show that learning brings about major changes in people's lives, and it also enables individuals and communities to "sustain what they are doing through preventing decay or collapse (at individual or community level) or consolidating a positive state of stability" (Schuller et al., 2004: 25). Whilst recognizing the cautions that are always needed in interpreting such data, the findings illustrate the impact that participation has on self-confidence, social capital, and learner identity and show the importance of providing good quality teaching and learning to enable this group of people to sustain and progress in their learning.

We must also remember that neither adult education in general, nor ALN learning in particular, operates in a politically neutral vacuum, cushioned from the social and economic struggles for power that are enacted in the contexts in which learning is located. Cervero and Wilson (2001: 11) maintain that this "requires us to recognize that adult learners exist in the structurally defined hierarchies of everyday life. Thus they enter this process marked by their location within larger systems of power and privilege that have shaped their experience." As Ecclestone argues, we want an education that "looks outwards to social change and genuine political consciousness" (2004: 131), and in recognizing the importance of learners' emotional as well as cognitive changes we are emphasizing their resilience and agency in bringing about change for themselves, their families, and their communities. With increased self-confidence and social capital, with a stronger sense of themselves as people and as learners, perhaps this first tentative step into learning will be the catalyst that enables them to fight back against existing power and privilege.

REFERENCES

Aldridge, F., & Tuckett, A. (2004). *Business as usual? NIACE survey on adult participation.* Leicester, UK: National Institute of Adult Continuing Education.

Baron, S., Field, J., & Schuller, T. (eds) (2000). *Social capital: Critical perspectives.* Oxford: Oxford University Press.

Bourdieu, P. (1990). *In other words.* Cambridge: Polity Press.

Bourdieu, P. (1997). Cultural reproduction and social reproduction. In J. Karabel & A. H. Halsey (eds.), *Power and ideology in education.* New York: Oxford University Press.

Bullen, P., & Onyx, J. (1998). *Measuring social capital in five communities in NSW.* Coogee, NSW, Australia: Management Alternatives.

Bynner, J., & Hammond, C. (2004). Benefits of adult learning: Quantitative insights. In T. Schuller et al. (eds.), *Wider benefits of learning* (pp. 161–78). London: Routledge Falmer.

Campbell, C., Wood, R., & Kelly, M. (1999). *Social capital and Health.* London: Health Education Authority.

Cervero, R., & Wilson, A. (2001). *Power in practice.* San Francisco: Jossey-Bass.

Chisman, F., & Campbell, W. (1990). Narrowing the jobs-skills gap: A focus on workplace literacy. In F. Chisman (ed.), *Leadership for literacy: The agenda for the 1990s.* San Francisco: Jossey-Bass.

Coffield, F. (1999). Breaking the consensus: Lifelong learning as social control. *British Educational Research Journal*, 25(4), 479–99.
Coleman, J. S. (1994). *Foundations of social theory*. Cambridge, MA: Belknap Press.
Croll, P. (2004). Families, social capital and educational outcomes. *British Journal of Educational Studies*, 52(4), 390–416.
Crowther, J., Hamilton, M., & Tett, L. (2001). *Powerful literacies*. Leicester, UK: NIACE.
Ecclestone, K. (2004). Learning or therapy? The demoralisation of education. *British Journal of Educational Studies*, 52, 2, 112–37.
Erwin, T. D., & Kelly, K. (1985). Changes in students' self-confidence in college. *Journal of College Studies Personnel*, 26, 359–400.
Falk, I. ., & Kilpatrick, S. (2000). What is social capital? A study of a rural community. *Sociologia Ruralis*, 1(40), 87–110.
Field, J. (2005). *Social capital and lifelong learning*. Bristol: Policy Press.
Gallacher, J., & Crossan, B. (2000). *What contribution can lifelong learning make to social inclusion?* Glasgow: Caledonian University.
Green, A., & Preston, J. (2001). Education and social cohesion: Re-entering the debate. *Peabody Journal of Education*, 76(3–4), 247–84.
Green, A., Preston, J., & Sabates, R. (2003). *Education, equity and social cohesion: A distributional model*. Research Report 7: Centre for Research on the Wider Benefits of Learning. London: Institute of Education and Birbeck College.
Hall, S. (1993). 'Cultural identity and diaspora' in Williams, P. and Chrisman, L. (eds.). *Colonial Discourse and Post Colonial Theory: A Reader*. London: Harvester Wheatsheaf.
Halman, L. (2001), *The European values study: A third wave*. Tilburg, The Netherlands: Tilburg University.
Hammond, C. (2004). Impacts on well-being, mental health and coping. In T. Schuller et al. (eds.), *Wider benefits of learning* (pp. 37–56). London: Routledge Falmer.
Lawrence, D. (1999). *Teaching with confidence: A guide to enhancing teacher self-esteem*. London: Paul Chapman.
Maclachlan, K., & Tett, L. (2005). *Diversity, difference and the power to decide in literacies learning*. SCUTREA 35th Annual Conference, University of Sussex.
Malcolm, I., & Field, J. (2005). *Researching learning/working lives: Issues of identity, agency and changing experiences of work* (pp. 235–42). Diversity and Difference in Lifelong Learning: Proceedings of the 35th Annual SCUTREA Conference, University of Sussex.
McGivney, V. (2001). *Fixing or changing the pattern?* Leicester, UK: NIACE.
Murtagh, B. (2002). Social activity and interaction in Northern Ireland. *Northern Ireland Life and Times Survey Research Update 10*. Belfast: ARK.
Norman, M., & Hyland, T. (2003). The role of confidence in lifelong learning. *Educational Studies*, 29(2–3), 261–72.
OECD. (2001). The *well-being of nations: The role of human and social capital*. Paris: OECD.
Owens, T. J. (1993). Accentuate the positive—and the negative: Rethinking the use of self-esteem, self-deprecation and self-confidence. *Social Psychology Quarterly*, 56(4), 288–99.
Parsons, S., & Bynner, J. (2002). *Basic skills and social exclusion*. London: Basic Skills Agency.
Preston, J. (2004). *Identity, learning and engagement: A qualitative enquiry using the NCDS.*
Research Report No 13. London: Centre for Research on the Wider Benefits of Learning.
Putnam, R. (1993). *Making democracy work*. Princeton, NJ: Princeton University Press.

Putnam, R. (2000). *Bowling alone: The collapse and revival of American community*. New York: Simon & Schuster.

Raudenbush, S., & Kasim, R. (2002). *Adult literacy, social inequality and the information economy: Findings from the National Adult Literacy Survey*. Ottawa and Hull: Statistics Canada and Human Resource Development Canada.

Rees, G., et al. (2000). Participating in the learning society: History, place and biography. In F. Coffield (ed.), *Differing visions of a learning society*. Bristol, UK: The Policy Press.

Schuller T., Brasset-Grundy, A., Green, A., et al. (2004). *Wider benefits of learning*. London: Routledge Falmer.

Scottish Executive. (2001). *Adult literacy and numeracy in Scotland report* (ALNIS). Edinburgh:

Stationery Office.

Strawn, C. (2005). Social capital influences on lifelong learning. In Gallacher et al. (eds.), *Proceedings of the 3rd CRLL conference* (pp. 550–58). Glasgow, CRLL.

Stuart, M. (1995). Education and self identity: A process of inclusion and exclusion. In M. Stuart & A. Thomson (eds.), *Engaging with difference: The 'other' in adult education*. Leicester, UK: NIACE.

Tett, L., Hall, S., Maclachlan, K., et al. (2006). *Evaluation of the Scottish Adult Literacy and Numeracy Initiative*. Edinburgh: Scottish Executive. Available at http://www.scotland.gov.uk/Publications/2006/03/20102141/0.

Tuijnman, A., & Boudard, E. (2001). *International adult literacy survey: Adult education participation in North America: International perspectives*. Ottawa: Statistics Canada.

Vincent, C. (ed.) (2003). *Social justice, education and identity*. London: Routledge Falmer.

Willms, J. D. (2003). Literacy proficiency of youth: Evidence of converging socioeconomic Gradients. *International Journal of Educational Research*, 39, 247–52.

16 Adults' Lives and Learning in Different Contexts

A View Over Time[1]

Yvon Appleby and David Barton

INTRODUCTION

In this chapter we examine the contribution that detailed qualitative studies can make to understanding language, literacy, and numeracy issues in people's lives. To do this we discuss a longitudinal study of adults called Adult Learners' Lives which we have been involved in. This ethnographically informed qualitative project looked in detail over time at the connections between literacy, language, and numeracy in people's everyday lives and the formal learning they participated in. The research took account of a range of factors that shaped literacy, language, and numeracy purposes and practices, examining how these changed over time. Informed by a social practices approach, it took account of differing contexts in people's lives where they routinely used and developed new literacy, numeracy, and language practices, including those in work, families, and communities. The research also looked at different formal learning environments and contexts, from discrete literacy, language, and numeracy classes in college settings to those delivered in what we have called "community settings." Organizations offering supported learning in these community settings included a drug rehabilitation center, a homeless center, and a domestic violence support group. The research provides rich data from different life and learning contexts: from inside and outside the classroom and within formal provision to informal learning. In this chapter we describe the research showing how a longitudinal approach enabled us to see how lives adapted and changed in response to changes in circumstances and new events. This included, for example, changes in people's physical and mental health, in family circumstances and employment. To understand these factors over time, we found it useful to conceive of people as having different careers in different areas of their lives, for example, employment careers, learning careers, and health careers. These occurred over time and brought out life patterns, which both moved forward and sometimes appeared to go backwards. We were able to see how different factors, as part of these careers, impacted upon how and why people participated in new literacy, numeracy, or language practices to manage their existing circumstances and to respond to change—both immediately and in the future. As people's life stories became more multilayered throughout the research, we developed a way of connecting histories, immediate circumstances, and practices with the future—showing how all of these aspects of a person are engaged in the learning process. The learning process is lifelong and life wide, existing both outside as well as inside the classroom. In the concluding discussion we consider what a longitudinal approach brings to this area of research and how this relates to both policy and practice in adult language, literacy, and numeracy.

THE RESEARCH

The study was carried out in the northwest of England. It was funded by the National Center for Research and Development in adult literacy and numeracy (NRDC) in England as part of a wide-ranging program of research. The NRDC was established to provide evidence for policy and develop professional practice to develop the national adult literacy, ESOL, and numeracy strategy, Skills for Life (SfL), which was introduced in England in 2001. The government's national strategy represents a significant change to the way that adult literacy, numeracy, and language had previously been organized, funded, and delivered (see Hamilton & Hillier, 2006). Significantly, it introduced for the first time national curricula and assessments by tests, as well as professional standards for teachers and target numbers for students, participation, and achievement. Research using a variety of methods has focused on different aspects of teaching and learning within the new infrastructure (see examples in other chapters in this volume). The research reported here was concerned with the perspective of the learners, or the people identified as learners by the strategy (see Appleby, 2008a). It looked at why people attend classes and how the learning provision links to their lives. It focused on looking in detail at the purposes, uses, and meanings that people had for attributes being identified within a policy context as "basic skills." The overall rationale for the research was to understand the social and economic effects of participation in language, literacy, and numeracy programs, investigating the relationship between learner-related and teaching-related factors. The study aimed to investigate links between people's practices and social issues, focusing on why people came to classes and what enabled them to succeed with learning—responding to the educational challenges of motivation and persistence. In summary it looked at:

- The significance of language, literacy, and numeracy in the lives of the adults who have difficulties.

- The literacy, language, and numeracy demands encountered by people in their everyday lives and specific workplaces.
- The link between basic skills difficulties and social issues such as homelessness, poverty, community participation, health and disability, family violence, and unemployment.
- Strategies that adults with poor basic skills use to function effectively in their lives.
- Learners' perceptions and experiences of the learning program directed at them and their motivations for attending and persisting in attending.
- The link between the ways in which people deal with difficulties and their classroom experiences and how everyday practices support, or impede, what is learned in the classroom and, conversely, how classroom activities impact on everyday life.
- The reasons for teachers' decisions about pedagogic practices; characteristics of learner-teacher interactions and texts; relations between classroom "input" and learner "uptake."
- The approaches to increasing participation in basic skills tuition that are likely to be effective.

There were two phases in the research looking at different types of provision and working with different learners. In the first, learners were accessed through adult literacy, numeracy, or ESOL (English for Speakers of Other Languages) classes at college that were part of discrete SfL provision. In the second, learners were reached through community and outreach provision where literacy, numeracy, and language learning was less discrete and often integrated within other life skills learning and ongoing support. Many of these people are in the SfL priority groups, which include homeless, asylum seekers, and unemployed people, and are part of wider policy concerns about social inclusion and social participation. Overall, 282 people participated in the research, 134 were students, and the remainder tutors, managers, and other support workers. The electronic database from the project consists of 403 data files, which include 198 recorded interviews. Where we worked in depth with people in learning programs, this ranged from carrying out several interviews over a six-month period to keeping in touch with the person and their learning for up to three years. More detailed information is provided in Ivanič, Appleby, Hodge, Tusting, and Barton (2006), which reports on the college classes' work, and Barton, Appleby, Hodge, Tusting, and Ivanič (2006), which covers the community settings.

Three research sites were chosen for their demographic, economic, social, and cultural diversity. The first, Blackburn, a medium-sized ex-mill town, has a large ethnically diverse population. It is characterized by higher than average unemployment, lower than average wages, and poor housing. The second, Liverpool, is a large thriving city containing a diverse population of many language communities, some well established, some newly arrived.

Although experiencing economic hardship and social fragmentation, after the decline of the docks, it is now in a process of regeneration and reinvigoration. The third site, Lancaster, is a small city with a large university population. It has a large migratory population of students coexisting with a fairly settled local population. There are visible areas of poverty and poor housing, but generally Lancaster represents an economically and socially stable population.

A sample of learners in both phases of the study became part of a developing longitudinal cohort. The longitudinal cohort is part of the overall data whilst being distinct as it provides more detailed information about some individual lives over time. The research questions, as can be seen above, were wide ranging. However, we also wanted to know in more depth what happened in people's lives over time and the everyday practices they used and developed independently of provision which connected to their experiences of formal learning. We also wanted to gain insight into what were the features of people's lives that supported their learning over a period of time and what acted to make engaging in formal learning difficult or impossible. The data reported here cover a three-year period, and we are currently extending this by returning to interview a subset of the participants.

THE LONGITUDINAL COHORT

The first group within the longitudinal cohort was accessed through literacy, numeracy, and ESOL classes in colleges in our three case-study sites. In discussion with the practitioner researchers, who were our research collaborators, individuals were invited to approach to become part of the longitudinal group. The practitioner researchers were literacy, numeracy, and ESOL tutors working in classrooms in our three sites who were given paid time to work with us. They provided detailed local knowledge and access to learners and contributed their professional insights to the fieldwork and analysis. As an organization, the NRDC supports various types of practitioner research, linking research and practice (see Hamilton, Davies, & James, 2008; Hamilton & Wilson, 2005). The second group was made up from learners in community settings in the three sites. As most of the organizations worked with adults who were dealing with other issues in their lives, including drug and alcohol use, homelessness, and violence, negotiations to become part of the study were carefully carried out with the support of the organizations and staff. Our first longitudinal cohort comprised twenty-six learners, fifteen of whom were women and eleven were men. They were between eighteen and seventy years old with fourteen attending a literacy class, eight an ESOL class, and four a maths class. In the second cohort there were twenty-nine learners—eighteen women and eleven men. They were between the ages of sixteen and mid-forties. Overall, fifty-five people participated in the longitudinal cohort; Table 16.1 shows how they were

Table 16.1 The Longitudinal Cohort

	Literacy	*Numeracy*	*ESOL*	*Total*
College classes	14	4	8	26
Community provision	11	18		29

distributed between literacy, numeracy, and ESOL and between community-based as opposed to college provision.

The people included in the longitudinal cohort were chosen to represent a wide spectrum of learners. This included taking account of gender, age, work histories, health issues, race/ethnicity, along with type of provision, the level being worked at, and people's reasons for attending. We responded to opportunities provided in the provision to which we had access rather than imposing a predetermined sampling frame. We were responsive to the environment that we worked in where the sample was representative of the people who attended the classes rather than being representative of the general adult population. This method allowed us to work in detail with a range of learners showing some of the complexity in their lives and learning. It is therefore insightful and illustrative.

DATA COLLECTION METHODS

Data were collected in a variety of ways depending on the type of provision. In the first year those included in the longitudinal cohort were part of the project focal on "classroom studies" that looked at the social aspects of learning. Data collection methods for this project included taking detailed notes of observations, photographs, video, and examples of learners' work, taped discussions with learners, tutors' notes, and in-depth interviewing. Not all methods were used in each classroom; they were selected for appropriateness and practicality. Repeated interviews, e-mails, phone, and face-to-face contact in the classroom, at home, and work were carried out over both years with this first group. Some people from the first year were "tracked" over nearly three years by using a range of "keeping in touch" methods such as dropping into classes or work, e-mailing or phoning after special events such as a job interview or an asylum hearing, and home visits where appropriate. In year two, working in community settings, a wide range of methods was used, but because of the nature of the provision, which was supporting adults dealing with complex issues in their lives, video recording was not used. Other context-sensitive and collaborative methods were developed which included working with photographic life stories and coresearch relationships. There were heightened issues of confidentiality to be negotiated, particularly when everyday practices frequently related to bureaucratic texts which individuals found positioned them as vulnerable and powerless (see Taylor, 1996, and Crowther, Hamilton, & Tett, 2001). Bureaucratic texts include such things as welfare forms, health records, insurance claims, school and educational forms, and legal documents. These texts require a high level of concentration and competency, or experience, in reading and responding to densely written information which often contains jargon and abbreviations not found in everyday language. These "official" texts often tend to "depersonalize" people, making responding to them difficult and challenging, particularly where their lives don't neatly fit into the categories and tick boxes provided. These texts were experienced by many people in our study as disempowering, invasive, and out of their control. It was more difficult to "track" this group as lives were frequently disrupted as people experienced turbulence (Reder, forthcoming) and sometimes were part of moving populations responding to being homeless or refugee or asylum seekers.

LOOKING AT LIVES OVER TIME

Using an ethnographic approach

An ethnographic approach is part of a general qualitative social science methodology that looks in detail at people's lives (Denzin & Lincoln, 2005; Silverman, 2001, 2005). A qualitative approach facilitates the study of complexity and change in people's lives by looking in detail at significant factors both in immediate surroundings such as home and family and at wider factors within communities, workplace, and identity networks. It takes account of the situatedness of individual lived experience that is mediated through the social construction of gender, race, and class. Within this overall research perspective an ethnographic approach involves detailed study of people's everyday practices, in particular those situated in communities, providing rich information about their lives (Atkinson & Hammersley, 1994; Tusting & Barton, 2005). This approach seeks to understand meanings and practices from the perspective of the participants, capturing both patterns that people have in common and the uniqueness of lives and situations. It provides a local view, often containing unpredictability, set within wider and potentially more stable contexts. For example, this might be how an individual negotiates the complexity and turbulence of his or her life whilst engaging in the more stable, and sometimes experienced as rigid, adult education system.

By looking in detail over time we found that we needed a way to represent people fully to explain what we were observing in the immediate context of their lives—how they operated locally. It needed to be three-dimensional because lives are complex and multifaceted. This phrase was used by one of our teacher researchers to explain that she frequently saw people in her math class only in a one-dimensional way, as a "learner." Our joint research (her practitioner project embedded within our larger study) found that people responded better when they were treated as whole persons who had a lives outside of the classroom. Through this means they became "three-dimensional": they had shape and form, relating to the specific context of their lives. He or she was a person first, someone who had come to learn, rather than simply being someone who was there to learn what was being taught. Each person we talked to came with a past; he or she was living in the present and had hopes and aspirations for a future. In response to this we developed a way of taking account of each person's history, current circumstances, everyday practices, and future. (This is developed in more detail in Barton et al., 2006.)

Each person has a history, which may be a combination of positive and negative experiences and influences. Many of these fit into what, drawing upon Goffman's work (1961), we describe as "life careers." He uses this concept to describe a social course through life. People have many careers within their lives, including learning, health, and work careers; these careers form their past and affect both their current circumstances and the future. People's histories shape their confidence, attitudes, values, expectations, identities, and sense of place. We found that people brought their history into the classrooms and different learning environments that we researched. Sometimes this was visible but often the more difficult aspects, both past and present, reminded hidden (see Horsman, 2000). The notion of career will be expanded further in Elizabeth's life following, showing how differing identities and contexts intersect, overlap, and interlink.

People's current practices and circumstances tell us much of what people do, how they identify themselves, and how others identify them. A social practices approach to literacy, numeracy, and language (Barton & Hamilton, 2000; Papen, 2005) looks at what people do in their everyday lives enabling them to be agentive and capable of change. It also enables recognition of the various resources people bring to formal learning environments illustrating the differing concepts of capital suggested by Bourdieu (1986). This approach takes account of socioeconomic and cultural factors that are significant in shaping individual lives, where people benefit across many areas of their life from learning opportunities, identified as "the wider benefits of learning" (Schuller et al., 2004). Significant factors in the lives of people in the longitudinal cohort included homelessness, poverty, chronic ill health, mental illness, disability, violence, drug or alcohol addiction, or abusive relationships. Reder (forthcoming) identifies these factors as contributing to turbulence in people's lives, and Taylor (1996) shows the sometimes-overwhelming nature of these circumstances affecting people's ability to attend learning provision regularly. For others in the longitudinal cohort, life was generally stable and circumstances were predicable and manageable.

People's dreams and aspirations for the future were a tangible part of their current circumstances—even when difficult to achieve they were significant in providing "hope for the future" and "something to aim for." Most people, whatever their circumstances, described very "ordinary" aspirations for the future such as a nice home, a good job, and more control in their lives. This was even the case when their circumstances were complex and difficult to manage. Some described just being healthy and not dependent on drugs as the most important thing in their future. Others described longer-term goals, specifically linked to learning, such as going on to higher education or of learning new skills to gain new employment opportunities. For most, the future was about taking more control of life, creating more opportunities for themselves and their families, and in gaining more social confidence. Although often linked to broader learning objectives and skills acquisition, this was not generally how people talked about it themselves.

History, current circumstances, current practices, and hopes for the future inform who people are and what they bring of this to learning—whether to a three-hour spelling class or numeracy provision embedded within a drug rehabilitation program. Our analysis looked at individual lives, using a life-history approach, as well as looking for overall themes across our data set. Within each cases-study site and across the data set as a whole we analyzed and interpreted data, drawing out themes and noting links and interrelationships that were emerging. We also noted particular areas of difference, for example, ESOL provision raised issues that were not found in literacy and numeracy classes. Overall we used a grounded theoretical approach, looking for emergent themes allowing us to capture general areas and concerns, as well as being able to systematically refocus and concentrate on particular issues. Examples of general issues included embarrassment at not being able to read or write as well as other people, of early school failure, and of finding individual ways of managing in different contexts. Particular issues that emerged were around work and health where specific literacy practices, and the difficulties of engaging with them, were described to us. By using an approach that worked with people over time, in some cases nearly three years, and which took account of their lives over time we were able to see some of the complex interrelationships and the changes that occurred in lives and circumstances. This approach enabled us to achieve a high level of validity in relation to the people we worked with as we were able to check our emerging findings with them. By using different types of data, from different aspects of people's lives (both inside and outside the classroom), we were able to build a richer picture by triangulating information in our analysis. This meant, for example, that we gained more detailed understanding of why Tommy attended a spelling class when he told us on the first time we met him that he was not working and wanted to improve his spelling. Rather than just seeing him as "unemployed" and in need of "basic skills" to improve his chance of finding work, we were able to see how physical disability, partly caused through heavy manual work, and a need to care for family members meant that full-time work was difficult for him. We were also able to see his literacy and numeracy practices "in action," seeing how he used the library, reading for many varied hobbies such as astronomy and deep-sea fishing, as well as using a home computer to organize his extensive CD and games collection. Tommy's circumstances changed over the time that we worked him, sometimes relating to things that had happened in his past and sometimes relating to things in his future. They related to both his life and his learning.

We used the notion of career to understand how events and circumstances influenced each other and how sometimes learning appeared to be progressing, and at other times it ceased or appeared to go backwards. Elizabeth's life story illustrates this. By using the concept of career it is possible to see points of transition and change that are influenced by the various roles taken in people's lives and their biographies (see Bloomer & Hodkinson, 2000, and also Hodkinson & Sparkes, 1997). Studies of youth transitions have shown the interplay and relationships between school and work, family, drug taking, and criminal careers (e.g. Cieslik & Simpson, 2006). From a social practice perspective this enables different roles and identities both inside and outside of education to be understood and taken into account.

Using multiple careers: A case study of Elizabeth

Elizabeth (her chosen pseudonym) was fifty years old at the beginning of the study when we met her in a spelling class. Over a period of two and half years we interviewed her about attending class, asking why she had come, and what she was gaining, as well as what was happening in her life outside the classroom. She participated in making a video illustrating classroom teaching and discussed with others in the class how this related to their learning. She contributed to a photography project on everyday literacies using a disposable camera to take photos of real and concrete literacy and numeracy practices in her life. Her photos showed texts and images from the supermarket, train timetables, and road signs; things she described as significant signs and symbols that shaped her daily life and allowed her to participate or not. Her motivation for coming to class was to improve her spelling so that she would not be embarrassed by her lack of skill either in seeking new employment or in managing her daily and domestic affairs.

Over the time we worked with her Elizabeth participated in repeated interviews as she completed her spelling course and moved on to another class. Interviews and home visits carried out after she left the course captured her everyday life, including her literacy and numeracy practices, as she took a trip to America to visit her sister, started a new job, and considered buying a car. To bring together her practices and uses and meanings of literacy and numeracy from inside and outside the classroom, we will use three types of careers in Elizabeth's life: her learning career, her work career, and her health career.

From interviews and conversations, Elizabeth's *learning career* showed that she identified herself as dyslexic at school and had an absolute "fear" of English language and spelling as a child. She explained how she had developed learning strategies for coping, but that as an adult wanted to learn "properly." She felt that she always worked hard but that she was just "no good" and therefore hated learning, as it was difficult and negative. Her return to adult education aged thirty, after the death of her husband and whilst a working single parent of two girls, was experienced as boring and childlike. The teaching in the adult evening spelling class used children's books, which she described as "the cat sat on the mat" type of learning. Frustrated, she left knowing that she couldn't spell and that adult learning wasn't much different from her earlier school experiences. The pattern of her learning career can be described as mainly flat and in a negative direction. It took almost forty years for her to eventually experience a positive learning experience in the spelling class where this pattern was broken. The timing of this transition related to her work career.

In comparison to a negative learning career, Elizabeth's' early *work career* was varied and could be seen as moving up and down. Early experiences of factory work, making brushes, and working in a toy factory provided Elizabeth with some money, and, as she described it, the fun of being part of a large social group of female workers. Like many women at the time, she stopped full-time work with the birth of her children. The early death of her husband, at just over thirty years old, meant that she returned to work at a time in the United Kingdom when the job market was shrinking and unemployment was increasing. After working in a pub, which didn't need any literacy skills and didn't pay well, she started working for an electrical retail group as a salesperson. Within this work she developed many strategies for managing her weak spelling in recording sales information. She also managed by "sheer hard work and determination" to become their number one salesperson. Whilst this gave her confidence and new skills, her success attracted hostility and even physical violence from some male members of the sales team. Under the repeated strain Elizabeth experienced a nervous breakdown followed by depression. Her work career contains both highs and lows, producing a pattern resembling waves moving up and down, in contrast to her flatter learning career. She started with positive experiences of factory work, which changed to become a downward movement as she became a young widow and took on jobs she could manage as the sole carer of two children. This changed, taking a positive turn upward as she not only held a job in retail but also excelled at it, only to change again in a downward turn as she left because of the sexist pressure and harassment.

Elizabeth's *health career* remained fairly stable throughout her early life; her factory work was "hard graft" but fun, while as a young woman she had energy and no domestic commitments. As a typically working-class woman, she had married young and had children soon afterwards. This she described as hard work, as there was so little money and everything had to be borrowed or gone without, but it was considered the norm in her close-knit community. The early and unexpected death of her husband from a heart attack, followed by the premature death of her father, also with a heart attack, changed Elizabeth's life course dramatically. Although she managed to recover, bringing up two children and eventually forging a successful retail career, she described living with an emotional time bomb. This exploded under the pressure of sexist harassment and bullying at work, resulting in a complete breakdown and the loss of her job, becoming clinically depressed and housebound. The doctor treating her suggested that she take up sport and some adult learning to get her out of the house, improve her confidence, and to mix with other people. These were significant factors in her motivation for attending the course. The trajectory of Elizabeth's health career started with an almost uninterrupted line but fell dramatically in her early thirties and thereafter has small highs and more lows between her thirties towards becoming fifty.

Taken together it is possible to see how the different careers in Elizabeth's life related to each other, also being affected by external events and factors. Her adult learning career was informed by her early negative experiences of learning, so that successful learning only occurred when the teaching enabled her to feel safe and supported as a vulnerable learner. Debbie, the teacher of the spelling class, created a learning environment where Elizabeth felt comfortable in explaining about her suspicion that she had dyslexia, and together she and the tutor were working on learning strategies to use in class and at home. The teacher created a learning environment where people's emotional selves were recognized as important, enabling Elizabeth to explain what had happened at work and the impact on her health. The teacher was able to recognize and respond to aspects of Elizabeth's history, her current circumstances and practices as well as her hopes for the future. She was able to see and support significant points of transition, for example, from a low point in Elizabeth's health career to her returning to adult learning.

In her everyday life Elizabeth used her gain in confidence and awareness of learning from her formal class to apply for a new job and visit her sister in the States, booking her tickets online. She found that she panicked less when she couldn't do something and used the "look, cover, and learn" method she used with Debbie at home. Her increase in confidence in having existing skills recognized in class as well as becoming an independent learner at home meant that she was able to reorganize the recording system in her new job of selling beds. Although the system was text based, it relied upon an accurate knowledge of the process and procedures of "making a sale"—something Elizabeth was expert in. Although each of her careers had different trajectories, and were located in different contexts within her life, they each contributed to the factors and events in her life that we observed in our study. They were fluid, crossing boundaries as her positive experience of learning showed. Where teaching was able to relate to people's histories, current circumstances, and practices, experienced across different careers, linking teaching to future aspirations, adult education worked well. This involves more than being "student centered," as we shall discuss next, as it takes account of a person's life across different contexts.

DISCUSSION

This ethnographic research, which looks in detail over time, produces rich or '"thick" data that show some of the factors, events, and context that shape people's lives. As such, it provides more than a situation-free "snapshot" of learning events as it recognizes and acknowledges the complex interrelationships within lives that people bring to learning. Although there are different types of ethnography (Hammersley, 1994; Silverman, 2005), as an overall methodological approach ethnography allows the study of phenomena that are not accessible elsewhere—that is, in different contexts and specific or local situations. This generates different information, insight, and meanings to purely quantitative research that predominantly focuses on relationships between specific variables as assessed for a representative sample of a specified population. As Silverman (2005) argues, this is because qualitative research uses naturally occurring data that can capture the "how" and the "what" of people's meanings in a specified context or contexts. Ethnographically informed qualitative research conducted over time is therefore important in showing the detail of lives which are situated and local as well as being located within different and sometimes wider contexts. In large-scale quantitative surveys, with a focus on relationships between variables, this detail is almost impossible to capture, although it can show significant large-scale trends and changes.

The rich detail reported here captures significant aspects of people's lives that we argue are integral to their practices, uses, and meanings of literacy, language, and numeracy. These relate to Silverman's "how" and "what" of, in our case, formal and informal adult learning. Whilst there is an internal integrity in this research and its findings, it is important to ask what its significance is. By producing rich detail it could be seen as providing potentially interesting case studies or examples to illustrate other studies in the field. This may of course happen, but as used here its strength lay in being able to ask questions and provide analysis from the perspective of people in the contexts of everyday lives, revealing the complexity generated from this grounded theoretical approach. This both challenges and adds to other research paradigms, providing both critique and complementarity. In terms of critique, the knowledge generated in accordance with this epistemological position provides a critical challenge to studies that do not take account of people's lives and the specificity of their material and cultural resources (Bathmaker, 2005; Hamilton & Barton, 1999). In terms of complementarity, it provides useful comparative data illuminating similarities and differences in lives and learning in different contexts. For example, it offers a comparative perspective complementing the findings of the NCSLL longitudinal study carried out in the different adult education system of the United States. Although employing different methods of research, the U.S. research focused on many of the same questions as our research (e.g., Bingman, Ebert, & Smith, 1991). Our research also complements and adds a comparative dimension to research of the Scottish adult literacy, language, and numeracy framework (Tett et al., 2006; Tett & Maclachlan, 2006). This can be illustrative aiding evaluation; with its focus on the perspective of the learners, the approach does not claim to produce generalizable findings. Rather, it generates critical questions, or issues, by which to evaluate "the general" system. Using different data across time, which is itself is analyzed by a team of researchers, adds a level of triangulation within the study enabling it to be valid and close to the experience of those who are being represented by the research. The ethnographically informed research reported here also has the potential to add an element of triangulation with existing large-scale longitudinal studies conducted in the United Kingdom. For example, the notion of history, current circumstances and practices, aspirations for the future, along with the concept of life careers, can help to elaborate explanations of findings of quantitative cohort studies showing the negative economic and social impact for people at the most basic levels (Bynner & Parsons, 2006).

Taking a closer and longer view of learning provision from the learners' perspective can inform policy of incremental benefits and gains in people's lives, things that are not always apparent in the short term or in more tightly designed evidence-based studies. Currently, funding mechanisms in England create a climate that focuses on short-term achievement and retention of what has been learnt, which often hides real learning gains and benefits experienced over time. Such short-termism is at variance with some of the wider and longer-term social inclusion aspirations of adult learning policies. It is vital, therefore, to produce critical research that examines learning and lives over time, showing some of the more subtle and complex relationships that exist. These need to be understood in a more finely grained way that supports learning delivery, teacher training, and continuing professional development. Such an anti-instrumental approach to policy research, which focuses on more than simple "policy outcomes" (see discussion in Hammersley, 1999), has an important critical purpose whilst also making it possible to address issues of power and representation in the research process itself. This can occur either within one study as in the project reported here or as part of other larger or ongoing studies. For example, other NRDC research, such as the Teachers and Learners, two linked studies of literacy, language, and numeracy (www.nrdc.org.uk), shows the potential of complementary perspectives (both qualitative and quantitative) in design, data collection, analysis, and their potential usefulness for the policy field.

As well as providing both critique and complementarity, this approach enables collaboration, not simply as a by-product but as an integral part of the research itself. For example, those involved in the research being able to take the findings back into their teaching practice and everyday life was a crucial aspect. Taking a longer, as well as detailed, view has enabled some of these findings to be collaboratively converted into guides for practice with the active engagement of practitioners (Appleby & Barton, 2008; Appleby 2008b). This has supported practitioner research (see Ivanič, 2004) and dissemination across many networks built over the life of the research itself. Not only is the product of the research the result of a longitudinal approach, but the research process is also developed over time; this leads to greater participation and collaboration, which increases both research validity and the relevance of findings. Working over time also supported collaboration with the learners across different context of their lives. This not only provided us as researchers with a richer and more detailed picture but also enabled those involved as learners to participate in the reflective process of telling their life stories and contributing to emerging analysis. Most of the learners involved found it a positive experience as they were able to reflect upon their past, their present, and their future, knowing this was contributing to a wider understanding of the possibilities and challenges of adult literacy, language, and numeracy learning and teaching, both now and in the future.

NOTES

1. The research reported here was part of the NRDC Adult Learners' Lives project (PG1.2), based at Lancaster University and directed by David Barton and Roz Ivanič and including Rachel Hodge and Karin Tusting. We are grateful to them as well as to the many practitioner researchers who contributed to the project.

REFERENCES

Appleby, Y. (2008a). (forthcoming) Who are the learners? In I. Schwab & N. Hughes (eds.), *SfL teachers' literacy handbook*. Milton Keynes, UK: Open University Press.

Appleby, Y. (2008b). *Bridges into learning for adults to find provision for hard to reach*. Leicester, UK: NIACE.

Appleby, Y., & Barton, D. (2008). *Responding to people's lives*. Leicester, UK: NIACE.

Atkinson, P., & Hammersley, M. (1994). Ethnography and participant observation. In N. Denzin & Y. Lincoln (eds.), *Handbook of qualitative research* (pp. 249–61). London: Sage.

Barton, D., Appleby, Y., Hodge, R., et al. (2006). *Relating adults' lives and learning: Issues of participation and engagement in different settings*. London: NRDC publications.

Barton, D., & Hamilton, M. (2000). *Local literacies: Reading and writing in one community*. London: Routledge.

Barton, D., Ivanič, R., Appleby, Y., et al. (forthcoming). *Literacy, lives and learning*. London: Routledge.

Bathmaker, A. (2005, September). *Achieving the SfL adult basic skills targets in England: What picture can we gain from available statistical data and what issues does this raise?* Paper presented at the British Educational Research Association Conference at the University of Glamorgan, Wales.

Bingman, M., Ebert, O., & Smith, M. (1991). *Changes in learners' lives one year after enrollment in literacy programs: An analysis from the Longitudinal Study of Adult Literacy participants in Tennessee*. NCSALL Report 11. Cambridge, MA: Harvard Graduate School of Education.

Bloomer, M., & Hodkinson, P. (2000). Learning careers: Continuity and change in young people's disposition to learning. *British Educational Research Journal*, 26(5), 528–97.

Bourdieu, P. (1986). The forms of capital. In J. G. Richardson (ed.), *Handbook of theory and research for sociology of education*. Oxford: Greenwood Publishing Group.

Bynner, J., & Parsons, S. (2006). *New light on literacy and numeracy* London: Institute of Education, NRDC.

Cieslik, M., & Simpson, D. (2006). SfL: Basic skills and marginal transitions from school-to-work. *Journal of Youth Studies*, 9(2), 213–29.

Crowther, J., Hamilton, M., & Tett, L. (eds.) (2001). *Powerful literacies*. Leicester, UK: NIACE.

Denzin, N., & Lincoln, Y. (2005). *The Sage handbook of qualitative research* (3rd ed.). London: Sage.

Goffman, E. (1961). *Asylums*. Harmondsworth, UK: Penguin.

Hammersley, M. (1994). Introducing ethnography. In D. Graddol, J. Maybin, & B. Stierer (eds.), *Researching language and literacy in social contexts* (pp. 1–17). Milton Keynes, UK: Open University Press.

Hammersley, M. (1999). Some reflections on the current state of qualitative research. *Research Intelligence*, 70, 16–18.

Hamilton, M., & Barton, D. (1999). *The International Adult Literacy Survey: What does it really measure?* Center for Language in Social Life Working Paper Series. Lancaster, UK: Lancaster University.

Hamilton, M., Davies, P., & James, K. (2008). Practictioners Leading Research. London: NRDC.

Hamilton, M., & Hillier, Y. (2006). *A history of adult literacy, numeracy and ESOL 197–2000*. Stoke on Trent, UK: Trentham Books.

Hamilton, M., & Wilson, A. (eds.) (2005). *New ways of engaging new learners: Lessons from round one of the practitioner-led research initiative*. London: NRDC.

Hodkinson, P., & Sparkes, A. (1997). Careership: A sociological theory of career decision making. *British Journal of Sociology of Education*, 18(1), 29–44.

Horsman, J. (2000). *Too scared to learn: Women, violence and education*. Mahwah, NJ, and London: Lawrence Erlbaum Associates.

Ivanič, R., Appleby, Y., Hodge, R., et al. (2006). *Linking learning and everyday life: A social perspective on adult language, literacy and numeracy classes*. London: NRDC Publications.

Ivanič, R. (ed.) (2004), with D. Beck, G. Burgess, K. Gilbert, et al. *Listening to learners: Practitioner research on the Adult Learners Project*. London: NRDC.

Papen, U. (2005). *Adult literacy as social practice: More than skills*. London: Routledge.

Reder, S. (forthcoming). *Dropping out and moving on: Life, literacy and development among high school dropouts*. Cambridge, MA: Harvard University Press.

Schuller, T., Preston, J., Hammond, C., et al. (2004). *The benefits of learning: The impact of education on health, family life and social capital*. London: Routledge Falmer.

Silverman, D. (2001). *Interpreting qualitative data: Methods for analysing talk, text and interaction*. London: Sage.

Silverman, D. (2005). Instances or sequences? Improving the state of the art of qualitative Research. *Forum: Qualitative Social Research*, 6(3), 30.

Taylor, D. (1996). *Toxic literacies: Exposing the injustice of bureaucratic texts*. Portsmouth, NH: Heinemann.

Tett, L., Hall, S., Maclaclan, K., et al. (2006). *Evaluation of the Scottish Adult Literacy and Numeracy (ALN) Strategy*. Edinburgh: Crown Copyright.

Tett, L., & Maclachlan, K. (2006). *Self-confidence, social capital and learner identity*. Paper presented at the NRDC International Conference, Nottingham, UK.

Tusting, K., & Barton, D. (2005). Community-based local literacies research. In R. Beach, J. Green, M. Kamil, et al. (eds.), *Multidisciplinary perspectives on literacy research* (2nd ed., pp. 243–64). Cresskill, NJ: Hampton Press Inc.

Contributors

Lul A. Admasachew has a BA in Economics from Addis Ababa University and MSc in Social Science Research Methodology from Nottingham Trent University. In the last six years, she has worked as a researcher at various universities in the United Kingdom, including Nottingham Trent University, Lincoln University, University of Birmingham, and King's College London. She is currently working as a research fellow at Queen Mary, University of London. Her research interests include education, children, young people's well-being, domestic violence, race, culture, and mental health.

Judith A. Alamprese is a principal associate in the Social and Economic Policy Division of Abt Associates Inc. Her research and evaluation studies have addressed a range of topics, including adult reading instruction, adult education program quality, interagency coordination, and transition from adult education to postsecondary education. These studies have employed a range of qualitative and quantitative methods. Alamprese has served on two committees on adult literacy for the National Academies' Board on Testing and Assessment and has published reports and articles in the fields of adult education, workplace and family literacy, and organizational change.

Yvon Appleby is Senior Lecturer in the Department of Education and Social Science at the University of Central Lancashire in the United Kingdom. Previously she was Research Fellow at Lancaster University in the Lancaster Literacy Research Centre. Work at the Centre included a range of ethnographically informed research projects with the National Research and Development Centre for adult literacy and numeracy (NRDC) which looked at adult learners' lives and their literacy practices both inside and outside of formal education. She is committed to linking research with practice, a position which informs recent writing in *Literacy, Lives and Learning* with David Barton and colleagues, and two practitioner guides, *Responding to People's Lives* and *Bridges Into Learning for Adults Who Find Provision Hard to Reach.*

David Barton is Professor of Language and Literacy in the Department of Linguistics at Lancaster University and Director of the Lancaster Literacy Research Centre. His current interests include: the changing nature of literacy in contemporary society; reading and writing on the Internet; literacy and social justice; and research methodologies. Main publications have been concerned with rethinking the nature of literacy (*Literacy*), carrying out detailed studies of everyday literacies *(Local Literacies*; *Situated Literacies*; *Letter Writing as a Social Practice*), and the relations of literacy and learning (*Beyond Communities of Practice*; *Literacy, Lives and Learning*).

Ann-Marie Bathmaker is Professor of Further Education and Lifelong Learning at the University of the West of England, Bristol. Her particular area of interest is the transformation of policy, as espoused, into policy, as experienced, especially in relation to the changing experience of learning, education and training, amongst those who do not follow traditional smooth learning trajectories. Recent projects include Impact of Skills for Life on Learners from 2003 to 2007 and the ERSC Further Higher Project, investigating widening participation, of which she was co-director. Recent publications include papers on *Dual Sector Further and Higher Education, Adult Basic Skills* and the *New Skills Agenda.*

Mary Beth Bingman is Senior Research Associate at the Center for Literacy Studies at the University of Tennessee, Knoxville. She coordinated work for the National Center for the Study of Adult Learning and Literacy (NCSALL) on Connecting Practice, Policy, and Research. She also develops training resources for Equipped for the Future and leads an initiative creating a model for standards-based professional development for adult basic education practitioners. Her research focuses on professional development and assessment and accountability in adult literacy. She is co-author of *Life at the Margins: Literacy, Language and Technology in Everyday Life* and articles, reports, and chapters in adult education.

Laura Brazier is a Research Officer in the Faculty of Policy and Society at the Institute of Education, University of London. Ms. Brazier has a background in psychology and applied social policy research with extensive experience of working with young people and practitioners in various youth justice contexts. Her profile of work includes improving the literacy and numeracy of young people involved in offending, developing inclusive learning in the secure estate, and phenomenological approaches to engaging young people in learning.

Greg Brooks is a (recently retired) Professor of Education at the University of Sheffield. His research interests encompass initial, family, and adult literacy, the assessment of oracy (speaking and listening) skills, phonological coding in silent reading, trends in levels of attainment in literacy and numeracy over time, the English spelling system, and the history of reading. He has pursued these interests in a range of studies that have used both quantitative and qualitative methods. His recent publications include *What Works for Pupils With Literacy Difficulties? The Effectiveness of Intervention Schemes*; *Effective Teaching and Learning: Reading*; and *Effective and Inclusive Practices in Family Literacy, Language and Numeracy: A Review of Programmes and Practice in the UK and Internationally*.

John Bynner is Emeritus Professor of Social Sciences in Education at the London Institute of Education. He is past Director of the Bedford Group for Life Course and Statistical Studies, the Centre for Longitudinal Studies, the Wider Benefits of Learning Research Centre, and was the first Director of the National Research and Development Centre for Adult Literacy & Numeracy. He has published widely on the transition to adulthood and the role of literacy and numeracy in the life course. Recent coedited books include *Changing Britain, Changing Lives,* and *Youth and Social Capital.*

Olga Cara is a Researcher at the Institute of Education, National Research and Development Centre for Adult Literacy and Numeracy. She has been doing research in the field of adult education for almost three years. She has been involved in numerous studies involving both quantitative and qualitative methodologies. One of her strengths is working with longitudinal data.

John Comings is a member of the faculty of the Harvard Graduate School of Education. From 1996 to 2007, he was Director of the National Center for the Study of Adult Learning and Literacy. He was one of the editors of the *Annual Review of Adult Learning and Literacy* and the author of *Building a Level Playing Field, Establishing an Evidence-Based Adult Education System*, and *New Skills for a New Economy.*

Larry Condelli is a managing director in the Education, Human Development and Workforce Division at the American Institutes for Research. His work includes research on adult ESL students, accountability, and conducting professional development and technical assistance for adult educators. His research area is the study of effective instruction for adult ESL literacy students, and he currently directs the Adult ESL Literacy Impact Study. He is also the project director for the National Reporting System for Adult Education and was instrumental in developing this national accountability system for federally funded adult education and literacy program.

Karen Evans is University Professor and Chair in Education (Lifelong Learning) at the Institute of Education, University of London. Her research and teaching interests focus on learning in life and work transitions throughout the life course, and on comparative and international studies in the field of lifelong learning. Karen Evans's previous publications include the books

Improving Workplace Learning; *Reconnection: Countering Social Exclusion Through Situated Learning*; *Working to Learn*; *Learning and Work in the Risk Society*; and numerous articles in the fields of learning and working life. She is also editor of the international journal *COMPARE*.

Stuart Hall is a Research Fellow at the SCRE Centre, Department of Education, at the University of Glasgow. His research interests include issues of social and educational inclusion and the use of action research in organizational development. He is currently involved in a number of action research projects funded by the Scottish Government.

Ursula Howard has worked as a teacher, organizer, and researcher in the field of adult literacy and adult education, most recently as Director of the National Research and Development Centre for Adult Literacy and Numeracy (NRDC). She has published a number of articles and chapters on adult literacy, focusing on policy and on the history of literacy, specializing on the acquisition of writing skills and writing practices in life and work. Her forthcoming book explores nineteenth-century adult literacy learning, practices, and meanings.

Jane Hurry is a Senior Lecturer in child and adolescent development at London University, Institute of Education. She conducts research in literacy and numeracy acquisition and has a particular interest in youth offenders' learning in these domains. Recent publications include: J. Hurry and K. Sylva (2007), "Long-Term Outcomes of Early Reading Intervention," *Journal of Reading Research,* 30, 227–48; J. Hurry and E. Doctor (2007), "Assessing Literacy in Children and Adolescents," *Child and Adolescent Mental Health,* 12, 38–45; J. Hurry, L. Brazier, M. Parker, and A.Wilson (2006), *Rapid Evidence Assessment of Interventions that Promote Employment for Offenders*, DfES Research Report 747.

Jenny Litster is a project officer at the National Research and Development Centre for Adult Literacy and Numeracy, where she has worked on a large-scale study of Skills for Life teachers and assisted with the NRDC's research on learner persistence. She gained her Ph.D. from the University of Edinburgh in 2001. Her doctoral research focused on the Canadian children's author L. M. Montgomery, on whom she has published a number of articles and book chapters.

Kathy Maclachlan is a senior lecturer and Head of Postgraduate Studies in the Department of Adult and Continuing Education at the University of Glasgow, Scotland. She has been extensively involved in adult literacy and numeracy research over the last decade, including the evaluation of Scotland's national literacies strategy with Lyn Tett, and developing practitioner research to engage new learners. She is currently leading a project on persistence, progress, and achievement in literacies' learning. She also teaches on the department's Postgraduate Certificate in Adult Literacy and Numeracy and their master's programs in Adult and Continuing Education.

Barbara McDonald is a lecturer in Psychology at San Diego State University in California where she teaches research methods and developmental psychology. She has conducted extensive research on adult learning, both technical training and basic skills in literacy. Her main interests lie in the emotional and motivational aspects of learning, including the family and life circumstances that influence the process of learning.

Pamela Meadows is a visiting fellow at the National Institute of Economic and Social Research in London. She specializes in the economics of the labor market and social policy, particularly issues related to worklessness, social exclusion, and disadvantage. She was previously Director of the Policy Studies Institute and before that spent nearly twenty years as a government economist in both the Home Office and the Department of Employment, where she was Chief Economist and Head of Economics, Research, and Evaluation.

Hillary Metcalf is Senior Research Fellow at the National Institute of Economic and Social Research in the United Kingdom. Her interests lie in disadvantage in the labor market and policy to address disadvantage. She has conducted a range of policy evaluations and recently led a five-year quantitative and qualitative evaluation of the impact of the English adult basic skills program Skills for Life.

T. Scott Murray is President of DataAngel Policy Research Inc. and an adjunct professor at the University of New Brunswick. His research interests are related to the development and application of evidence in policy making. Mr. Murray has specialized in the assessment of skills in student and adult populations and shedding light on how skill influences outcomes. Mr. Murray retired from the post of Director General, Institutions and Social Statistics at Statistics Canada in 2007.

Samantha Parsons is a Research Fellow in both the Centre for Longitudinal Studies and the National Research and Development Centre for Adult Literacy and Numeracy at the Institute of Education, University of London. Much of her research has involved working with two of Britain's longitudinal birth cohorts, the 1958 National Child Development Study (NCDS) and the 1970 British Cohort Study (BCS70). Her research has focused on the antecedents and consequences of adult literacy and numeracy difficulties. Recent publications include *Illuminating* *disadvantage: Profiling the experiences of adults with entry level literacy or numeracy over the lifecourse* and *Does Numeracy Matter More?*, both co-authored with John Bynner.

Stephen Reder is University Professor and Chair of the Department of Applied Linguistics at Portland State University. His research and teaching interests focus on processes of literacy and language development during adulthood. He has pursued these interests in a range of studies that have used both qualitative and quantitative methodologies. His previous publications include *The State of Literacy in America; Learning Disabilities, Literacy, and Adult Education* (coedited with Susan Vogel); and numerous articles and chapters in the fields of adult literacy, adult education, and applied linguistics.

Stephen Rose is a nationally recognized labor economist who has been doing innovative research and writing about social class in America for the last thirty years. His *Social Stratification in the United States*, originally published in 1978, is now in its sixth edition. He has worked for the Joint Economic Committee of Congress, as an advisor to Secretary of Labor Robert Reich, and in various research organizations in Washington, DC. He is currently working on a book, *Mythonomics: Ten Things That You Think You Know About the Middle Class That Are Wrong.*

Patricia Scollay is Professor Emeritus in Psychology at San Diego State University where she taught experimental and developmental psychology. She has conducted research using both quantitative and qualitative methodologies and is interested in finding ways to integrate them. Her main interests lie in cross-cultural differences in life change. She currently teaches and mentors graduate students at Walden University.

Lyn Tett is Professor of Community Education and Lifelong Learning at the University of Edinburgh. Her research interests lie within the broad area of community education and lifelong learning and have involved an investigation of the factors such as class, gender, and disability that lead to the exclusion of adults from postcompulsory education and of the action that might be taken to promote social inclusion. Her most recent books include *Community Education, Lifelong Learning and Social Inclusion* (2nd Edition) and *Adult Literacy, Numeracy and Language: Policy, Practice and Research*, coedited with Mary Hamilton and Yvonne Hillier.

John Vorhaus is Associate Director, Research, at the National Research and Development Centre in Adult Literacy and Numeracy (NRDC) at the Institute of Education. He is also Director of the Centre for

Research on the Wider Benefits of Learning. John has directed numerous projects on adult literacy, language and numeracy, many of these focused on teaching and learning practices, and also on disadvantaged groups such as offenders, thenic minority groups and people who are not in education, training or employment. Ongoing research is taken up with persons with profound and multiple learning difficulties and disabilities: their political status, the question of whether and how they are shown respect, and an examination of the teaching and learning practices best fitted to their needs and abilities. John has taught philosophy at the University of Bristol and London, and also in prison, adult and further education, and he continues to publish in the areas of political philosophy, philosophy of law and philosophy of education.

Edmund Waite is a researcher at the Department of Lifelong and Comparative Education, Institute of Education, London University. His research interests and publications relate to the study of adult literacy in the United Kingdom and international contexts as well as the anthropological study of education in Muslim societies. He is currently working on the Enhancing Skills for Life: Adult Basic Skills and Workplace Learning research project (funded by the ESRC and NRDC).

Anita Wilson is a prison ethnographer, currently based in the Literacy Research Centre, at Lancaster University, United Kingdom. Her work focuses on the ways in which communicative practice impacts on the sustainability of social identity in prison and the place of formal, informal, and nonformal education within such practice. She takes a transdisciplinary, qualitative approach and has adapted her conceptualization of "thirdspace" literacies to health studies, human geography, correctional education, and various criminal justice settings. She has published widely in the fields of literacy studies, nonformal education, and the creative arts.

Heide Spruck Wrigley is a senior researcher with Literacywork International, a small independent research firm focusing on education and training for vulnerable youth and low-income adults. She is a fellow with the Center for Immigrant Integration Policy and has been the subject matter expert for a number of national studies that have examined effective approaches and promising practices in adult ESL literacy. She has been involved in all aspects of adult ESL, and her recent publications include *Capturing What Counts: Language and Literacy Assessment for Bilingual Adults*; *Beyond the Life Boat: Improving Language, Citizenship and Training Services for Immigrants and Refugees*; and *We Are the World: Serving Language Minority Adults in Family Literacy Programs*.

Kwang Suk Yoon is Principal Research Scientist at American Institutes for Research in Washington, DC. His research focuses on school reform efforts, teacher professional development, motivation, program evaluation, and quantitative research methodologies. Yoon's recent publications include *What Works Study for Adult ESL Literacy Students* (with Larry Condelli) and *Reviewing the Evidence on How Teacher Professional Development Affects Student Achievement.*

Author Index

Subject Index

G

H

I

J

Made in the USA
Middletown, DE
27 July 2020

13754716R00230